Elements of Life and Death

Elements of Life and Death

By

Lars Olof Björn

Cambridge Scholars Publishing

Elements of Life and Death

By Lars Olof Björn

This book first published 2023

Cambridge Scholars Publishing

Lady Stephenson Library, Newcastle upon Tyne, NE6 2PA, UK

British Library Cataloguing in Publication Data
A catalogue record for this book is available from the British Library

ISBN (10): 1-5275-1261-4
ISBN (13): 978-1-5275-1261-0

TABLE OF CONTENTS

INTRODUCTION

Ninety-eight percent of the mass of the human body is made up of six elements: oxygen, carbon, hydrogen, nitrogen, sulfur, and phosphorus. The present book will be concerned with the importance of more than 40 other elements for the human body and for animals, plants, fungi, and bacteria. Hydrogen and oxygen are the elements of water, without which life is hard to imagine. They, and also carbon, nitrogen, sulfur, and phosphorus are necessary for proteins and nucleic acids, which are also components required by all life on our planet.

Some other elements are also required for all life forms – humans, animals, plants, fungi, bacteria, and archaea. They are K, Ca, Mg, Fe, Zn, Mo, Cu, Mn, and Ni. Well, not quite true. There are actually bacteria which do not need Fe (iron), and Ni (nickel) is probably not universally required either. And we and other animals need Na. We shall return to these elements later.

But some organisms require other elements, or their essentiality is controversial. This will be the elements that we shall focus on in this book, such as As, B, V, Cr, Cd, Si, Co, I, F, Br, Sb, Se, Te, La, Ce, Pr, Nd, Gd. We shall also have something to say about biological uses of some other elements, as Cl and Fe, which are extremely important in some ways that most people are unaware of. And we cannot abstain from Agatha Christies favourite poison, thallium (see https://en.wikipedia.org/wiki/Thallium_poisoning, https://www.nytimes.com/1977/06/24/archives/agatha-christie-booksaves-an-infants-life.html "The pale horse", https://en.wikipedia.org/wiki/The_Pale_Horse, and https://www.imdb.com/title/tt10485750/)."

We shall also include in our account several other elements which are not necessary for any organisms, but anyway are of biological interest.

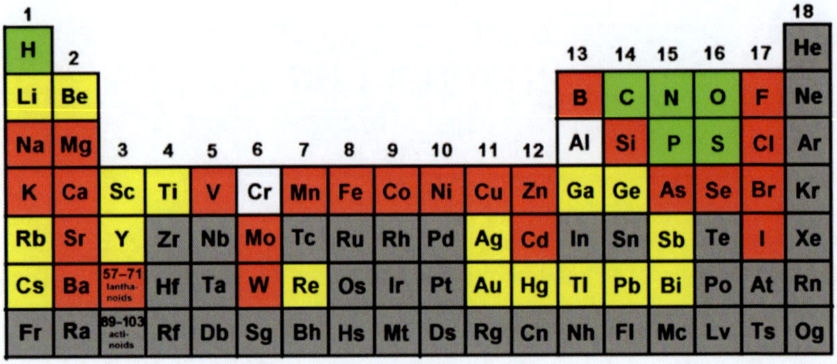

Overview of the periodic system of elements. Those with green background are major constituents of all organisms, and are not treated in this book. Those on grey background are of no known direct biological use, and are not treated here either. The elements on red background are required by at least some organisms, or are parts of known enzymes. The status of elements on white background, chromium and aluminium is disputed, and will be discussed. Note the numbers on top of the columns, which are used for numbering the following chapters (except that there is no chapter 18).

At the beginning of each chapter the periodic system is shown as above, but with the symbols for elements treated in the chapter surrounded by a light blue frame.

CHAPTER ONE

ALKALI METALS

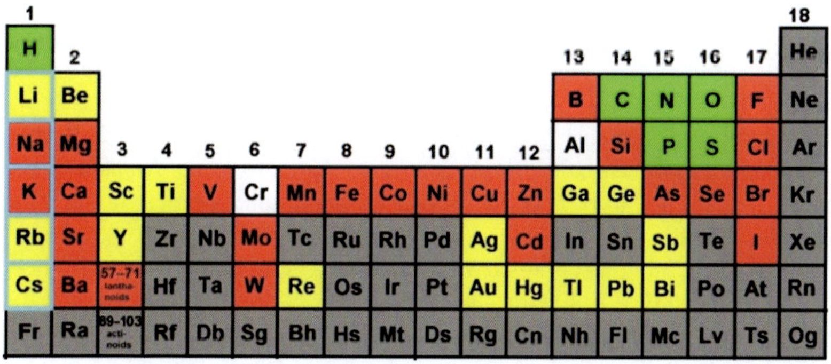

1.1 Lithium – the metal that was "in the beginning"

In the beginning was the Big Bang. The first elements that followed were hydrogen, helium and lithium (^{7}Li). Here we shall concentrate on biological aspects of this, the first metal to come into being. Present-day natural lithium also consists of a small amount of ^{6}Li. For further information about creation and destruction of lithium, see Lodders (2020).

Some plants are considered to be lithium hyperaccumulators (Jiang et al. 2014, Elektorowicz & Keropian 2015, Kawanagh et al. 2018, Qiao et al. 2018, Anguilano et al. 2022). Even closely related plants may show great differences in their accumulation ability. For instance, for plants grown in a soil amended with 20 mg Li/kg, the Li concentration in leaves of *Brassica napus* amounted to 7.8 mg per kg dry matter; for *Brassica oleracea* it was 1423 mg per kg dry matter (Kawanagh et al. 2018). There are lithium accumulators in several plant families: *Brassica oleracea, Brassica juncea* and *Cardamine hirsuta* in Brassicaceae, *Helianthus annuus* in Asteraceae (Elektorowicz & Keropian 2015, Kavanagh et al. 2018), *Apocynum venetum* in Apocynaceae (Jiang et al. 2014, Qiao et al. 2018), and *Agrostis tenuis* in Poaceae (Anguliano et al. 2022). At least in *Apocynum venetum* the lithium is deposited mostly in the vacuoles (Qiao et al. 2018). Since high amounts of lithium is a poison for many organisms, one can assume that lithium accumulation is a way for plants to protect themselves from herbivores and parasites, corresponding to what has been demonstrated for nickel accumulating plants.

Also some bacteria are able to accumulate high amounts of lithium. Of 70 organisms (bacteria, fungi, and actinomycetes) the most efficient accumulators (from a solution containing 72.0 μ*M* Li^+) are the bacteria *Arthrobacter nicotianae* (Micrococcaceae) 125.8 μmol/(g dry weight) and *Brevibacterium helovolum* (Brevibacteriaceae) 98.1 μmol/(g dry weight), both in the order Micrococcales (Tsuruta 2005). In the cyanobacterium *Arthrospira platensis* the concentration of lithium after three days reached 2500 μg/(g d.w.), i.e. 360 μmol/g d.w.) at a concentration of 500 mg/L (72 m*M*) (Cepoi et al. 2021). Bruna et al. (2022) studied bacterial strains from the Atacama desert that grew well at Li^+ concentrations up to 700 m*M* (*Pseudomonas rhodesiae*, *Planomicrobium koreense*, and *Pseudomonas sp*.). They seemed to protect themselves from lithium by precipitating it as lithium sulfide.

Also some fungi accumulate lithium; one species that has been studied is *Lentinus crinitus*, Polyporaceae (Faria et al. 2019, 2022). These authors also list data for a number of other fungi.

Lithium has been used as medicine for a number of medical problems. Most importantly, it has been used for a long time (Cade 1949, Ferensztajn-Rochowiak et al. 2021) for alleviating the symptoms of bipolar disorder, but, depending on genetic constitution, is effective only for about half of the patients, and only 20–30% are "excellent responders" (Papiol et al. 2022). The specific neurobiological mechanisms through which lithium exerts its effects on brain function are not fully understood; current knowledge is reviewed by Bergamelli et al. (2021). Bipolar disorder patients exhibit abnormally high cytosolic Ca^{2+} levels and protein kinase C hyperactivity that can be downregulated by long-term Li^+ treatment. Protein kinase C is a protein with three calcium ions in the molecule, and calculations show that lithium ions can displace one of the calcium ions, and thereby inhibit the enzymatic activity, in agreement with experiments (Grauffel et al. 2021). Rubidium and cesium ions do not have this effect on protein kinase C, and do not have the same effect on bipolar disorder as lithium ions have. It should be pointed out that several other explanations for the lithium effect on bipolar disorder have been proposed. For instance, Espanhol & Vieira-Coelho (2022) report that lithium medication has a positive effect on the brain's white matter. Different explanations may not be mutually exclusive. Rybakowski & Ferensztajn-Rochowiak (2022) are of the opinion that e potassium channels in the plasma membrane, with the cell in the state, are open, potassium ions can diffuse in both directions, and a state develops, with the inside 70 mV negative compared to the e. The cell membrane also contains sodium channels, but in the resting of the cell these are closed.

hen the nerve cell is stimulated in some way (by a signal from a bouring nerve cell, or by something else), some sodium channels open, n "action potential" is triggered. As sodium ions flow into the cell, the ntial on the inside changes from -70 mV to -55 mV. This causes the rest e sodium channels to open and the potassium channels to close, and the de potential to change from -55 to +30 mV (Fig. 1.2.2). As this potential eached the sodium channels close, and the potassium channels open ain, which causes a return to the resting potential.

The process just described takes place at one point of the long extension, e axon, of a nerve cell. But seen in a larger perspective, a pulse of positive side potential moves away from the point where the process started, i.e., e nerve impulse is transmitted along the axon.

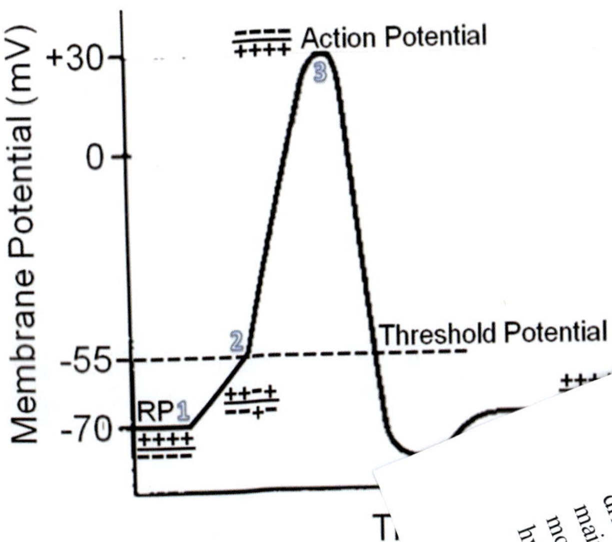

Fig. 1.2.2. The potential changes at one poin potential, 2 potential for triggering of mass open potential. Reproduced from the Vancouver Co (https://learningcentre.vcc.ca/media/vcc-library/co worksheets/by-coursex2fprogram/life-sciences/Nerv a CC by 4.0 (https://creativecommons.org/licenses/by

lithium is underused for treatment of bipolar disorder considering the good results obtained with it.

A question that has been very hotly debated is whether lithium medication lowers the risk of suicide or not, and there have been several suggestions that lithium should be added to drinking water in order to protect people from suicide (see Araya et al. 2022).

Lithium has occasionally been used or suggested for treating a number of other medical problems: Cancer (Ge & Jakobsson 2019), sepsis-induced skeletal muscle atrophy (Lee et al. 2021, glaucoma (Vallée et al. 2021), osteoarthritis (Todd et al. 2018), psoriasis (Grželj & Sollner Dolenc 2020), Alzheimer's disease (Wallace 2014, Ishii et al. 2021), and pulmonary injury caused by *Actinobacillus pleuropneumoniae* (Zhang et al. 2022). Presently we do not have data for verification of these claims. An inverse relation between lithium in drinking water and incidence of Alzheimer's disease is very weak, and can be shown only for men (Muronaga et al. 2022).

Lithium medication also has unwanted effects, such as kidney damage (Hayes et al. 2021) and endocrine dysfunction (Lerena et al. 2022) and must therefore be used with caution. In some places in Afghanistan the lithium content is high, and in vitro experiments point to a risk that this may induce malignant transformation of human cells (Sudo et al. 2020). In vitro experiments with cell cultures indicate that lithium ions at a concentration of more than 100 mg/L are toxic to neurons and may generate fibrils of amyloid-beta peptide, reminiscent of Alzheimer's disease (Wang et al. 2014).

Lithium ions affect the activities of a number of magnesium-dependent enzymes, including some in the respiratory chain. They also have an influence on the transcription of certain genes (reviewed by Roux & Dosseto 2017).

The total human mobilization of Li from the Earth's crust, >1,000 x 10^9 g/year, is much larger than Li mobilized by the natural processes of chemical and mechanical weathering, 94 and 132 x 10^9 g/year, respectively (Schlesinger et al. 2021). This means that there is a great risk for contamination of the environment.

Natural lithium consists of two isotopes whose natural relative abundance is 7.59% for ^{6}Li and 92.41% for ^{7}Li. It has been suggested that tritium (^{3}H) fusion should be used for future environment-friendly energy production, and that the tritium, which is rare in nature, should be synthesized using ^{6}Li + neutron → ^{3}H + ^{4}He. Thus there is need for fractionation of lithium to get ^{6}Li, and this one should use its enrichment by certain algae (López-Rodas et al. 2021). Also the ^{7}Li would find use in the nuclear energy industry.

1.2 Sodium and potassium

1.2.1 Roles in enzyme function

Many enzymes are activated by potassium ions, fewer by sodium ions, and some are less selective, and can be more or less activated by sodium, potassium, or other similar ions. Already long ago it was noticed that some enzymatic processes are stimulated by potassium ions. An example is the study by Boyer et al. (1942) of phosphorylations in muscle (Fig. 1.2.1). Kachmar and Boyer (1953) discovered that an enzyme, pyruvate kinase, required potassium for activity (although not all pyruvate kinases require potassium ions; Ramírez-Silva et al. 2022). Pyruvate kinase is a crucial enzyme of the glycolytic pathway that catalyzes production of the second of two ATP molecules by converting phosphoenolpyruvate and ADP to pyruvate and ATP. With a saturating concentration of pyruvate, the reaction rate is 400 times as high with (90 m*M*) as without potassium ions; further information on the activation of this enzyme by potassium is provided by Oria-Hernández et al. (2005).

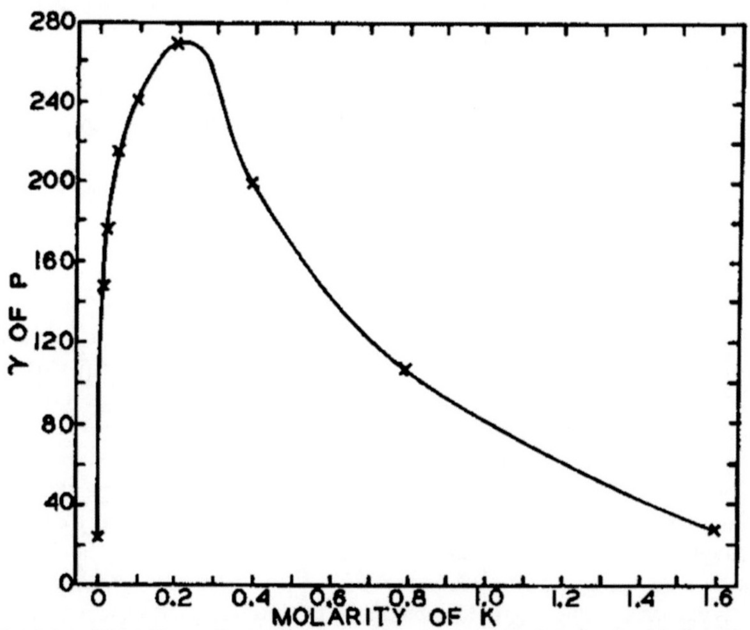

Fig. 1.2.1. The effect, of potassium ion concentration on the transfer of phosphate from 3-phosphoglycerate to creatine in rat muscle (Boyer et al. 1942). Reproduced under CC by 4.0 (https://creativecommons.org/licenses/by/4.0/) license.

Nerve cells have extensions of two kinds, shorter dendrites, and very long axons. In vertebrates the axons are equipped with myelin sheaths of lipid material (except at places called nodes of Ranvier) which isolate them from the surroundings. This speeds up signal transmission, since signals "jump" across the isolated segments, from one node to another node. Invertebrate axons generally lack myelin sheaths, which results in slower signal transmission.

1.2.3 Role of potassium ions in stomatal regulation

Regulation of the stomatal aperture in plants is very complicated, and not exactly the same in all plants. It has evolved to let carbon dioxide in to enable photosynthesis without losing too much water by transpiration – in many cases a precarious balance. With some exceptions stomata are closed in the dark, when no photosynthesis can take place anyway (but some plants keep stomata open at night to absorb carbon dioxide in an unreduced form for later use). The guard cells surrounding the stomatal apertures are so constructed that the stomata open when their inner pressure increases due to increased osmotically active substances. One of the most important of these substances is potassium ion (Sawhney & Zelitch 1969, Humble & Raschke 1971, Andrés et al. 2014). Because plant differ in the details, it would carry to far here to describe all the intricacies of stomatal regulation, in which red and blue light as well as carbon dioxide concentration and water status all play different important roles.

The uptake of potassium ions into the guard cells is electrically driven, and takes place if the inside of the cell has a negative potential of ca 70 mV in relation to the outside (Roelfsema et al. 2001); see also Roux & Leonhardt 2008).

Sodium is not required by all plants, but in case of insufficient potassium sodium can to some extent serve instead, but not completely replace it (seemingly *Salicornia bigelovii* Torr. is an exception; see below (Wakeel et al. 2011, Erel et al. 2020). Some plants need sodium as a micronutrient, such as several species having the C4 photosynthetic pathway (Brownell & Crossland 1972). Sodium chloride has been shown to alleviate drought stress in several species (He et al. 2019, Guo et al. 2020, Huihui et al. 2021).

According to Yamada et al. (2016) *Salicornia bigelovii* Torr. grows best with Na^+ but no K^+ in the root medium (Fig. 1.2.3). However, if the plants, before being transferred to the test medium, were pre-grown in a solution containing, in addition to other nutrients, 10 mM NaCl, the plants contained some sodium ions at the start of the treatment.

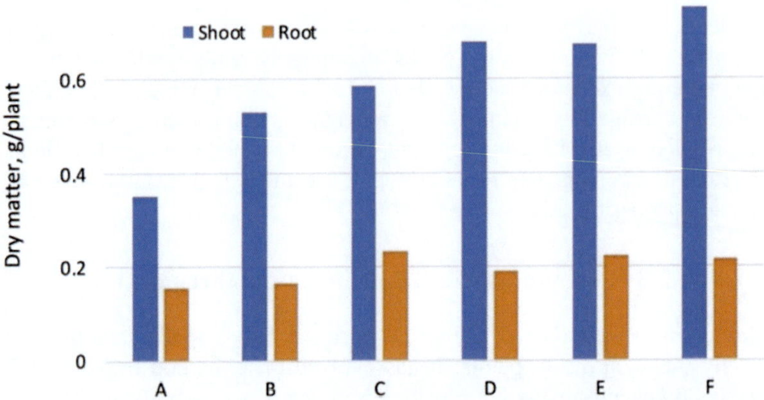

Fig. 1.2.3. Effect of shifting Na^+/K^+ ratio on growth of dwarf glasswort (*Salicornia bigelovii* Torr.) after growing for 1 week in nutrient solutions with the sum of sodium and potassium ion concentration 100 mM, but with varying concentrations of each of these ions: In A Na^+ 0, K^+ 100; in B Na^+ 20, K^+ 80; in C Na^+ 40, K^+ 60; in D Na^+ 40, K^+ 40; in E Na^+ 80, K^+ 20, and in F Na+ 100, K^+ 0 mM. Because the plants contained Na and K before the start of the differential treatment, the final amounts of the elements, in mol per kg dry matter were as follows: A 1.60 Na, 2.09 K; B 1.88 Na, 2.04 K; C 2.14 Na, 1.80 K; D 2.62 Na, 1.61 K; E 2.96 Na, 1.10 K; F 3.38 Na, 0.52 K. Data from Yamada et al. (2016).

As one moves upwards in the food chain, the concentration of sodium in organisms increases. Kaspari (2020) found the concentration to mostly range 10–100 ppm in forbs, grasses (and nectar as well), 100–1000 ppm in invertebrate herbivores, and it was above 3000 ppm in all of 33 invertebrate predators tested.

Some plants living under conditions with high sodium chloride are able to rid themselves of salt by salt glands on the leaves. Examples are *Atriplex canescens* (Guo et al. 2022), *Tamarix* species (Wei et al. 2020, Duan et al. 2022, Hussain et al. 2022), *Limonium bicolor* (Zhao et al. 2023), and some mangrove trees, e.g., *Avicennia* species (Tan et al. 2013).

1.3 Rubidium

According to Osretkar & Krauss (1965) *Chlorella pyrenoidosa* Chick can be grown in a medium with rubidium instead of potassium, although the growth rate was lower than with potassium. Baum (1965) tested several algae for their ability to use rubidium instead of potassium. Most species could not, but six species could. One of them, *Chlamydomonas reinhardtii*, showed the most pronounced response. With a contaminant level of 5.48 mg

K per liter, it initially grew equally well with 160.5 mg/L of KCl, and with the corresponding amount of RbCl, but reached a higher maximum cell density with KCl. MacLeod & Snell (1948) found that rubidium ion replaced K^+ completely for growth of *Lactobacillus casei* and *Streptococcus faecalis*, partially for *L. mesenteroides* 9135 and *L. arabinosus*, and not at all for *Leuconostoc mesenteroides* 8042.

1.4 Cesium

Cesium has in many cases a negative effect on organisms. For instance, it inhibits the growth of a mutant of *Arabidopsis thaliana* by reducing the uptake of potassium (Adams et al. 2019). It decreases the growth of *Chlorella emersonii* when the alga is grown photoautotrophically, but not when it is grown chemoheterotrophically, also by interfering with potassium (Avery et al. 1992). Also the growth of a *Streptomyces* species was inhibited by cesium, because it competitively inhibited potassium uptake (Kuwahara et al. 2011). But in many potassium-deficient microorganisms cesium can promote growth by partly replacing the functions of potassium. In the absence of potassium, Kup-mediated Cs^+ uptake partially supported cell growth of *Escherichia coli*, however, at a much lower rate than with sufficient K^+. In K^+-limited medium (0.1 mM), the presence of Cs^+ (up to 25 mM) in the medium enhanced growth as much as control medium containing 1 mM K^+ (Tanudjaja et al. 2017). Kato et al. (2017) observed a positive correlation between intracellular Cs^+ concentrations and growth yields under K^+-deficient conditions by constructing an *Escherichia coli* strain that can be induced to expresses a Cs^+-transport system.

References

Adams, E., Miyazaki, T., Saito, S., Uozumi, N. & Shin, R. 2019. Cesium inhibits plant growth primarily through reduction of potassium influx and accumulation in *Arabidopsis*. Plant Cell Physiol. 60, 63–76. https://doi.org/10.1093/pcp/pcy188

Andrés, Z., Pérez-Hormaeche, J., Leidi, E.O., Schlücking, K., Steinhorst, L., McLachlan, D.H., Schumacher, K., Hetherington, A.M., Kudla, J., Cubero, B. & Pardoa, J.M. 2014. Control of vacuolar dynamics and regulation of stomatal aperture by tonoplast potassium uptake. Proc. Natl Acad. Sci. USA 111, E1806–E1814.

Anguilano, L., Onwukwe, U., Aryani, D., Ojeda, J.J., Lingua, G., Gianotti, V. & Devoto, A. 2022. Characterisation of hyperaccumulators for lithium recovery from ancient mine soils. *In* A. Lazou et al. (eds.),

REWAS 2022: Developing tomorrow's technical cycles (Volume I), The Minerals, Metals & Materials Series, pp. 149–153. https://doi.org/10.1007/978-3-030-92563-516

Araya, P., Martínez, C. & Barros, J. 2022. Lithium in drinking water as a public policy for suicide prevention: Relevance and considerations. Front. Public Health 10:805774 (18 pp.). https://doi.org/10.3389/fpubh.2022.805774

Avery, S.V., Codd, G.A. & Gadd, G.M. 1992. Replacement of cellular potassium by caesium in *Chlorella emersonii*: differential sensitivity of photoautotrophic and chemoheterotrophic growth. J. Gen. Microbiol. 138, 69–76.

Baum, L.S. 1965. A study of the inter-relationship of potassium and rubidium in various species of fresh water algae. Dissertation.

Bergamelli, E., Del Fabro, L., Delvecchio, G., D'Agostino, A. & Brambilla, P. 2021. The Impact of lithium on brain function in bipolar disorder: An updated review of functional magnetic resonance imaging studies. CNS Drugs 35, 1275–1287.

Boyer, P.D., Lardy, H.A., Phillips, P.H. 1942. The role of potassium in muscle phosphorylations. J. Biol. Chem. 146, 673–681.

Brownell, P.F. & Crossland, C.J. 1972. The requirement for sodium as a micronutrient by species having the C4 dicarboxylic photosynthetic pathway. Plant Physiol. 49, 794–797. https://doi.org/10.1007/s40263-021-00869-y

Bruna, N., Galliani, E., Oyarzún, P., Bravo, D., Fuentes, F. & Pérez-Donoso, J.M. 2022. Biomineralization of lithium nanoparticles by Li-resistant *Pseudomonas rodhesiae* isolated from the Atacama salt flat. Biol. Res. 55:12 (13 pp.). https://doi.org/10.1186/s40659-022-00382-6

Cade, J.F.J. 1949. Lithium salts in the treatment of psychotic excitement. Med. J. Aust. 2, 349–351.

Cepoi, L., Zinicovscaia, I., Rudi, L., Chiriac, T., Pantelica, A., Apostol, A., Djur, S., Ganea, L. & Decebal, Iancu, D. 2021. Biomass of *Arthrospira platensis* enriched with lithium by bioaccumulation and biosorption process. Food Bioscience 41: 100950 (12 pp.). https://doi.org/10.1016/j.fbio.2021.100950

Cohn, M. & Monod, J. 1951. Purification et propriétés de la β-galactosidase (lactase) d'*Escherichia coli*. Biochim Biophys Acta 7, 153–174.

Di Cera, E. 2008. Thrombin. Molecular Aspects of Medicine 29, 203–254. https://doi.org/10.1016/j.mam.2008.01.001

Duan, Q., Zhu, Z., Wang, B. & Chen, M. 2022. Recent progress on the salt tolerance mechanisms and application of tamarisk. Int. J. Mol. Sci. 23: 3325 (13 pp.). https://doi.org/10.3390/ijms23063325

Díaz-Alejo, H.M, López-Rodas, V., Camino García-Balboa,C., Tarín, F., Barrado, A.I., A.I., Conde, E. & Costas, E. 2021. The upcoming ^{6}Li isotope requirements might be supplied by a microalgal enrichment process. Microorganisms 9: 1753 (14 pp.). https://doi.org/10.3390/microorganisms9081753

Elektorowicz, M. & Keropian, Z. 2015. Lithium, vanadium and chromium uptake ability of *Brassica juncea* from lithium mine tailings. Int. J. Phytoremediat. 17, 521–528. https://doi.org/10.1186/s40659-022-00382-6

Erel, R., Ben-Gal, A., Dag, A., Schwartz, A. & Yermiyahu, U. 2014. Sodium replacement of potassium in physiological processes of olive trees (var. Barnea) as affected by drought. Tree Physiol. 34, 1102–1117. https://doi.org/10.1093/treephys/tpu081

Espanhol, J.C.L. & Vieira-Coelho, M.A. 2022. Effects of lithium use on the white matter of patients with bipolar disorder – a systematic review. Nordic J. Psychiatry 76, 1–11. https://doi.org/10.1080/08039488.2021.1921264

Faria, M.G.I., Avelino, K.V., do Valle, J.S., da Silva, G.J., Gonçalves Jr., A.C., Dragunski, D.C., Colauto, N.B. & Linde, G.A. 2019. Lithium bioaccumulation in *Lentinus crinitus* mycelial biomass as a potential functional food. Chemosphere 235, 538–542. https://doi.org/10.1016/j.chemosphere.2019.06.218

Faria, M.G.I., Avelino, K.V., Philadelpho, B.O., dos Santos Bomfim, R., do Valle, J,S., Gonçalves Júnior, A.C., Dragunski, D.C., de Souza Ferreira, R., de Souza, C.O., Ferreira Ribeiro, C.D., Colauto, N.B. & Linde, G.A. 2022. Lithium bioaccumulation in *Lentinus crinitus* mycelia grown in media with different lithium sources and pH values. Environm. Sci-Pollut. Res. https://doi.org/10.1007/s11356-022-21893-5

Ferensztajn-Rochowiak, E., Chlopocka-Woz´niak, M. & Rybakowski, J.K. 2021. Braz. J. Psychiatry 43, 407–413. https://doi.org/10.1590/1516-4446-2020-1111

Ge, W. & Jakobsson, E. 2019. Systems biology understanding of the effects of lithium on cancer. Front. Oncol. 9:296 (8 pp.). https://doi.org/10.3389/fonc.2019.00296

Grauffel, C., Weng, W.-H., Dudev, T. & Lim, C. 2021. Trinuclear calcium site in the C2 Domain of PKCα/γ is prone to lithium attack. ACS Omega 6, 20657−20666. https://doi.org/10.1021/acsomega.1c02882

Grželj, J. & Sollner Dolenc, M. 2020. The role of xenobiotics in triggering psoriasis. Arch. Toxicol. 94, 3959–3982. https://doi.org/10.1007/s00204-020-02870-8

Guo, H., Cui, Y.-N., Pan, Y.-K., Wang, S.-M. & Bao, A.-K. 2020. Sodium chloride facilitates the secretohalophyte *Atriplex canescens* adaptation to drought stress. Plant Physiol. Biochem. 150, 99–108. https://doi.org/10.1016/j.plaphy.2020.02.018

Hawrylak-Nowak, B., Kalinowska, M. & Szymańska, M. 2012. A study on selected physiological parameters of plants grown under lithium supplementation. Biol. Trace Elem. Res., 149, 425–430.

Hayes, J.F., Osborn, D.P.J., Francis, E., Ambler, G., Tomlinson, L.A., Boman, M., Wong, I.C.K., Geddes, J.R., Dalman, C. & Lewis, G. 2021. Prediction of individuals at high risk of chronic kidney disease during treatment with lithium for bipolar disorder. BMC Medicine 19:99 (16 pp.). https://doi.org/10.1186/s12916-021-01964-z

He, F.-L., Bao, A.-K., Wang, S.-M. & Jin, H.-X. 2019. NaCl stimulates growth and alleviates drought stress in the salt-secreting xerophyte *Reaumuria soongorica*. Environm. Exp. Bot. 162, 433–443. https://doi.org/10.1016/j.envexpbot.2019.03.014

Humble, G.D. & Raschke, K. 1971. Stomatal opening quantitatively related to potassium transport. Plant Physiol. 48, 447–453.

Hussain, T., Asrar, H., Li, J., Feng, X., Gul, B. & Liu, X. 2022. The presence of salts in the leaf exudate improves the photosynthetic performance of a recreto-halophyte, *Tamarix chinensis*. Environm. Exp. Botany 199: 104896 (10 pp.). https://doi.org/10.1016/j.envexpbot.2022.104896

Ishii, N., Terao, T. & Hirakawa, H. 2017. The present state of lithium for the prevention of dementia related to Alzheimer's dementia in clinical and epidemiological studies: A critical review. Int. J. Environ. Res. Public Health 2021, 18, 7756. https://doi.org/10.3390/ijerph18157756

Jasper, P. 1978. Potassium transport system of *Rhodopseudomonas capsulata*. J. Bacteriol. 133, 1314–1322.

Jiang, L., Wang, L., Mu, S.-Y. & Tian, C.-Y. 2014. *Apocynum venetum*: A newly found lithium accumulator. Flora 209, 285–289. http://dx.doi.org/10.1016/j.flora.2014.03.007

Kachmar, J.F. & Boyer, P.D. 1953. Kinetic analysis of enzyme reactions. II. The potassium activation and calcium inhibition of pyruvic phosphoferase. J Biol Chem 200, 669–682.

Kato, S., Kanata, Y., Kitagawa, W., Sone, T., Asano, K. & Kamagata, Y. 2017. Restoration of the growth of *Escherichia coli* under K^+-deficient conditions by Cs^+ incorporation via the K^+ transporter Kup. Sci. Rep. 7: 1965 (6 pp.). https://doi.org/10.1038/s41598-017-02024-4

Kaspari, M. 2020. The seventh macronutrient: how sodium shortfall ramifies through populations, food webs and ecosystems. Ecol. Lett. 23, 1153–1168. https://doi.org/10.1111/ele.13517

Kavanagh, L., Keohane, J., Cabellos, G.G., Lloyd, A. & Cleary, J. 2018. Induced plant accumulation of lithium. Geosci. 8, 56 (18 pp.). https://doi.org/10.3390/geosciences8020056

Krem, M.M. & Di Cera, E. 2001. Molecular markers of serine protease evolution. EMBO J 20: 3036–3045.

Kuwahara, C., Fukumoto, A., Nishina, M., Sugiyama, H., Anzai, Y. & Kato, F. 2011. Characteristics of cesium accumulation in the filamentous soil bacterium *Streptomyces sp*. K202. J. Environm. Radioactivity 102, 138–144. https://doi.org/10.1016/j.jenvrad.2010.11.004

Lee, J.-H., Kim, S.-W., Kim, J.-H., Kim, H.-J., Um, JI. Jung, D.-W. & Williams, D.R. 2021. Lithium chloride protects against sepsis-induced skeletal muscle atrophy and cancer cachexia. Cells 20 10, 1017 (17 pp.). https://doi.org/10.3390/cells10051017

Lerena, V.S., León, N.S., Sosa, S., Deligiannis, N.G., Danilowicz, K. & Rizzo, L.F.L. 2022. Lithium and endocrine dysfunction. Medicina (Buenos Aires) 82–130–137.

Lodders, K. 2020. The cosmic lithium story. Elements 16, 241–246. https://doi.org/10.2138/gselements.16.4.241

MacLeod, R.A. & Snell, E.E. 1948. The effect of related ions on the potassium requirement of lactic acid bacteria. J. Biol. Chem. 176, 38–52.

Muronaga, M., Terao, T., Kohno, K., Hirakawa, H., Izumi, T. & Etoh, M. 2022. Lithium in drinking water and Alzheimer's dementia: Epidemiological findings from National Data Base of Japan. Bipolar Disorders. 24, 788–794. https://doi.org/10.1111/bdi.13257

Osretkar, A. & Krauss, R.W. 1965. Growth and metabolism of *Chlorella pyrenoidosa* Chick during Substitution of Rb for K. Phycol. 1, 23–33.

Oria-Hernández, J., Cabrera, N., Pérez-Montfort, R. & Ramı´rez-Silva, L. 2005. Pyruvate Kinase Revisited. The activating effect of K^+. J. Biol. Chem. 280, NO. 45, 37924–37929.

Page, M.J. & Di Cera, E. 2006. Role of Na and K in Enzyme Function. Physiol Rev 86: 1049–1092. doi:10.1152/physrev.00008.2006.

Papiol, S., Schulze, T.G. & Heilbronner, U. 2022. Lithium response in bipolar disorder: Genetics, genomics, and beyond. Neurosci. Lett. 785: 136786 (6 pp.). https://doi.org/10.1016/j.neulet.2022.136786

Qiao, L., Tanveer, M., Wang, L. & Tian, C. 2018. Subcellular distribution and chemical forms of lithium in Li-accumulator *Apocynum venetum*. Plant Physiol. Biochem. 132, 341–344. https://doi.org/10.1016/j.plaphy.2018.09.022

Ramírez-Silva, L., Hernández-Alcántara, G., Guerrero-Mendiola, C., González-Andrade, M., Rodríguez-Romero, A., Rodríguez-Hernández, A., Lugo-Munguía, A., Gómez-Coronado, P.A., Rodríguez-Méndez, C.

& Vega-Segura, A. 2022. The K^+-dependent and -independent pyruvate kinases acquire the active conformation by different mechanisms. Int. J. Mol. Sci. 23:1347 (22 pp.). https://doi.org/10.3390/ijms23031347

Roelfsema, M. R. G. & Hedrich, R. 2005. In the light of stomatal opening: New insights into "The Watergate". New Phytol. 167, 665–691. https://doi.org/10.1111/j.1469-8137.2005.01460.x.

Roux, B. & Leonhardt, N. 2018. The regulation of ion channels and transporters in the guard cell. Ch. 7 *in* Advances in Botanical Research 87, 171–214. Elsevier. https://doi.org/10.1016/bs.abr.2018.09.013

Roux. M. & Dosseto, A. 2017. From direct to indirect lithium targets: a comprehensive review of omics data. Metallomics 9, 1326--1351. https://doi.org/10.1039/c7mt00203c

Rybakowski, J.K. & Ferensztajn-Rochowiak, E. 2022. Mini-review: Anomalous association between lithium data and lithium use. Neuroscience Lett. 777: 136590 (6 pp.). https://doi.org/10.1016/j.neulet.2022.136590

Sawhney, B.L. & Zelitch, I. 1969. Direct determination of potassium ion accumulation in guard cells in relation to stomatal opening in light. Plant Physiol. 44, 1350–1354.

Schlesinger, W.H., Klein, E.M., Wang, Z. & Vengosh, A. 2021. Global biogeochemical cycle of lithium. Global Biogeochem. Cycles, 35, e2021GB006999 (18 pp.). https://doi.org/10.1029/2021GB006999

Sudo, M., Hashimoto, K., Yoshinaga, M., Azimi, M.D., Fayaz, S.H., Hamajima, N., Kondo-Ida, L., Yanagisawa, K. & Kato, M. 2020. Lithium promotes malignant transformation of nontumorigenic cells in vitro. Sci. Total Environm. 744: 140830 (7 pp.). https://doi.org/10.1016/j.scitotenv.2020.140830

Tan, W.-K., Lin, Q., Lim, T.-N., Kumar, P. & Loh, C.-S. 2013. Dynamic secretion changes in the salt glands of the mangrove tree species *Avicennia officinalis* in response to a changing saline environment. Plant, Cell and Environm. 36, 1410–1422. https://doi.org/10.1111/pce.12068

Tanudjaja, E., Hoshi, N., Su, Y.-H., Hamamoto, S., & Uozumi, N. 2017. Kup-mediated Cs^+ uptake and Kdp-driven K^+ uptake coordinate to promote cell growth during excess Cs^+ conditions in *Escherichia coli*. Sci. Rep. 7: 2122 (9 pp.). https://doi.org/10.1038/s41598-017-02164-7

Todd, T., Lu, Z., Zhao, J., Cline, B., Zhang, W., Chen, H., Kumar, A., Wen Jiang, W., West, F., Franklin, S., Zheng, L. & Xie, J. 2018. LiF@SiO2 nanocapsules for controlled lithium release and osteoarthritis treatment. Nano Research 11, 5751–5760. https://doi.org/10.1007/s12274-018-2061-5

Tomioka, N., Uchiyama, H. & Yagi, O. 1994. Cesium accumulation and growth characteristics of *Rhodococcus erythropolis* CS98 and *Rhodococcus sp*. strain CS402. Appl. Environ. Microbiol. 60, 2227–2231.

Tsuruta, T. 2005. Removal and recovery of lithium using various microorganisms. J. Biosci. Bioeng. 100, 562–566. https://doi.org/10.1263/jbb.100.562

Vallée, A., Vallée, J.-N. & Lecarpentier, Y. 2021. Lithium and atypical antipsychotics: The possible WNT/b pathway target in glaucoma. Biomedicines 9: 473 (20 pp.). https://doi.org/10.3390/biomedicines9050473

Wakeel, A., Farooq, M., Qadir, M. & Schubert, S. 2011. Potassium substitution by sodium in plants. Crit. Revs Plant Sci. 30, 401–413. https://doi.org/10.1080/07352689.2011.587728

Wallace, J. 2014. Calcium dysregulation, and lithium treatment to forestall Alzheimer's disease – a merging of hypotheses. Cell Calcium 55, 175–181. http://dx.doi.org/10.1016/j.ceca.2014.02.005

Wang, Y., Zhang, Y., Zhang H., Liu D., Liu, Z., Cheng, X., Lu, Y., Jia J., Liu, J. & Pan, F. 2014. The toxic effect of lithium ion on neurons (PC12 cells) and Aβ42 molecules. Biol. Trace Elem. Res. 159, 410–415. https://doi.org/10.1007/s12011-014-9949-z

Wei, X., Yan, X., Yang, Z., Han, G., Wang, L., Yuan, F. & Wang, B. 2020. Salt glands of recretohalophyte *Tamarix* under salinity: Their evolution and adaptation. Ecol. Evolut.10, 9384–9395. https://doi.org/10.1002/ece3.6625

Yamada, M., Kuroda, C. & Fujiyama, H. 2016. Function of sodium and potassium in growth of sodium-loving Amaranthaceae species. Soil Sci. Plant Nutr. 2016, 20–26. http://dx.doi.org/10.1080/00380768.2015.1075365

Zhang, H., Huo, Y., Guo, K., Xu, Z., Liu, S., Wang, Q., Wang, X., Xu, N., Wu, Y. & Sun, G. 2021. Na^+ accumulation alleviates drought stress induced photosynthesis inhibition of PSII and PSI in leaves of *Medicago sativa*, J. Plant Interact. 16:1, 1–11. https://doi.org/10.1080/17429145.2020.1866091

Zhang, Y., Xu, W., Yang, T. & Huanf, F. 2022. Protective effect of lithium chloride on pulmonary injury caused by *Actinobacillus pleuropneumniae* inhibition of GSK-3b-NF-kB-dependent pathway. Polish J. Vet. Sci. 25, 35–44. DOI: 10.24425/pjvs.2022.140838

Zhao, B., Zhou, Y., Jiao, X., Wang, X., Wang, B. & Yuan, F. (2023) Bracelet salt glands of the recretohalophyte *Limonium bicolor*: Distribution, morphology, and induction. Bracelet salt glands of the recretohalophyte *Limonium bicolor*: Distribution, morphology, and induction. J. Integr. Plant Biol. 00: 17 pp. (early access). https://doi.org/10.1111/jipb.13417

CHAPTER TWO

ALKALINE EARTHS

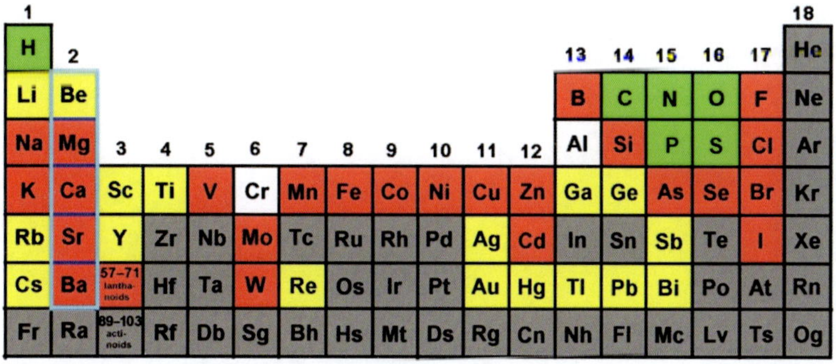

2.1 Beryllium

Beryllium is an element that is toxic to plants (Ali et al. 2018, Sheteiwy et al. 2022) as well as animals including humans. The initiation and progress of chronic beryllium disease are described by MacMurdo et al. (2020). Between 134,000 and 800,000 workers are exposed to beryllium in the workplace within the United States (beryllium ceramics, machining, rod and wire production, production of computers and other electronic devices), and more than one million people are exposed to beryllium dust in the United States annually (Infante et al. 2004), with 2–20% of them developing beryllium sensitization. But there are also other sources of beryllium poisoning. Many coals contain considerable amounts of beryllium, and burning such coal can result in health risk (Petanová & Bencko 2020). The beryllium in coal is to a small extent due to beryllium-uptake in the coal-forming plants, but most of the beryllium is derived from volcanic ash (Yang 2007). Concrete dust may sensitize people against beryllium (MacMurdo et al. 2021). Beryllium does not enter the body only through mouth and nose: Also uptake through the skin by contact with metallic objects has been documented (Magnano et al. 2022).

Twelve isotopes (^{5}Be to ^{16}Be) of beryllium are known, most of them radioactive, with half-lives from 5×10^{-21} s (^{6}Be) to 1.39×10^{6} years (^{10}Be);

only ^{9}Be is stable. An important radioactive isotope is ^{7}Be (Fig, 2.1.1), with a half-life of about 53.22 days (but somewhat dependent on the environment). Despite its short life, it is important because it is continuously formed by reaction between cosmic rays and nuclei of oxygen and nitrogen in the Earth's atmosphere. Arnold & Al-Salih (1955) predicted that this would happen, and found ^{7}Be in rainwater and snow.

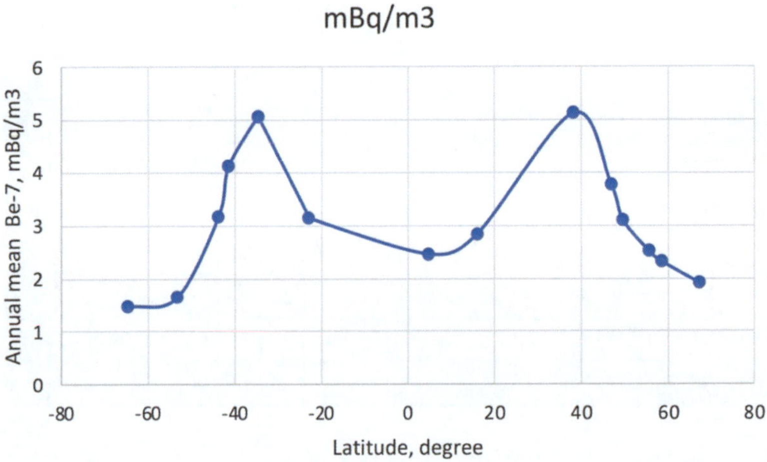

Fig. 2.1.1. The average distribution of ^{7}Be radioactivity in air (one outlying value deleted). The combined effect of production rate and air currents cause the ^{7}Be maxima to be located at ca 35°S and 35°N. Redrawn from data of Villarreal et al. (2022).

^{7}Be has been useful for studying air currents in the atmosphere, surface soil particle migration, and interception of particles in the air by plant canopies.

^{7}Be decays through electron capture (i.e., electron transferred to the atom's electron cloud to the nucleus, mainly from the K shell, and to a few percent from the L shell), a process whose rate to a slight extent depends on the atom's environment, e.g., whether it occurs as pure beryllium metal, in an alloy, or in an organic compound (Gholamian et al. 2021, Ray et al. 2020, Tkalya et al. 2021). Inside a C36 fullerene cage the half-life is about 2.5% larger than outside. The capture rate also to some extent depends on electrical field and pressure. A gamma photon is emitted when an electron falls from an outer orbital to the empty place in the K (or L) shell.

Chronic beryllium disease is associated with DNA methylation and gene expression changes (Yang et al. 2019). It is often mistaken for beryllium-unrelated sarcoidosis (Müller-Quernheim 2006).

2.2 Magnesium

Magnesium, as free Mg^{2+} ions, or bound to organic compounds, are essential for all organisms. We shall start by mentioning its presence in all forms of chlorophyll, since without chlorophylls and photosynthesis in which they play a crucial role, this planet would be very different, and not inhabited by us. Magnesium is inserted into the chlorophyll precursor protoporphyrin IX by the enzyme magnesium chelatase with the help of ATP (Fig. 2.2.1). Further steps leading to chlorophylls (including bacteriochlorophyll) are described by Willows (2019).

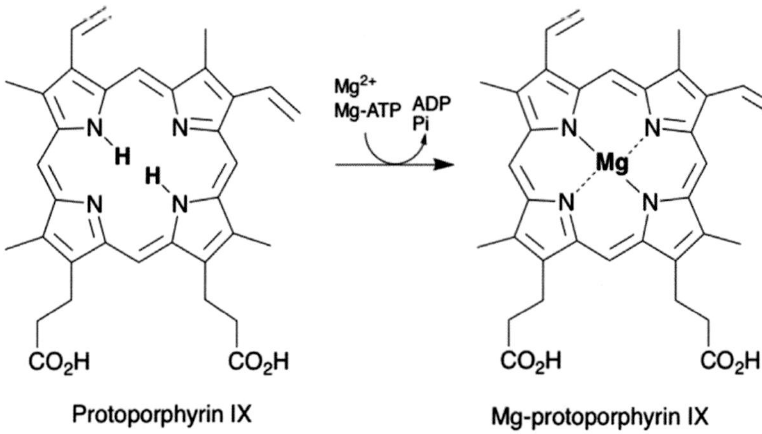

Fig. 2.2.1. The insertion of magnesium into protoporphyrin IX.

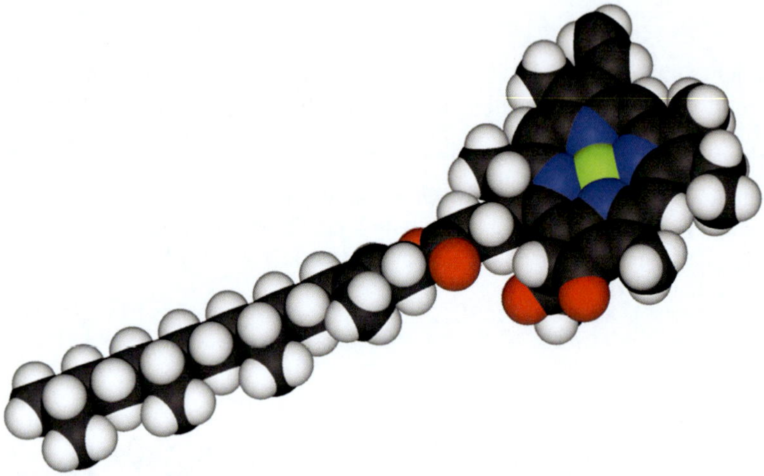

Fig. 2.2.2. Model of a chlorophyll *a* molecule. Carbon black, hydrogen light grey, nitrogen blue, oxygen red, magnesium green. The long extension to the left is the alcohol phytol, forming an ester with the acid chlorophyllide *a*, to the right. Wikipedia (https://upload.wikimedia.org/wikipedia/commons/9/92/Chlorophyll-a-3D-vdW.png).

What is commonly referred to as ATP, is, in most cases, a complex between ATP and a magnesium ion (Fig. 2.2.3). The magnesium ion can be coordinated either with two or three of the phosphate groups. Buelens et al. (2021) found some preference for two groups, although according to Dudev et al. (2017) Mg^{2+} prefers to bind ATP tridentately to each of the three phosphate groups, and Liao et al. (2004) found both arrangements equally probable. However, in free solution with pH<6, magnesium ions are displaced by hydrogen ions, with far-reaching effects (Wolf & Cittadini 2003).

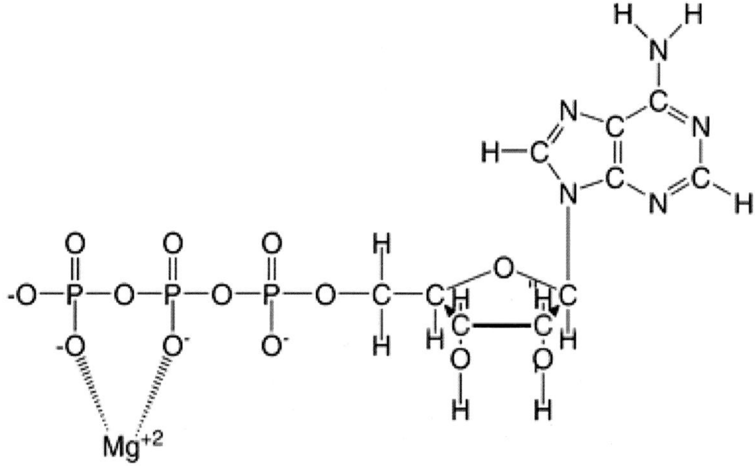

Fig. 2.2.3. ATP-Mg with the magnesium ion coordinated with the two terminal phosphate groups, the most common configuration. The magnesium ion can also be coordinated with all three phosphate groups.

Mg^{2+} ions act as activator of over 300 different enzymes (Pasternak et a. 2010). See Table 2.2.1 for the major classes. The magnesium ions do not function in the same way for different enzymes. As shown in Table 1, the ions sometimes bind to ATP or GTP, in others to the enzyme, or to the substrate.

Table 2.2.1 Major classes of Mg-requiring enzymes, adapted from Williams, R.J.P. (1993) by Wolf & Cittadini (2003).

Enzyme	Requirement
Kinases	ATP–Mg phosphate transfer
G-proteins	GTP–Mg phosphate transfer
Adenylate-cyclase	ATP–Mg substrate
ATPases	ATP–Mg substrate
Alkaline phosphatase	Mg binds and activates the enzyme
Enolase	Mg binds the enzyme
Isocitrate lyase	Mg–isocitrate substrate
Glutamine synthase	Mg in the enzyme
Methyl aspartase	Mg binds the enzyme
Pyrophosphatase	Mg bound to substrate
Xylose isomerase	Mg binds the enzyme
Ribulosephosphate carboxylase	Mg in the enzyme

In some cases Mg^{2+} to some extent can be replaced by certain other divalent cations, in particular by Mn^{2+}, and to a lesser degree by Co^{2+} (Table 2.2.2).

However, magnesium often behaves in another way than does manganese or cobalt. For glucose/xylose isomerase from *Streptomyces* cobalt is a more efficient activator than magnesium at low concentrations of the ion, but at saturating ion concentration magnesium is twice as activating as cobalt (Kasumi et al. 1982). For isocitrate lyase from *Phycomyces* both the magnesium and manganese activate the enzyme at pH 7.3, but at pH 8.5 manganese still acts as an activator, while magnesium inhibits the reaction (Rúa et al. 1997).

Table 2.2.2 Effect of metal cations on isocitrate activity (Data from Hoyt et al. 1988)

Salt in medium, 5 mM	Relative isocitrate lyase activity
None	0
$MgCl_2$	100
$MnCl_2$	54
$CoCl_2$	17
$NiCl_2$	7
$SrCl_2$	3
$CaCl_2$	0
$ZnCl_2$	0
$CuCl_2$	0
$FeCl_2$	0
$FeCl_3$	0

A large part of the intracellular Mg^{2+} is bound to ribosomes or polynucleotides (Wolf & Cittadini 2003).

In the ocean magnesium is not a limiting nutrient, but on land there are areas where animals as well as plants may suffer from magnesium deficiency if it is not ameliorated by human intervention. The magnesium supply for humans has improved much since 1992, but was still low 2011 in some countries, e.g., Colombia, Madagascar, and Thailand (Kumssa et al. 2015). Even in some countries where the magnesium status in the population in general is good, some groups may be undernourished with respect to magnesium. Such people can be found among alcoholics (Vanoni et al. 2021), and old people are also thought to be another risk group (Veronese & Barbagallo 2021). One reason for this is that low serum magnesium level is associated with risk of dementia (Kieboom et al. 2017, Alam et al. 2020). Fig. 4 shows a map of magnesium availability in soil.

Fig. 2.2.5, from Pelczýnska et al. (2022), based on Medical Art collection, http://smart.servier.com/, shows the normal intake, distribution, and excretion of magnesium in adult men and women. Insufficient intake or absorption of magnesium is not uncommon. For instance, Mejía-Rodríguez et al. (2013) found that in Mexico 36.3% of the women and and 31.0% of the men had too little magnesium, and Mexico is no exception in this respect. However, diagnosing magnesium deficiency is not easy. Normal plasma magnesium concentration does not exclude magnesium depletion (e.g., Ismail et al. 2010). Several other tests to probe magnesium status have been proposed, and especially analysis of erythrocyte (red blood corpuscle) magnesium content is one common method, but also this has been criticized (Basso et al. 2000). Muscle magnesium appears to be a good marker (Arnaud 2008), but need for biopsy limits the usefulness of such testing.

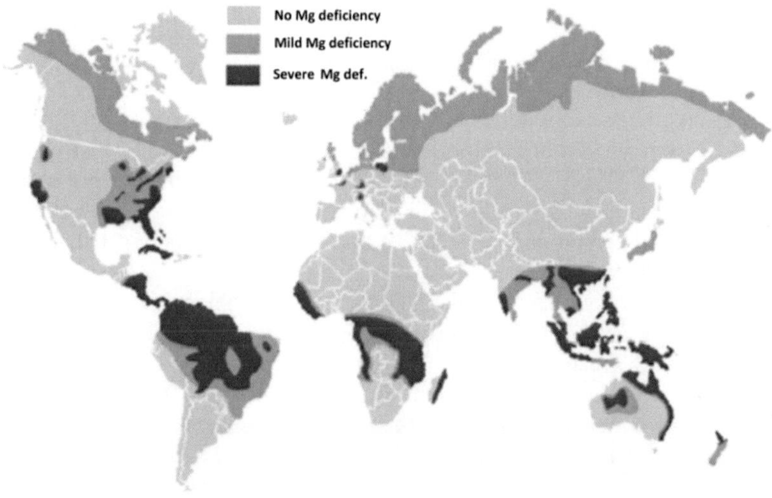

Fig. 2.2.4. This map depicts the magnesium availability in the soil. It is not the same as the distribution of the nutritional magnesium status of the human population. The darker the more deficiency. Adapted from https://mobile.twitter.com/DanakaliLtd/status/1483037486507073539/photo/1

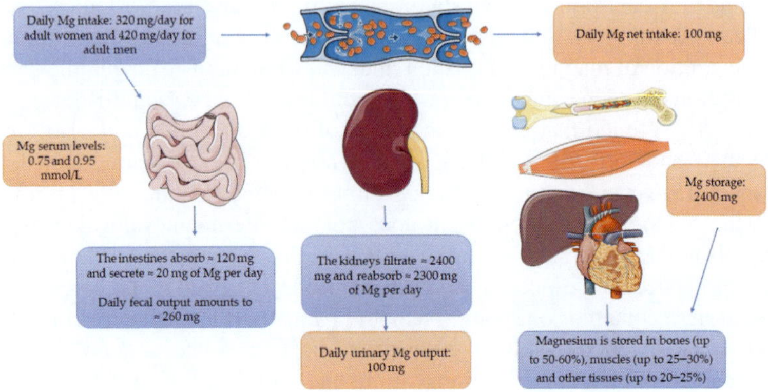

Fig. 2.2.5. Typical intake, distribution, and excretion of magnesium in men and women. From Pelczýnska et al. (2022), based on Medical Art collection, http://smart.servier.com/. Reproduced under the Creative Commons Attribution (CC BY) license (https://creativecommons.org/licenses/by/4.0/).

Human magnesium deficiency is associated with many medical issues: Risk of systemic inflammation and cardiovascular mortality (Fan et al. 2021), obesity and type 2 diabetes mellitus (Piuri et al. 2021, Zahra et al. 2021), arterial hypertension with cardiovascular and cerebrovascular complications (Dominguez et al. 2020), and dementia (Kieboom et al. 2017, Alam et al. 2020).

Magnesium deficiency in plants is a common problem (see Fig. 4). As one can suspect from the composition of chlorophyll, deficiency becomes visible as light green or yellow coloration of the leaves. Also other mineral deficiencies or lack of sufficient light result in this, but typical for magnesium deficiency (although this feature is shared with, e.g., iron deficiency) is that the parenchymatic cells next to the veins remain green (Fig. 2.2.6).

Fig. 2.2.6. Mineral deficient fig (*Ficus carica*) leaves. (a) Control (full nutrition); (b) N deficiency; (c) P deficiency; (d) K deficiency; deficiency; (e) Ca deficiency; (f) Mg deficiency. From Garza-Alonso et al. (2019) under Creative Commons Attribution (CC BY) license (http://creativecommons.org/licenses/by/4.0/).

2.3 Calcium

Calcium has so many biological functions that they cannot even be enumerated here. We shall first mention one of its functions, on which much of life on Earth depends, but which is perhaps not generally known: Its role in the photosynthetic water-oxidizing centre, also known as the oxygen-evolving centre, OEC (Fig. 2.3.1).

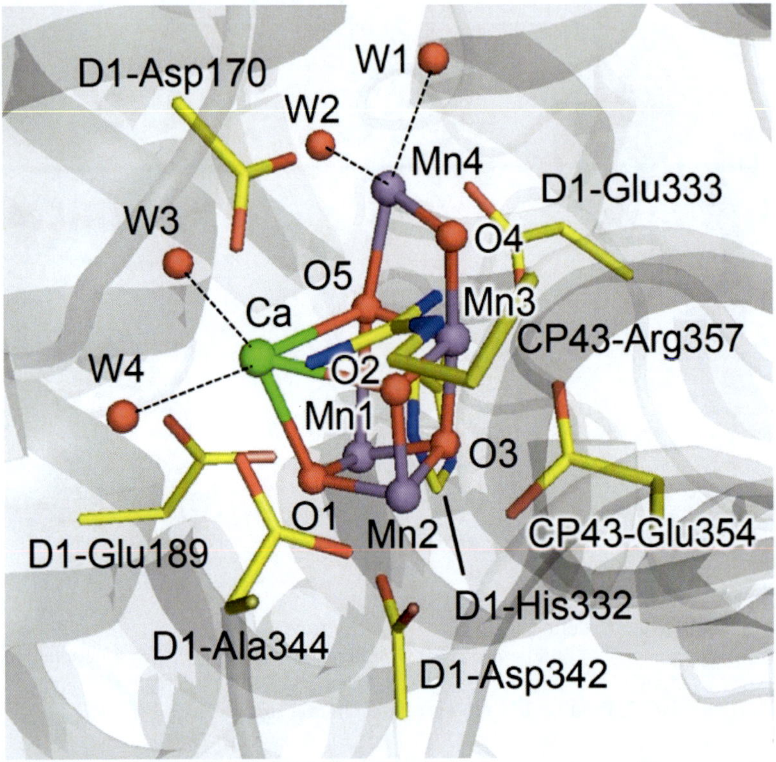

Fig. 2.3.1. The oxygen-evolving centre in an organism with oxygenic photosynthesis. Reproduced from Mandal et al. (2022) under the Creative Commons Attribution 4.0 License (https://creativecommons.org/licenses/by-nc-nd/4.0/).

Many enzymes and other proteins contain calcium, where it has two kinds of functions. It can keep the protein in a certain configuration, but it can also have a more direct role in the catalytic function of an enzyme.

Calcium ions can also have a signaling function. Often, they exert this function together with the protein calmodulin, which is present in practically all cells of all eukaryotes. The calcium-calmodulin has many functions. For instance, it is involved in muscle contraction, in the conversion of glycogen to glucose, and in various memory functions.

Finally, calcium forms part in many kinds of skeleton. Human and other vertebrate skeletons consist, to a large part, of calcium phosphate (calcium hydroxyapatite, $Ca_{10}(PO_4)_6(OH)_2$, to be more exact). Ruben & Bennett (1987) suggest that, for vertebrates, calcium phosphate was preferred as skeletal material over calcium carbonate because it is resistant to low pH.

Vertebrates produce pH-lowering lactic acid during bursts of muscle activity. However, calcium carbonate is used by some other organisms, such as in crustacean mandibles, frequently in the form of crystalline fluorapatite (Bentov et al. 2016).

Many other skeletons, such as those of stony corals, consist mainly of calcium carbonate, or forms a considerable part of the skeleton, as in crustaceans. Certain red algae, so called coralline algae (Corallinaceae, Corallinales) are among the the first multicellular organisms forming calcium carbonate skeletons. As science progresses, the record of coralline algae seems to stretch further and further back in time. Tomás et al. (2007) describe Late Hauterivian (i.e. early Cretaceous, ca 130 million years old) coralline algae from Spain. Aguirre et al. (2010) use various methods for constructing a phylogenetic tree with a similar origin (138.40 Ma ago) for the group. Teichert et al. (2019) describe fossils of coralline algae from my home island (Gotland) that are about 300 million years older, from the Silurian. Chen et al. (2020) write about reef-building red algae from the Permian. Finally, Zhang & Zhang (2022) describe a fossil as probably being a coralline alga, and date it to the Ediacaran, i.e. more than 541 million years ago.

But formation of calcium carbonate is carried out also by procaryotes, e.g., by the thermophilic soil bacterium *Geobacillus stearothermophilus* DSM 13240 (Simon et al. 2022). Sponges are an ancient animal group in which many species contain structures of calcium carbonate (or silica, or both silica and calcium carbonate). Hypercalcified sponges were ubiquitous Silurian —Devonian reef builders (Balthasar et al. 2021). There is circumstantial evidence that the structure in the ancient sponges had an original composition of high-Mg calcite. Uriz et al. (2012) as well as Garate et al. (2017) describe calcifying bacteria living in association with sponges, and suspect that there may be a bacterial origin of calcification in metazoans. Sakowski et al. (2020) suspect that bacteria are involved in the calcification of oyster shells.

2.4 Strontium and barium

2.4.1. Barium in plants and animals

For mammals barium is very toxic, because it blocks potassium channels (e.g., Bhoelan et al. 2014). "To a K^+ channel, the alkaline earth metal Ba^{2+} is like a divalent K^+ ion. Barium's size allows it to fit into the selectivity filter, but its charge apparently causes it to bind too tightly. When it is bound in the filter, Ba^{2+} prevents the rapid flow of K^+." (Jiang & MacKinnon 2000). The

LD50 values for barium chloride in rats range from 132 to 277 mg barium/kg (Borzelleca et al. 1988; Tardiff et al. 1980). Because of their beautiful light emission when heated, both strontium (red light) and barium (green light) have been used in fireworks and for signalling. Because of its toxicity, barium (main emission ≈ 560 nm) has, for this use, to some extent been replaced by boron.

The best-known biological role of barium is as barium sulfate statoliths, which sense the direction of gravity in Charophyceae, a group of algae (Schröter et al. 1975, Steudle et al. 1978, Sievers & Schmitz 1982, Braun 2001) and perhaps other algae (Brook et al. 1980, 1988) that also accumulate this element, and perhaps also strontium is used by Cnidaria for this purpose (Toyokawa et al. 2018).

Crystals of barium sulfate occur also in other algae. In *Spirogyra* they are localized in the cytoplasm, not in the cell wall (Kreger & Boeré 1969). *Ankistrodesmus sp.* and *Selenastrum* accumulate barium several thousand times from their media. Brook et al. (1990) found particles containing barium in cells of almost all members of Chlorophyta; *Spirogyra sp.* and *Ulothrix sonata* are significant barium accumulators (Mann & Fyfe 1984). Following the disaster at the Fukushima Daiichi nuclear power plant in 2011, it was found that *Bryopsis maxima* accumulated (among other radioactive nuclides) ^{85}Sr and ^{140}Ba. The total content of strontium (stable nuclide) in this alga is 15.9 g and of barium 3.79 g per kg dry matter, while the content of calcium is only 12.5 g per kg dry matter. The rate of uptake of ^{85}Sr was determined to be 0.13 g per kg dry matter per day (Takahashi et al. 2015). The concentrations of strontium and barium in seawater vary, but the concentration of strontium is several hundred times, according to Bowen (1956) more than a thousand times that of barium. Thus *B. maxima* concentrates barium to a higher degree than strontium. In an investigation (Bowen 1956), not including the exceptional *B. maxima*, the ratio between strontium and barium by weight was, on average, for brown algae about 80, for red algae about 40, and for marine green algae about 90. So the preference for barium (in relation to the concentrations in seawater) appears to be common in marine algae.

Xenophyophorea (Fig. 2.4.1) are a group of deep sea foraminifera with intracellular barium sulphate crystals (Hopwood et al. 1997).

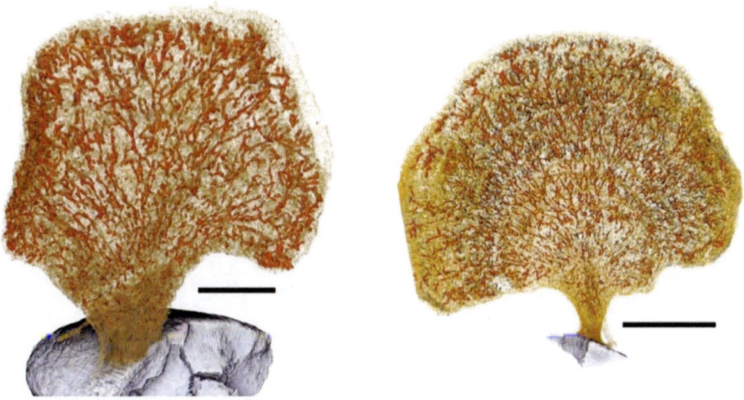

Figure 2.4.1. Two tests (skeletons) of *Psammina aff. limbata*, belonging to Xenophyophora, Foraminifera. From Gooday et al. (2018), Creative Commons Attribution 4.0 International License, http://creativecommons.org/licenses/by/4.0/

2.4.2. Strontium in animals

Strontium has been studied intensively, partly because of the nuclide ^{90}Sr that is generated in nuclear power plants and by use of nuclear weapons. ^{90}Sr has a physical half-life of 28.79 years (10516 ± 21 days; Woods & Lucas 1996). It is taken up in the skeleton (mistaken for calcium), where it has a biological half-life of 13.5 ± 1.0 years (Froidevaux et al. 2010). It is one of the most radiotoxic nuclides produced by nuclear fission. It undergoes β-decay to yttrium-90 (^{90}Y), which in turn undergoes β-decay to stable zirconium with a half-life of 64 hours. ^{90}Sr causes bone cancer, but since administration of stable zirconium citrate reduces the incidence (Zander-Principati & Kuzma 1964), much of the cancer seems to be caused by β-particles from ^{90}Y. Strontium is transferred to offspring via mother's milk (e.g., Della Rosa et al. 1996).

In mammals stable strontium is much less toxic than barium, but in high concentrations it can cause chromosomal aberrations in bone marrow cells (Ghosh et al. 1990). It has a number of applications in medicine (e.g., Wu et al. 2021, Song et al. 2022) and dentistry (e.g., Dai et al. 2021, Dhivya et al. 2021). Alloys containing strontium have been recommended for certain medical implants (Jiang et al. 2018).

Acantharia (Acantharea) (Fig. 2.4.2, Fig. 2.4.3) are the best investigated type of organism for which strontium is an essential element. They are a kind of radiolaria, microscopic protozoa, and the main component in their skeletons is strontium sulfate (Müller 1858, Odum 1951, Brass 1980, Fujimaki et al. 2022). Another group of radiolaria, the Polycystinea, have silica skeletons, but the reproductive swarmers accumulate strontium and precipitate crystals of strontium sulfate (Anderson et al. 1990).

In corals, on the contrary, strontium is to about 40% present as strontianite, a form of strontium carbonate; the rest being together with calcium in aragonite, also a carbonate (Greegor et al. 1997).

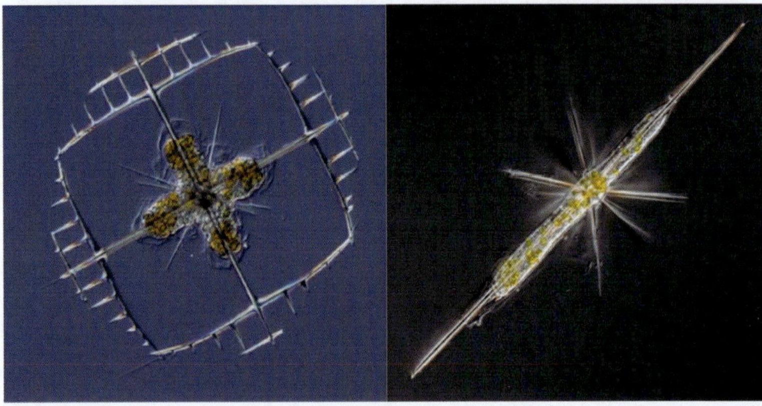

Figure 2.4.2. Two members of Acantharia: Left, *Lithoptera fenestra*; right *Amphilonche elongata*. From Biard (2022), Creative Commons Attribution-Non-Commercial License.

2.4.3. Strontium and barium in the food chain

As elements pass through the food-chain, calcium is preferentially retained over strontium and barium; Balter (2004) gives several examples of this. Values for strontium, originally obtained from Elias et al. (1982), are shown in Fig. 2.4.4. The preference for calcium over barium is even greater.

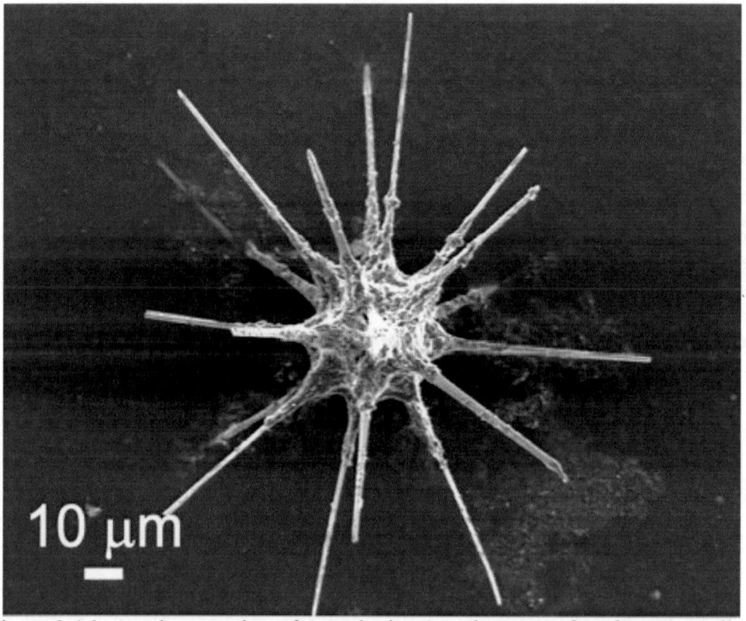

Figure 2.4.3. Another member of Acantharia: *Acanthometra cf. multispina* Müller. From Fujimaki et al. (2022), Creative Commons Attribution 4.0 International License http://creativecommons.org/licenses/by/4.0/ .

2.4.4. Importance of strontium for mammal and human bone and teeth

In the middle of the past century researchers started to pay attention to the importance of strontium for skeleton and teeth, e.g., Shorr & Carter (1952), Sawchuck (1953). The older literature is reviewed by Nielsen (2004). Strontium ranelate strongly stimulates human cartilage matrix formation in vitro (Henrotin et al., doi:10.1016/j.bone.2004.04.0262001). Due to low fracture toughness and poor strength of calcium phosphate (a main component of bone), strontium compounds are added to improve the properties of bone (review by Demirel & Kaya 2020). At the start of this century strontium ranelate, the strontium salt of ranelic acid, began to be used for delivering strontium. Since then the literature has expanded rapidly: In mid-2022 more than 2000 articles on strontium ranelate, with more than 26000 citing articles, are listed in the Web of Science core

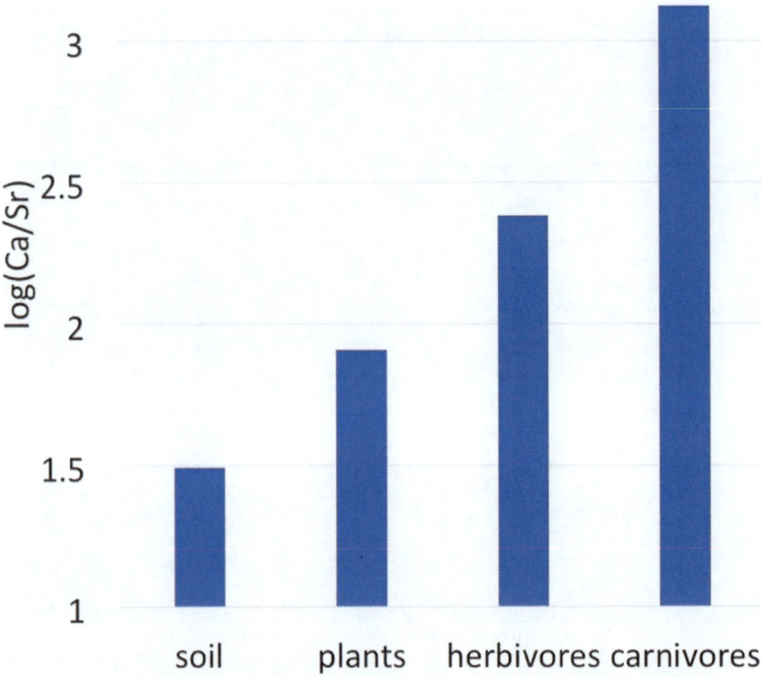

Figure 2.4.4. Calcium to strontium molar ratios for soil, plants, herbivores, and carnivores from a remote site in Yellowstone National Park, USA. Data originally from Elias et al. (1982), extracted from Balter (2004).

collection. Martín-del-Campo et al. (2019) recommend the use of strontium folate instead of strontium ranelate.

Strontium ions act on two types of bone cells: (1) Osteoblasts that construct bone by producing collagen type I, proteoglycans, osteocalcin, osteonectin, osteopontin, and sialoproteins, and (2) osteoclasts that dissolve bone by secreting hydrolytic enzymes. Strontium has the dual function of stimulating osteoblast formation and function, inhibiting formation of osteoclasts (e.g., Bonnelye et L. 2008, Marx et al. 2020, Kołodziejska et al. 2021), and induces osteoclast apoptosis (Hurtel-Lemaire et al. 2009). It thus stimulates the formation and maintenance of bone. These results have been obtained in vitro. However, it is not clear that the same holds in vivo, and if strontium is administrated orally it may not hold (Chavassieux et al. 2013, Blake & Fogelman 2013). One possibility for misinterpretation when results are evaluated by use of x-rays is that the presence of strontium renders specimens more opaque to x-rays even if the strength of the bone has not improved. Anyway, from a systematic review and meta-analysis Yan et al.

(2022) conclude that strontium supplementation is advantageous in terms of promoting new bone formation and accelerating material degradation. These effects of strontium are explained by the effects on calcium-sensing receptor of osteoclasts and osteocytes (see Stepan 2013 for literature).

Overdosage of strontium may induce induce rickets and defective bone mineralization, possibly by reducing 25-hydroxyvitamin D α-hydroxylase activity (Omdahl et al. 1972).

Many studies have been done on the effect of strontium on teeth. A majority of studies suggest that strontium has cariostatic properties, predominantly in the presence of fluoride. Because no randomized, controlled investigations have been carried out, no certain connection has been established (Lippert & Hara 2013).

2.5 Radium

All radium isotopes are radioactive and have no known biological function. The cyanobacterium *Gleomargarita lithophora* is able to accumulate radium, and has been proposed as a means for removal of radium pollution (Mehta et al. 2019, 2022). In plants (ryegrass and clover) radium is taken up to a fifth the amount of barium under the same conditions (Vandenhove et al. 2005). In people radium is retained in the skeleton, its chronic accumulation results in an ongoing source of internal ionising radiation, increasing the long-term risks of cancers, such as bone sarcomas (Pearson et al. 2022; for further information and references see this article).

References

Aguirre, J., Perfectti, F. & Braga, J.C. 2010. Integrating phylogeny, molecular clocks, and the fossil record in the evolution of corallinealgae (Corallinales and Sporolithales, Rhodophyta. Paleobiol. 36, 519–533. https://doi.org/10.1666/09041.1

Alam, A.B., Lutsey, P.L., Gottesman, R.F., Tin, A. & Alonso, A. 2020. Low serum magnesium is associated with incident dementia in the ARIC-NCS cohort. Nutrients 12, 3074 (11 pp.). https://doi.org/10.3390/nu12103074

Ali, S., Jin, R., Gill, R.A., Mwamba, T.M., Zhang, N., Hassan, Z. ul, Islam, F., Ali, S. & Zhou, W. 2018. Beryllium stress-induced modifications in antioxidant machinery and plant ultrastructure in the seedlings of black and yellow seeded oilseed rape. BioMed. Res. Internat. 2018: 1615968, (14 pp.). https://doi.org/10.1155/2018/1615968

Amicosante, M. & Fontenot, A.P. 2006. T cell recognition in chronic beryllium disease. Clin. Immunol. 121, 134–43.

Anderson, O.R., Perry, C.C. & Hughes, N.P 1990. Transmission and scanning electron microscopic evidence for cytoplasmic deposition of strontium sulphate crystals in colonial radiolaria. Phil. Trans. R. Soc. Lond. B 329, 81–86.

Arnaud, M.J. 2008. Update on the assessment of magnesium status. British J. Nutrit, 99, Suppl. 3, S24–S36. https://doi.org/10.1017/S000711450800682X

Arnold, J.R. & Al-Salih, H.A. 1955. Beryllium-7 produced by cosmic rays. Science 121, 451–453.

Balthasar, U., Kershaw, S., Da Silva, A.-C., Seuss, B., Cusack, M., Eichenseer, K. & Chung, P. 2021. Palaeozoic stromatoporoids and chaetetids analysed using electron backscatter diffraction (EBSD); implications for original mineralogy and microstructure. Facies 67:8 (18 pp.). https://doi.org/10.1007/s10347-020-00618-5

Balter, V. 2004. Allometric constraints on Sr/Ca and Ba/Ca partitioning in terrestrial mammalian trophic chains. Oecologia 139, 83–88. https://doi.org/10.1007/s00442-003-1476-0

Basso, L.E., Ubbink, J.B. & Delport, R. 2000. Erythrocyte magnesium concentration as an index of magnesium status: a perspective from a magnesium supplementation study. Clinica Chimica Acta 291, 1–8.

Bentov, S., Aflalo, E.D., Tynyakov, J., Glazer, L. & Sagi, A. 2018. Calcium phosphate mineralization is widely applied in crustacean mandibles. Sci. Rep. 6:22118 (10 pp.). https://doi.org/10.1038/srep22118

Berger, R.J.F., Håkansson, P. & Mera-Adasme, R. 2020. A consistent model for the key complex in chronic beryllium disease. Z. Naturforsch. 75b, 413–419. https://doi.org/10.1515/znb-2020-0010*h*

Bertazzo, S.; Bertran, C. A. 2006. "Morphological and Dimensional Characteristics of Bone Mineral Crystals". *Key Engineering Materials*. 309–311: 3–6. doi:10.4028/www.scientific.net/KEM.309-311.3.

Bhoelan, B.S., Stevering, C.H., van der Boog, A.T.J. & van der Heyden, M.A.G. 2014. Barium toxicity and the role of the potassium inward rectifier current. Clin. Toxicol. *52*, 584–593. https://doi.org/10.3109/15563650.2014.923903

Biard, T. 2022. Diversity and ecology of Radiolaria in modern oceans. Environm. Microbiol. 24, 2179–2200.

Blake, G.M. & Fogelman, I. 2014. Strontium ranelate does not have an anabolic effect on bone. Nature Revs Endocrinol. *9*, 696–697.

Bonnelye, E., Chabadel, A., Saltel, F. & Jurdic, P. 2008. Dual effect of strontium ranelate: Stimulation of osteoblast differentiation and

inhibition of osteoclast formation and resorption in vitro. Bone 42, 129–138.
Borzelleca, J.F., Condie, L.W. Jr & Egle, J.L. Jr. 1989. Short-term toxicity (one- and ten-day gavage) of barium chloride in male and female rats. J. Am. Coll. Toxicol. 7, 675–685.
Bowen, H.J.M. 1956. Strontium and barium in sea water and marine organisms. J. Mar. Biol. Ass. U.K. 35, 451–460.
Brass, G.W. 1980. Trace elements in acantharian skeletons. Limnol. Oceanogr. 25, 146–149.
Braun, M. 2001. Gravisensing in single-celled systems: Characean rhizoids and protonemata. Adv. Space Res. 27, 1031–1039.
Brook, A.J., Fotheringham, A., Bradly, J. & Jenkins, A. 1980. Barium accumulation by desmids of the genus *Closterium* (Zygnematophyceae). Br. Phycol. J. 15, 261–264.
Brook, A.J., Grime, G.W. & Watt, F. 1988. A study if barium accumulation in desmids using the Oxford scanning proton microprobe (SPM). Nuclear Instrum. Meth. Physics Res. B30, 372–377.
Brook, A.J., Watt, F. & Grime, G.W. 1990. A contribution to the study of biogeochemical cycling in an English lake by PIXE: The flux of barium and strontium in the phytoplankton. Nuclear Instrum. Meth. Physics Res. B49, 481–484.
Buchner, M.R. 2020. Beryllium-associated diseases from a chemist's point of view. Z. Naturforsch. 75b, 405–412. https://doi.org/10.1515/znb-2020-0006
Buelens, F.P., Leonov, H., de Groot, B.L. & Grubmüller, H. 2021. ATP−magnesium coordination: Protein structure-based force field evaluation and corrections. J. Chem. Theory Comput. 17, 1922−1930. https://dx.doi.org/10.1021/acs.jctc.0c01205
Chavassieux, P., Meunier, P.J., Roux, J.P., Portero-Muzy, N., Pierre, M. & Chapurlat, R. 2014. Bone histomorphometry of transiliac paired bone biopsies after 6 or 12 months of treatment with oral strontium ranelate in 387 osteoporotic women: Randomized comparison to Alendronate. J. Bone Mineral Res. 29, 618–628. https://doi.org/10.1002/jbmr.2074
Chen, W., Wang, Y., Huang, Y., Wang, T., Yi, Z. & Kiessling, W. 2020. Reef-building red algae from an uppermost Permian reef complex as a fossil analogue of modern coralline algal ridges. Facies 66:22 (14 pp.). https://doi.org/10.1007/s10347-020-00606-9
Dai, L.-L., Mei, M.-L., Chu, C.-H., Zhao, I.S. & Lo, E.C.-M. 2021. Effect of strontium-doped bioactive glass on preventing formation of demineralized lesion. Materials 14: 4645 (10 pp.).

https://doi.org/10.3390/ma14164645
Della Rosa, R.J., Wolf, H.G. & Goldman, M. 1968. Translocation of ^{90}Sr from maternal skeleton to progeny during gestation and lactation. Proc. First. Internat. Congr. Radiation Prot. 201–204. https://doi.org/10.1016/B978-1-4832-8312-8.50036-9
Demirel, M. & Kaya, A.I. 2020. Effect of strontium-containing compounds on bone grafts. J. Mater. Sci. 55, 6305–6329. https://doi.org/10.1007/s10853-020-04451-7
Dhivya, V., Mahalaxmi, S., Rajkumar, K., Premkumar, V.V., Saravanakarthikeyan, B., Karpagam, R., Priyatharshini, R., Sakthipandi, K., Saikumari, V., Vijay, N. & Rajkumar, G. 2021. Effects of strontium-containing fluorophosphate glasses for enhancing bioactivity and enamel remineralization. Materials Characteriz. 181: 111496 (12 pp.). https://doi.org/10.1016/j.matchar.2021.111496
Ding, J., Lin, L., Hang, W. & Yan, X. 2009. Beryllium uptake and related biological effects studied in THP-1 differentiated macrophages. Metallomics 1, 471–478.
Dominguez, L.J., Veronese, N. & Barbagallo, M. 2020. Magnesium and hypertension in old age. Nutrients 13, 139. https://doi.org/10.3390/nu13010139
Dudev, T., Grauffel, C. & Lim, C. 2017. How native and alien metal cations bind ATP: Implications for lithium as a therapeutic agent. Sci. Rep. 7:42377 (10 pp.). https://doi.org/10.1038/srep42377
Elias R.W., Hirao, Y. & Patterson, C. 1982. The circumvention of the natural biopurification of calcium along nutrient pathways by atmospheric inputs of industrial lead. Geochim. Cosmochim. Acta 46, 2561–2580.
Froidevaux, P., Bochud, F. & Haldimann, M. 2010. Retention half times in the skeleton of plutonium and 90Sr from above-ground nuclear tests: A retrospective study of the Swiss population. Chemosphere 80, 519–524. https://doi.org/10.1016/j.chemosphere.2010.04.049
Fujimaki, R., Suzuki, N., Kimoto, K., Nagai, Y., Oaki, Y., Shimode, S., Toyofuku, T. & Imai, H. 2022. Micro- and nanometric characterization of the celestite skeleton of acantharian species (Radiolaria, Rhizaria). Scientific Rep. 12:2848 (9 pp.). https://doi.org/10.1038/s41598-022-06974-2
Garate, L., Sureda, J., Agell, G. & Uriz, M.J. 2017. Endosymbiotic calcifying bacteria across sponge species and oceans. Sci. Rep. 7:43674 (14 pp.). https://doi.org/10.1038/srep43674
Garza-Alonso, C.A., Olivares-Sáenz, E., Gutiérrez-Díez, A., Vázquez-Alvarado, R.E. & López-Jiménez. A. 2019. Visual symptoms,

vegetative growth, and mineral concentration in fig tree (*Ficus carica* L.) under macronutrient deficiencies. Agronomy 9: 787 (12 pp.). https://doi.org/10.3390/agronomy9120787

Ghosh, S., Talukder, G. & Sharma, A. 1990. Clastogenic activity of strontium chloride on bone marrow cells in vivo. Biol. Trace Elem. Res. 25, 51–56.

Gooday, A.J., Sykes, D., Góral, T., Zubkov, M.V. & Glover, A.G. 2018. Micro-CT 3D imaging reveals the internal structure of three abyssal xenophyophore species (Protista, Foraminifera) from the eastern equatorial Pacific Ocean. Scientific Rep. 8:12103 (12 pp.). https://doi.org/10.1038/s41598-018-30186-2

Greegor, R.B., Pingitore Jr., N.E. & Lytle, F.W. 1997. Strontianite in coral skeletal aragonite. Science 275, 1452–1454.

Henrotin, Y., Labasse, A., Zheng, S.X., Galais, P.; Tsouderos, Y., Crielaard, J.M. & Reginster, J.Y. 2001. Strontium ranelate increases cartilage matrix formation. J. Bone Mineral Res. 16, 299–308.

Hopwood, J.D, Mann, S. & Gooday, A.J. 1997. The crystallography and possible origin of barium sulphate in deep sea rhizopod protists (Xenophyophorea). J. Mar. Biol. Assoc. U.K. 77, 969–987.

Hoyt, J.C., Robertson, E.F., Berlyn, K.A. & Reeves, H.C. 1988. Escherichia coli isocitrate lyase: properties and comparisons. Biochim. Biophys. Acta, 966, 30–35.

Hurtel-Lemaire, A.S., Mentaverri, R., Caudrillier, A., Cournarie, F., Wattel, A., Kamel, S., Terwilliger, E.F., Brown, E.M. & Brazier, M. 2009. The calcium-sensing receptor is involved in strontium ranelate-induced osteoclast apoptosis. New insights into the associated signaling pathways. J. Biol. Chem. 284, 575–584.

Infante, P.F. & Newman, L.S. 2004. Beryllium exposure and chronic beryllium disease. Lancet 363, 415–416.

Ismail, Y., Ismail, A.A. & Ismail, A.A.A. 2010. The underestimated problem of using serum magnesium measurements to exclude magnesium deficiency in adults; a health warning is needed for ‘‘normal’’ results. Clin. Chem. Lab. Med. 48, 323–327.

Jiang, W., Cipriano, A.F., Tian, Q., Zhang, C., Lopez, M., Sallee, A., Lin, A., Cortez Alcaraz, M.C., Wu, Y., Zheng, Y. & Liu, H. 2018. In vitro evaluation of MgSr and MgCaSr alloys via direct culture with bone marrow derived mesenchymal stem cells. Acta Biomaterialia 72, 407–423. https://doi.org/10.1515/CCLM.2010.077

Jiang, Y. & MacKinnon, R. 2000. The barium site in a potassium channel by X-ray crystallography. J. Gen. Physiol. 115, 269–272. http://www.jgp.org/cgi/content/full/115/3/269

Kieboom, B.C.T., Licher, S., Wolters, F.J., Ikram, M.K., Hoorn, E.J., Zietse, R., Stricker, B.H. & Ikram, M.A. 2017. Serum magnesium is associated with the risk of dementia. Neurology 89, 1716–1722. DOI: https://doi.org/10.1212/WNL.0000000000045

Kołodziejska, B., Stepien´, N. & Kolmas, J. 2021. The influence of strontium on bone tissue metabolism and its application in osteoporosis treatment. Int. J. Mol. Sci. 22, 6564 (13 pp.). https://doi.org/10.3390/ijms22126564

Kreger, D.R. & Boere, H. 1969. Some observations on barium sulphate in *Spirogyra*. Acta Bot. Neerl. 18, 143–151.

Kumssa, D.B., Joy, E.J.M., Ander, L., Watts, M.J., Young, S.D., Rosanoff, A., White, P.J., Walker, S., Broadley, M.R 2015. Global magnesium supply in the food chain. Crop & Pasture Sci. 66, 1278–1289. http://dx.doi.org/10.1071/CP15096

Liao, J.-C., Sun, S., Chandler, D. & Oster, G. 2004. The conformational states of Mg·ATP in water. Eur. Biophys, J. 33, 29–37. https://doi.org/10.1007/s00249-003-0339-2

Lippert, F. & Hara, A.T. 2013. Strontium and caries: A long and complicated relationship. Caries Res. *47*, 34–49. DOI: 10.1159/000343008

MacMurdo, M.G. & Raed A. Dweik, R.A. 2021. Cluster of beryllium-sensitized individuals may be related to environmental exposure to beryllium-containing concrete dust. Chest 159, 900–901. https://doi.org/10.1016/j.chest.2020.09.242

MacMurdo, M.G., Mroz, M.M., Culver, D.A., Dweik, R.A. & Maier, L.A. 2020. Chronic beryllium disease – Update on a moving target. Chest 158, 2458–2466.

Magnano, G.C., Marussi, G., Adami, G., Crosera, M. & Larese Filon, F. 2022. Assessment of dermal absorption of beryllium and copper contained in temple tips of eyeglasses. Toxicol. Lett. 361, 64–71. https://doi.org/10.1016/j.toxlet.2022.04.001

Mandal, M., Saito. K. & Ishikita, H. 2022. Release of electrons and protons from substrate water molecules at the oxygen-evolving complex in photosystem II. J. Phys. Soc. Japan 91: 091012 (9 pp.). http://dx.doi.org/10.1071/CP15096

Mann, H. & Fyfe, W,S. 1984. An experimental study of algal uptake of U, Ba, V, Co and Ni from dilute solutions. Chem. Geol. 44, 385–398.

Martín-del-Campo, M., Sampedro, J.G., Flores-Cedillo, M.L., Rosales-Ibañez, R. & Rojo, L. 2019. Bone regeneration induced by strontium folate loaded biohybrid scaffolds. Molecules 24: 1660 (21 pp.). https://doi.org/10.3390/molecules24091660

Marx, D., Yazdi, A.R., Papini, M. & Towler, M. 2020. A review of the latest insights into the mechanism of action of strontium in bone. Bone Reps 12: 100273 (15 pp.).

Mehta, N., Benzerara, K., Kocar, B. D. & Chapon, V. 2019. Sequestration of radionuclides radium-226 and strontium-90 by cyanobacteria forming intracellular calcium carbonates. Environ. Sci. Technol. 53, 12639−12647. https://doi.org/10.1021/acs.est.9b03982

Mehta, N., Bougoure, J., Kocar, B.D., Duprat, E. & Benzerara, K. 2022. Cyanobacteria accumulate radium (^{226}Ra) within intracellular amorphous calcium carbonate inclusions. ACS EST Water 2, 616−623. https://doi.org/10.1021/acsestwater.1c00473

Mejía-Rodríguez, F.; Shamah-Levy, T., Villalpando, S., García-Guerra, A., Méndez-Gómez Humarán, I. 2013. Iron, Zinc, Copper and Magnesium Deficiencies in Mexican Adults from the National Health and Nutrition Survey 2006. Salud Publica Mex. 55, 275–284.

Müller, J. 1858. Über die Thalassicollen, Polycystinen und Acanthometren des Mittelmeeres. Abhandl. Preuss. Akad. Wiss. 1–62, mit 11 Tafeln (with 11 plates).

Nielsen, S.P. 2004. The biological role of strontium. Bone 35, 583– 588. https://doi.org/10.1016/j.bone.2004.04.026

Odum, H.T. 1951. Notes on the strontium content of sea water, celestite radiolaria, and strontianite snail shells. Science 114, 211–213.

Omdahl, J.L. & DeLuca, H.F. 1972. Rachitogenic activity of dietary strontium. I. Inhibition of intestinal calcium absorption and 1,25-dihydroxycholecalciferol synthesis. J. Biol. Chem. 247; 5520–5526.

Pasternak, K., Kocot, J. & Horecka, A. 2010. Biochemistry of magnesium. J. Elementol. 15, 601–616.
https://doi.org/10.5601/jelem.2010.15.3.601-616

Pearson, A.J., Gaw, S., Hermanspahn, N., Glover, C.N. & Anderson, C.W.N. 2022. Radium in New Zealand agricultural soils: Crop uptake and estimation of current and future ionising radiation dose. J. Environm. Radioactivity 244–245: 106808 (8 pp.).
https://doi.org/10.1016/j.jenvrad.2021.106808

Petanová, J. & Bencko, V. 2020. Health aspects of exposure to emissions from burning coal of high beryllium content: Interactions with the immune system. Cent. Eur. J. Public Health 28, 198–201.
https://doi.org/10.21101/cejph.a5851

Piuri, G., Zocchi, M., Della Porta, M., Ficara, V., Manoni, M., Zuccotti, G.V., Pinotti, L.; Maier, J.A. & Cazzola, R. 2021. Magnesium in obesity, metabolic syndrome, and type 2 diabetes. Nutrients 13, 320. https://doi.org/10.3390/nu13020320

Rúa, J., Soler, J., Busto, F. & de Arriage, D. 1997. Effect of pH on the role of Mg^{2+} and Mn^{2+} on *Phycomyces* isocitrate lyase kinetics. Biochimie 79, 179–186.

Ruben, J.A. & Bennett, A.A. 1987. The evolution of bone. Evolution, 41, 1187–1197.

Sakowski, E.G., Wommack, K.E. & Polson, S.W. 2020. Oyster calcifying fluid harbors persistent and dynamic autochthonous bacterial populations that may aid in shell formation. Mar. Ecol. Prog. Ser. 653, 57–75. https://doi.org/10.3354/meps13487

Sawchuck, W.B. 1953. Effects of strontium and fluoride on bone repair. J. Dental Res. 32, 680–681.

Schröter, K., Läuchli, A. & Sievers, A. 1975. Mikroanalytische Identifikation yon Bariumsulfat-Kristallen in den Statolithen der Rhizoide yon *Chara fragilis*, *Desv.* Planta 122, 213–225.

Sheteiwy, M.S., El-Sawah, A.M., Korany, S.M., Alsherif, E.A., Mowafy, A.M., Chen, J., Jo´sko, I., Selim l, S. & AbdElgawad, H. 2022. Arbuscular mycorrhizal fungus "*Rhizophagus irregularis*" impacts on physiological and biochemical responses of ryegrass and chickpea plants under beryllium stress. Environm. Pollut. 315: 120356 (13 pp.) https://doi.org/10.1016/j.envpol.2022.120356

Shorr, E. & Carter, C. 1982. The usefulness of strontium as an adjuvant to calcium in the remineralization of the skeleton in man. Bull. Hosp. Joint Dis. 13, 59–66.

Sievers, A. & Schmitz, M. 1982. Röntgen-Mikroanalyse von Barium, Schwefel und Strontium in Statolithen-Kompartimenten von *Chara*-Rhizoiden, Ber. Dtsch. Bot. Ges. *95*, 353–360.

Simon, P., Pompe, W., Gruner, D., Sturm, E., Ostermann, K., Matys, S., Vogel, M. & Rödel, G. 2022. Nested formation of calcium carbonate polymorphs in a bacterial surface membrane with a graded nanoconfinement: An evolutionary strategy to ensure bacterial survival. ACS Biomater. Sci. Eng. 8, 526−539. https://doi.org/10.1021/acsbiomaterials.1c01280

Song, T., Yang, J., Liu, P., Liu, M., Li, D., Xiao, Y., Wang, Y. & Zhang, X. 2022. Icariin self-crosslinked network functionalized strontium-doped bioceramic scaffolds synergistically enhanced the healing of osteoporotic bone defects. Composites Part B 235: 109759 (13 pp.). https://doi.org/10.1016/j.compositesb.2022.109759

Stepan, J.J. 2013. Strontium ranelate: in search for the mechanism of action. J. Bone Miner. Metab. 31, 606–612. https://doi.org/10.1007/s00774-013-0494-1

Steudle, E., Läuchli, A. & Sievers, A. 1978. X -R ay Microanalysis of barium and calcium in plant material: Significance for the analysis of statoliths. Z. Naturforsch. 33 c, 444—446.

Takahashi, S., Aizawa, K., Nakamura, S., Nakayama, K., Fujisaki, S., Watanabe, S. & Satoh, H. 2015. Accumulation of alkaline earth metals by the green macroalga *Bryopsis maxima*. Biometals 28, 391–400. https://doi.org/10.1007/s10534-015-9843-y

Tardiff, R.G., Robinson, M.; & Ulmer, N.S. 1980. Subchronic oral toxicity of $BaCl_2$ in rats. J. Environ. Pathol. Toxicol. 4, 267-–275.

Teichert, S., Woelkerling, W. & Munnecke, A. 2019. Coralline red algae from the Silurian of Gotland indicat that the order Corallinales (Corallinophycidae, Rhodophyta) is much older than previously thought. Palaeontol. 62. 599–613. https://doi.org/10.1111/pala.12418

Tkalya, E.V., Avdeenkov, A.V., Bibikov, A.V., Bodrenko, I.V. & Nikolaev, A.V. 2012. Electron capture β decay of 7Be located inside and outside the C_{36} fullerene. Physical Rev. 86, 014608 (7 pp.). https://doi.org/10.1103/PhysRevC.86.014608

Tomás, S., Aguirre, J., Braga, J.C. & Martín-Closas, C. 2007. Late Hauterivian coralline algae (Rhodophyta, Corallinales) from the Iberian Chain (E. Spain). Taxonomy and the evolution of multisporangial reproductive structures. Facies 53, 79–95. https://doi.org/10.1007/s10347-006-0088-7

Toyokawa, M., Shirai, K., Takahata, N., Ishida, A. & Sano, Y. 2018. Growth and formation of statoliths in *Aurelia coerulea* examined by using ^{34}S- and Sr-labels. J. Plankton Res. 40, 619–626. https://doi.org/10.1093/plankt/fby034

Uriz, M.J., Agell, G., Blanquer, A., Turon, X. & Casamayor, E.O. 2012. Endosymbiotic calcifying bacteria: A new cue to the origin of calcification in metazoan? Evolution 66, 2993–2999. https://doi.org/10.1111/j.1558-5646.2012.01676.x

Vandenhove, H., Eyckmans, T. & Van Hees, M. 2005. Can barium and strontium be used as tracers for radium in soileplant transfer studies? J. Environm. Radioactivity 81, 255–267. https://doi.org/10.1016/j.jenvrad.2004.01.039

Vanoni, F.O., Milani, G.P., Agostoni, C., Treglia, G., Faré, P.B., Camozzi, P., Lava, S.A.G., Bianchetti, M.G. & Janett, S. Magnesium 2021. Metabolism in chronic alcohol-use disorder: Meta-analysis and systematic review. Nutrients 13, 1959 (11 pp.). https://doi.org/10.3390/nu13061959

Veronese, N, & Barbagallo, M. 2021. Magnesium and Micro-Elements in Older Persons. Nutrients 13, 847 (3 pp.).

https://doi.org/10.3390/nu13030847
Villarreal, R.E., Arazi, A. & Fernández Niello, J.O. 2022. Correlation between the latitudinal profile of the ^{7}Be air concentration and the Hadley cell extent in the Southern Hemisphere. J. Environm. Radioact. 244-245: 106760 (8 pp.). https://doi.org/10.1016/j.jenvrad.2021.106760
Willows, R.D. 2019. The Mg branch of chlorophyll synthesis: Biosynthesis of chlorophyll *a* from protoporphyrin IX. Ch. 5 in Adv. Botanical Res. 90, 141–182.
Wolf, F.I. & Cittadini, A. 2003. Chemistry and biochemistry of magnesium. Molecular Aspects of Medicine 24, 3–9.
Woods, M. & Lucas, S.E.M. 1996. Half-life of ^{90}Sr - mcasurement and critical review. Nuclear Instr. Meth. Phys. Res. A 369, 534–538.
Wu, T., Liu, W., Huang, S., Chen, J., He, F., Wang, H., Zheng, X., Li, Z., Zhang, H., Zha, Z., Lin, Z. & Chen, Y. 2021. Bioactive strontium ions/ginsenoside Rg1–incorporated biodegradable silk fibroin-gelatin scaffold promoted challenging osteoporotic bone regeneration. Materials Today Bio. 12: 100141 (15 pp.). https://doi.org/10.1016/j.mtbio.2021.100141
Yan, M.-D., Ou, Y.-J, Lin, Y.-J., Liu, R.-M., Fang, Y., Wu, W.-L., Zhou, L., Yao, X. & Chen, J. 2022. Does the incorporation of strontium into calcium phosphate improve bone repair? A meta-analysis. BMC Oral Health 22:62 (21 pp.). https://doi.org/10.1186/s12903-022-02092-7
Yang, I.V., Konigsberg, I., MacPhail, K., Li L, Davidson, E.J., Mroz, P.M., Hamzeh, N., Gillespie, M., Silveira, L.J., Fingerlin, T.E. & Maier, L.A. 2019. DNA methylation changes in lung immune cells are associated with granulomatous lung disease. Am. J. Respir. Cell Mol. Biol. 60, 96–105.
Yang, J. 2007. Modes of occurrence and geological origin of beryllium in coals from the Pu'an Coalfield, Guizhou, Southwest China. Acta Geologica Sinica 81, 470–476.
Zahra, H., Berriche, O., Mizouri, R., Boukhayatia, F., Khiari, M., Gamoudi, A., Lahmar, I., Ben Amor, N., Mahjoub, F., Zayet, S. & Jamoussi, H. 2021. Plasmatic magnesium deficiency in 101 outpatients living with Type 2 diabetes mellitus. Clin. Pract. 11, 791–800. https://doi.org/10.3390/clinpract11040095
Zander-Principati, G.E & Kuzma, J.F. 1964. Reduction of strontium-90 bone cancer by zirconium citrate. Int. J. Rad. Biol. Rel. Stud. Phys. Chem. Med. 8, 427–437. https://doi-org.ludwig.lub.lu.se/10.1080/09553006414550541

Zhang, Y. & Zhang, X. 2022. A possible erect coralline alga from the Ediacaran Dengying Formation in the Zhenba area of South China. J. Paleontol. 96, 1209–1222. https://doi.org/10.1017/jpa.2022.33

CHAPTER THREE

RARE EARTHS

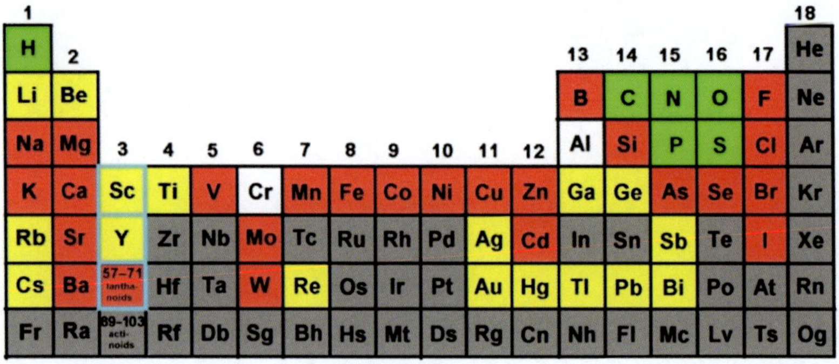

3.1 Scandium

As far as is known, scandium is not a necessary element for the life of any organism. However, it has a number of interesting effects. It stimulates the synthesis of a number of compounds in various bacteria. In several species of *Streptomyces* it stimulates the synthesis of various antibiotics. Thus, in *Streptomyces coelicolor*, and *Streptomyces lividans*, scandium (as scandium chloride, $ScCl_3$) enhances production of actinorhodin, a deep blue-pigmented polyketide antibiotic, and its precursor guanosine 5′-diphosphate 3′-diphosphate (Kawai et al. 2007). The optimum concentration of scandium chloride is 30 μM; with higher concentration the accumulation declines. Similarly, in *Streptomyces griseus*, scandium stimulates the production of the antibiotic streptomycin.

Scandium, as $ScCl_3$, and also lanthanum, $LaCl_3$, activate, ranging from 2.5- to 12-fold, the expression of nine genes belonging to nine secondary metabolite–biosynthetic gene clusters of *Streptomyces coelicolor* A3(2) (Tanaka et al. 2010).

In *Bacillus subtilis* scandium stimulates the production of amylase 3.5-fold at a concentration of 10 mg/L, and also the production of the antibiotic bacilysin (Inaoka et al. 2011).

In cultures of rat mesenchymal stem cells, scandium (as $ScCl_3$) promotes the proliferation of rat bone mesenchymal cells. $ScCl_3$ at concentrations of $1x10^{-8}$ –1x 10^{-6} mol/L can enhance levels of osteogenic differentiation (differentiation of bone cells) in a dose-dependent manner and reduce adipogenic differentiation (differentiation of fat cells) (Ren et al. 2022).

Gómez-Aracena et al. (2002) found that higher scandium content in men's toenails was associated with a slightly (but statistically significant) lower risk of myocardial infarction. Toxicology and Industrial Health 18, 353–360.

3.2 Yttrium

On the island of Resarö in the archipelago outside Stockholm is a small village called Ytterby (meaning "Outer Village"). As a child I often passed the place on my way to our summer resort further out in the archipelago, without knowing that it is a famous place. Once there was a mine there, in which a black kind of stone was found. The chemist Johan Gadolin (after whom the element gadolinium has its name) discovered that the stone contained something that did not fit any previous description. Other chemists later found no less than seven new chemical elements in this stone (later called gadolinite). Since there was not much around to name so many new elements after, some of them got names that are perturbations of the village name: Yttrium (Y), ytterbium (Yb), terbium (Tb), and erbium (Er). One, scandium (Sc), was named after Scandinavia, another one, holmium (Ho), after Stockholm, and the final one, thulium (Tm), after the name that the Romans had for the unknown far North, Ultima Thule. We shall here concentrate on yttrium (Y). Together with the lanthanoids, which we treat in another chapter, and scandium, it was earlier included in the group "rare earths".

As far as we know, yttrium is not an essential element for any organism. However, it is taken up by plants as well as by various microorganisms. As can be seen from Fig. 3.2.1, yttrium and iron become localized to the outer part of the root, in contrast to the other elements. Already 0.1 mM Y^{3+} results in a marked decline of plant dry weight, water content, height, leaf area, concentration of chlorophylls *a* and b and carotenoids in young sunflower plants (Maksimovic et al. 2012).

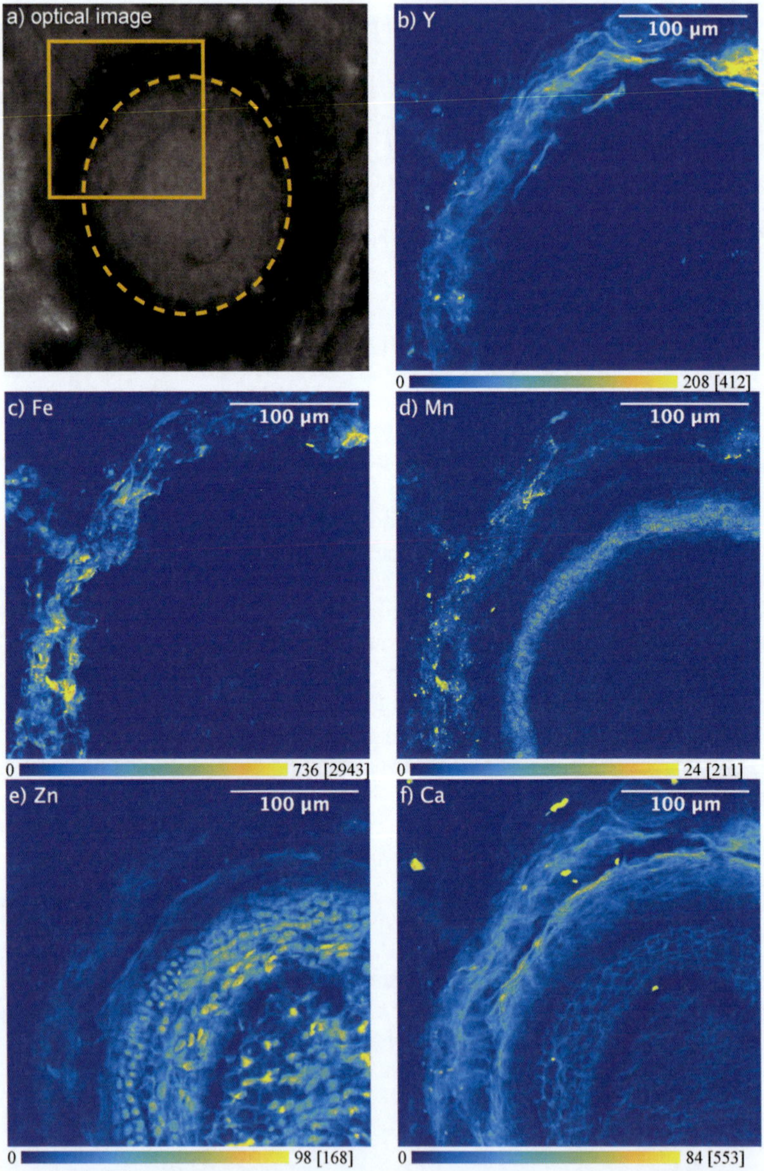

Fig. 3.2.1. Distribution of elements, imaged by synchrotron-based micro X-ray fluorescence, in a root of *Saxifraga paniculata*, which has been grown in yttrium-doped soil. Reproduced from Fehlauer et al. 2022 under a Creative Commons Attribution 4.0 International License (http://creativecommons.org/licenses/by/4.0).

Both because yttrium is a valuable metal with many uses, and because it is poisonous to many organisms, effort has been expended to collect it from mine seep-water and wastewater. One of the organisms that has been tried is the bacterium *Serratia marcescens* (Yersiniaceae). It adsorbs (by a combination of ion exchange, electrostatic attraction and complexation) rather than absorbs yttrium to more than a tenth of its own weight (Liang & Shen 2022).

The growth of a fungal strain, *Penicillium* sp. ZD28, is markedly increased by Y^{3+} at low concentrations (Fig. 3.2.2) (Wang et al. 2020).

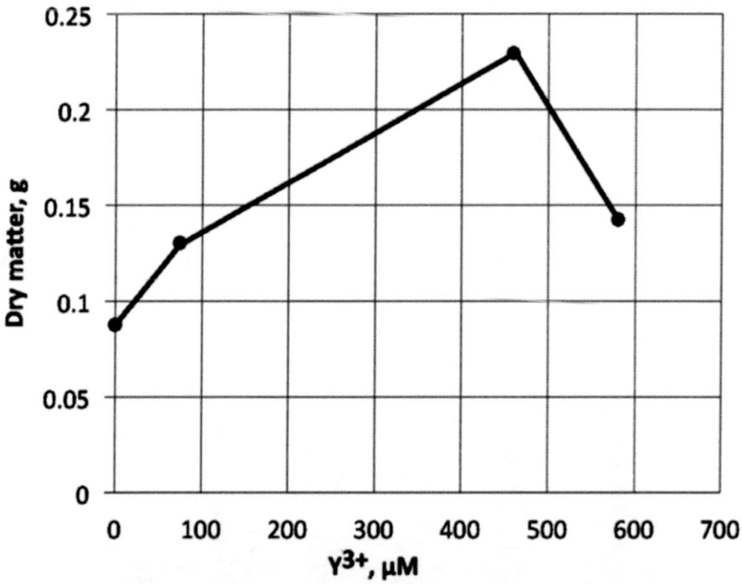

Fig. 3.2.2. Dry matter of the fungus *Penicillium* sp. ZD28 as a function of the concentration of yttrium ions in the medium. Data from Wang et al. (2020).

Yttrium is a major component in the fluorescent material used in fluorescent lamps. According to Pinto et al. (2022) yttrium can be recovered from fluorescent lamp waste using the green alga *Ulva* sp. The highest fraction (50%) recovered was obtained using a low salinity (10 g/kg), Y^{3+} concentration = 20 mg/L, and 9 g/L algae (fresh weight).

3.3 Lanthanides

Lanthanides form a group of 14 elements from lanthanum (atomic number 57) to luthetium (atomic number 71). Together with scandium and yttrium they are sometimes referred to as "rare earth metals", although they are not so rare in rocks. Cerium is the 25th most common element in the Earth's crust. However, lanthanides occur at exceedingly low levels in most natural waters (picomolar to nanomolar range) with the exception of volcanic hot springs (Glass et al. 2020). Since lanthanide atoms have the same number of electrons in their outermost orbitals (but differ by the numbers of electrons in orbitals further in) they have very similar chemical properties.

Enzymes having a lanthanide as cofactor seem to prefer the lighter lanthanides (lanthanum, cerium, neodymium and praseodymium). Bacteria having them seem to grow more slowly in media having only higher atomic weight lanthanides. However, Good et al. (2022) found a strain of *Methylorubrum extorquens* that is satisfied with gadolinium, which has a higher atomic weight than six of the other lanthanides. With *Methylacidiphilum fumariolicum* SolV from a volcanic mud volcano Pol et al. (2013) got the growth rates that are shown in Figure 3.3.1.

The best explored lanthanide enzymes are methanol dehydrogenases. They have a pyrrolquinolin quinone (PQQ) in their active centers. Methanol dehydrogenase type MxaF1 contains calcium, another type (XoxF) one of the lighter lanthanides (Pol et al. 2014, Fig. 3.3.2). Some organisms are equipped with both types, others have one only.

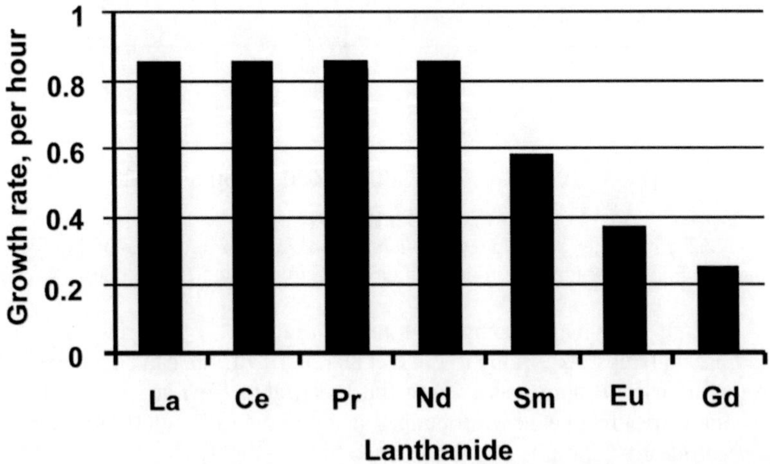

Fig. 3.3.1. Relative growth rate of the bacterium *Methylacidiphilum fumariolicum* SolV in a growth medium containing various metal ions (concentration 66 nanomolar). Under natural conditions only lanthanum and cerium are of any importance for the organism, since the other lanthanoids occur in much lower concentrations in the volcanic springs where it lives. Values from Pol et al. (2014).

Bacteria using lanthanides live in very different places. *Methylobacterium extorquens* PA1, a bacterium living on plant shoots and having genes for Ca-dependent as well as for lanthanide-dependent methanol dehydrogenases, seems to prefer the lanthanide-dependent ones if light lanthanides are available, and in this case repress the synthesis of the Ca-dependent enzyme (Ochsner et al. 2019). The proportions of various lanthanides did not differ significantly from those in the medium (Fig. 3.3.3). Bayon et al. (2020), on the other hand, found enrichments of lanthanum and cerium in the oxic part of a siboglinid worm tube at a cold methane seep at the ocean bottom, with lanthanide distribution patterns that depart significantly from the ambient seawater signature. This was interpreted as reflecting the signature of lanthanide-dependent methanotrophic activity. Daszczyn´ska et al. (2022) found many bacterial species carrying out lanthanide-dependent methanol oxidation in a mine.

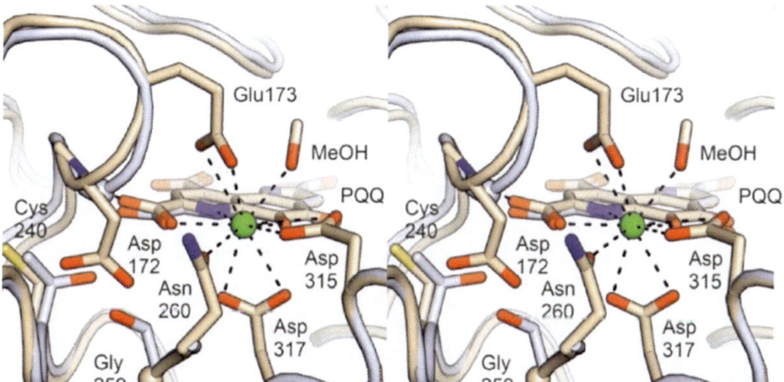

Fig. 3.3.2. A composite stereo-view of the central parts of two lanthanide proteins. Beige, neodymium-containing XoxF1 dimer from *Methylacidimicrobium thermophilum* AP8 methanol dehydrogenase; light blue, cerium-containing XoxF2 of *Methylacidiphilum fumariolicum* SolV methanol dehydrogenase. The green sphere is the lanthanide ion. Reproduced from Schmitz et al. (2021) under the Creative Commons Attribution 4.0 International license.

Bacteria using lanthanides live in very different places. *Methylobacterium extorquens* PA1, a bacterium living on plant shoots and having genes for Ca-dependent as well as for lanthanide-dependent methanol dehydrogenases, seems to prefer the lanthanide-dependent ones if light lanthanides are available, and in this case repress the synthesis of the Ca-dependent enzyme (Ochsner et al. 2019). The proportions of various lanthanides did not differ significantly from those in the medium (Fig.

3.3.3). Bayon et al. (2020), on the other hand, found enrichments of lanthanum and cerium in the oxic part of a siboglinid worm tube at a cold methane seep at the ocean bottom, with lanthanide distribution patterns that depart significantly from the ambient seawater signature. This was interpreted as reflecting the signature of lanthanide-dependent methanotrophic activity. Daszczyn´ska et al. (2022) found many bacterial species carrying out lanthanide-dependent methanol oxidation in a mine.

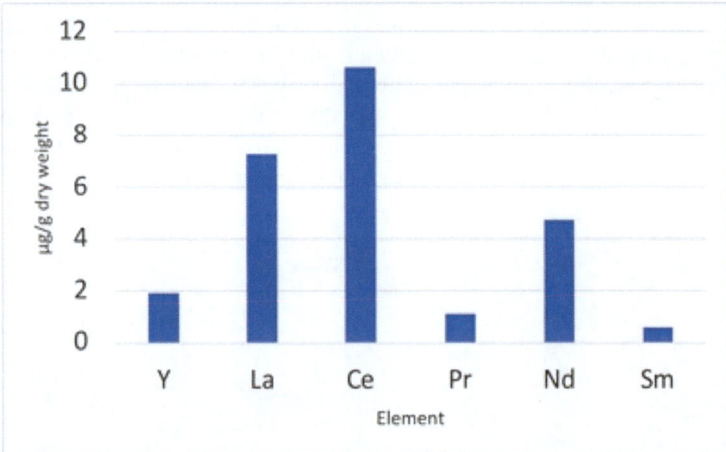

A

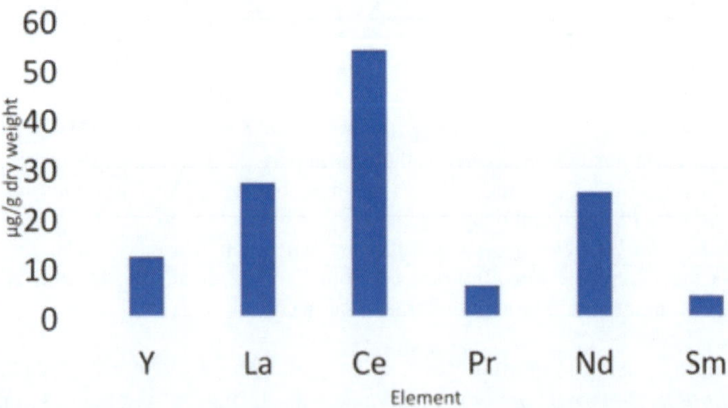

B

Fig. 3.3.3. (A) Lanthanide and yttrium distribution in the shoot of *Arabidopsis thaliana* grown on a medium with the lanthanide distribution shown in (B). Replotted from Ochsner et al. (2019).

XoxF enzymes are diverse and can be divided into at least five clades with potentially different biochemical properties (Keltjens et al. 2014; Huang et al. 2018). Schmitz et al. (2021) compare the active site structures of XoxF1 and XoxF2 types, with neodymium and cerium, respectively. Huang et al. (2018) study types XoxF4 and XoxF5, and divide them further into subtypes. XoxF4-1 is the dominant XoxF4 subtype in nature while other XoxF4 subtypes appear to be auxiliary, found only in Methylophilaceae (a family in Betaproteobacteria). XoxF5 enzymes are widespread among Alphaproteobacteria (e.g., the family Beijerinckiaceae, Wegner et al. 2020), Beta-, and Gammaproteobacteria. Huang et al. (2018) found that two XoxF4 subtypes from *Methylotenera mobilis* and one XoxF5 subtype from another *Methylotenera* species showed broad specificities toward alcohols and aldehydes and strict dependence on lighter lanthanides.

The bacteria in the experiment in Fig. 3.3.1 were cultured at pH 2.7 and with methane as carbon source, and with the lanthanides as trivalent cations. They were cultured I plastic flasks, since it turned out that too much lanthanide (probably mainly cerium) was dissolved in glass flasks. The concentration used in the experiment shown in Fig. 3.3.1 is far below saturating.

XoxF-type methanol dehydrogenase genes can be found in bacteria that are not believed to be methylotrophs. An example of this is *Bradyrhizobium sp.* strain Ce-3, a nitrogen-fixing legume symbiont belonging to the *Bradyrhizobium elkanii* superclade (Pastawan et al. 2020).

As mentioned above, organisms seem to prefer the lighter lanthanoids. However, the maximum reaction rates and Michaelis constants do not always follow this pattern (Fig. 3.3.4A, 3.3.4B).

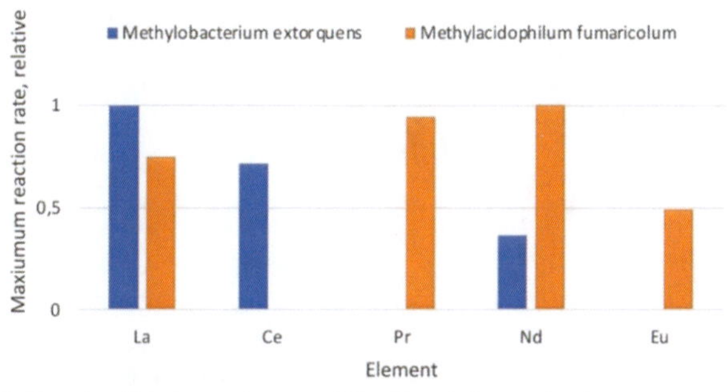

Fig. 3.3.4A. Lanthanide dependence of XoxF activity varies between organisms. A) Comparison of the V_{max} values for *M. extorquens* AM1 (blue) and *M. fumaricolum* SolV (red) XoxFs, metalated with different lanthanides, as determined by an artificial dye-linked assay. For each enzyme, the metalated enzyme with highest activity was normalized to 1. Redrawn from data of Featherston et al. (2021).

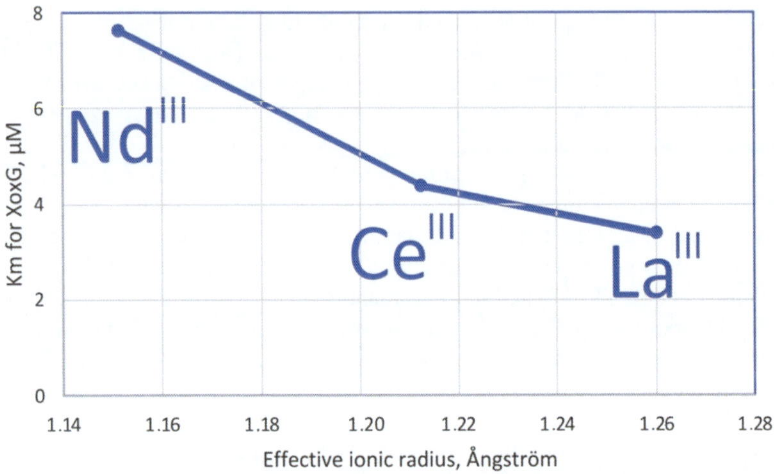

Fig. 3.3.4B. In *M. extorquens* AM1, the Km of XoxF for its physiological electron acceptor XoxG increases significantly from La-XoxF to Nd-XoxF. Redrawn from data of Featherston et al. (2021).

Dicranopteris linearis is a fern that hyperaccumulates lanthanum and cerium. The distribution of the lanthanides within the plant was studied using synchrotron-based X-ray fluorescence microscopy (Liu et al. 2020). Dinh et al. (2022) have a table with many other data on rare element accumulation in *Dicranopteris linearis* and other plants. Lanthanides occur also in zooplankton (Lobus et al. 2019), but no function in animals is known.

Lanmodulin is a lanthanid binding protein present in many bacteria. The structure is shown in Cook et al. (2019). Surprisingly it binds an actinide, americium, more strongly than lanthanides, although no biologic function for americium is known (Singer et al. 2022).

References

Bayon, G., Lemaitre, N., Barrat, J.-A., Wang, X., Feng, D. & Duperron, S. (2020) Microbial utilization of rare earth elements at cold seeps related to aerobic methane oxidation. Chemical Geology 555: 119832 (13 pp.). https://doi.org/10.1016/j.chemgeo.2020.119832

Cook, E.C., Featherston, E.R., Showalter, S.A. & Cotruvo, Jr., J.A. (2019) Structural basis for rare earth element recognition by *Methylobacterium extorquens* lanmodulin. Biochemistry 58, 120–125. https://doi.org/10.1021/acs.biochem.8b01019

Al Danaf, N., Kretzschmar, J., Jahn, B., Singer, H., Pol, A., Op den Camp, H.J.M., Steudtner, R., Lamb, D.C., Drobot, B. & Daumann, L.J. (2022) Studies of pyrroloquinoline quinone species in solution and in lanthanide-dependent methanol dehydrogenases. Phys. Chem. Chem. Phys. 24, 15397–15405. https://doi.org/10.1039/d2cp00311b

Daszczyn´ska, A., Krucon', T., Stasiuk, R., Koblowska, M. & Matlakowska, R. (2022) Lanthanide-dependent methanol metabolism of a *Proteobacteria*-dominated community in a light lanthanide-rich deep environment. Int. J. Mol. Sci. 23, 3947 (15 pp.). https://doi.org/10.3390/ijms23073947

Dinh, T., Dobo, Z. & Kovacs, H. (2022) Phytomining of rare earth elements – A review. Chemosphere 297: 134259 (13 pp.). https://doi.org/10.1016/j.chemosphere.2022.134259

Cotruvo, J. A., Jr., Featherston, E. R., Mattocks, J. A., Ho, J. V. & Laremore, T. N. (2018) Lanmodulin: A highly selective lanthanide-binding protein from a lanthanide-utilizing bacterium. J. Am. Chem. Soc. 140, 15056−15061. https://doi.org/10.1021/jacs.8b09842

Crombie, A.T. (2022) The effect of lanthanum on growth and gene expression in a facultative methanotroph. Environm. Microbiol. 24, 596–613. https://doi.org/10.1111/1462-2920.15685

Featherston, E.R. & Cotruvo Jr, J.A. (2021) The biochemistry of lanthanide acquisition, trafficking, and utilization. BBA - Molecular Cell Research 1868: 118864 (11 pp). https://doi.org/10.1016/j.bbamcr.2020.118864

Featherston, E.R., Rose, H.R., McBride, M.J., Taylor, E.M., Boal, A.K. & Cotruvo, Jr., J.A (2019) Biochemical and structural characterization of XoxG and XoxJ and their roles in lanthanide-dependent methanol dehydrogenase activity. ChemBioChem 20, 2360–2372.

Fehlauer, T., Collin, B., Angeletti, B., Negahi, M.M., Dentant, C., Chaurand, P., Lallemand, C., Levard, C. & Rose. J. (2022) Multiscale imaging on *Saxifraga paniculata* provides new insights into yttrium uptake by plants. Sci. Rep. 12:18268 (13 pp). https://doi.org/10.1038/s41598-022-23107-x

Glass, J.B., Cowan, E.T. & Go´mez-Aracena, J., Martin-Moreno, J.M., Riemersma, R.A. et al. (2012) Association between toenail scandium levels and risk of acute myocardial infarction in European men: the EURAMIC and Heavy Metals Study.

Good. N.M., Lee, H.D., Hawker, E.R., Su, M.Z., Gilad, A.A. & Martinez-Gomez, N.C. (2022) Hyperaccumulation of gadolinium by *Methylorubrum extorquens* AM1 reveals impacts of lanthanides on cellular Processes beyond methylotrophy. Front. Microbiol. 13:820327 (15 pp.). https://doi.org/10.3389/fmicb.2022.820327

Huang, J., Yu, Z. & Chistoserdova, L. (2018) Lanthanide-dependent methanol dehydrogenases of XoxF4 and XoxF5 clades are differentially distributed among methylotrophic bacteria and they reveal different biochemical properties. Front. Microbiol. 9:1366 (13 pp.). https://doi.org/10.3389/fmicb.2018.01366

Huang J, Yu Z, Groom J et al. Rare earth element alcohol dehydrogenases widely occur among globally distributed, numerically abundant and environmentally important microbes. ISME J 2019;13:2005–17.

Inaoka, K. & Ochi, K. (2011) Scandium stimulates the production of amylase and bacilysin in *Bacillus subtilis*. Appl. Environm. Microbiol. 77, 8181–8183. https://doi.org/10.1128/AEM.06205-11

Johannesson, K.H. (2020) Lanthanide rarity in natural waters: implications for microbial C1 metabolism. FEMS Microbiology Letters, 367: fnaa165 (4 pp.). https://doi.org/10.1093/femsle/fnaa165

Kawai, K., Wang, G., Okamoto, S. & Ochi, K. (2007) The rare earth, scandium, causes antibiotic overproduction in *Streptomyces* spp. FEMS Microbiol. Lett. 274, 311–315.

Li, Z., Yao, Q., Guo, X., Crits-Christoph, A., Mayes, M. A., Hervey, W. J. I. V., Lebeis, S. L., Banfield, J. F., Hurst, G. B., Hettich, R. L. & Pan, C. (2019) Genome resolved proteomic stable isotope probing of soil microbial communities using $^{13}CO2$ and ^{13}C-methanol. Front. Microbiol., 10, 2706.

Liang, C.-L. & Shen, J.-L. (2022) Removal of yttrium from rare-earth wastewater by *Serratia marcescens*: biosorption optimization and mechanisms studies. Sci. Rep. 12:4861 (14 pp.). https://doi.org/10.1038/s41598-022-08542-0

Liu, W.-S., van der Ent, A., Erskine, P.D., Morel, J.L., Echevarria, G., Spiers, K.M., Montargès-Pelletier, E., Qiu, R.-L. & Tang, Y.-T. (2020) Spatially Resolved Localization of Lanthanum and Cerium in the Rare Earth Element Hyperaccumulator Fern Dicranopteris linearis from China. Environ. Sci. Technol. 54, 2287−2294. https://dx.doi.org/10.1021/acs.est.9b05728

Lobus, N.V., Arashkevich, E,G. & Flerova, E.A. (2019) Major, trace, and rare-earth elements in the zooplankton of the Laptev Sea in relation to community composition. Environm. Sci. Pollut. Res. 26, 23044–23060. https://doi.org/10.1007/s11356-019-05538-8

Maksimovic, I., Kastori, R., Putnik-Delic, M. & Zeremski, T. (2012) Yttrium – accumulation, translocation and distribution in young sunflower plants (*Helianthus annuus* L.) Fresenius Envirom. Bull. 21, 11–18.

Ochsner, A.M., Hemmerle, L., Vonderach, T., Nüssli, R., Bortfeld-Miller, M., Hattendorf, B. & Vorholt, J.A. (2019) Use of rare-earth elements in the phyllosphere colonizer *Methylo- bacterium extorquens* PA1. Molecular Microbiol. 111, 1152–1166. https://doi.org/10.1111/mmi.14208

Pastawan, V., Suganuma, S., Mizuno, K., Wang, L., Tani, A., Mitsui, R., Nakamura, K., Shimada, M., Hayakawa, T., Fitriyanto, N.A. & Nakagawa, T. (2020) Regulation of lanthanide-dependent methanol oxidation pathway in the legume symbiotic nitrogen-fixing bacterium *Bradyrhizobium sp.* strain Ce-3. J. Biosci. Bioeng. 130, 582–587. https://doi.org/10.1016/j.jbiosc.2020.07.012

Pinto. J., Colónia, J., Viana, T., Ferreira, N., Tavares, D., Jacinto, J., Abdolvasei, A., Monteiro, F.L.S., Henriques, B. & Pereira, E. (2022) Potential of the macroalga *Ulva* sp. for the recovery of yttrium obtained from fluorescent lamp waste. J. Cleaner Prod. 369: 133299 (11 pp.). https://doi.org/10.1016/j.jclepro.2022.133299

Pol, A., Barends, T.R.M., Dietl, A., Khadem, A.F., Eygensteyn, J., Jetten, M.S.M. & Op den Camp, H.J.M. (2014) Rare earth metals are essential for methanotrophic life in volcanic mudpots. Environm. Microbiol. 16, 255–264. https://doi.org/10.1111/1462-2920.12249

Ren, N., Yu, X., Wang, A., Liang, N., Feng, Z. & Sun, C. (2022) Effects of scandium chloride on osteogenic and adipogenic differentiation of mesenchymal stem cells. J. Rare Earths 40, 161–168. https://doi.org/10.1016/j.jre.2020.11.007

Schmitz, R.A.,Picone, N., Singer, H., Dietl, A., Seifert, K.-A, Pol, A., Jetten, M.S.M., Barends, T.R.M., Daumann, L.J. & Op den Camp, H.J.M. (2021) Neodymium as metal cofactor for biological methanol oxidation: Structure and kinetics of an XoxF1-type methanol dehydrogenase. mBio 12:e01708-21 (14 pp.). https://doi.org/10.1128/mBio.01708-21.

Singer, H., Drobot, B., Zeymer, C., Steudtner, R. & Daumann, L.J. (2022) Americium preferred: lanmodulin, a natural lanthanide-binding protein favors an actinide over lanthanides. Chem. Sci. 12, 15581–15587.

Tanaka, Y., Hosaka, T. & Ochi, K. (2010) Rare earth elements activate the secondary metabolite–biosynthetic gene clusters in *Streptomyces coelicolor* A3(2). J. Antibiotics, 63, 477–481. https://doi.org/10.1038/ja.2010.53

Vu, H. N., Subuyuj, G. A., Vijayakumar, S., Good, N. M., Martinez-Gomez, N. C. & Skovran, E. (2016). Lanthanide-dependent regulation of methanol oxidation systems in *Methylobacterium extorquens* AM1 and their contribution to methanol growth. J. Bacteriol. 198, 1250–1259. https://doi.org/10.1128/JB.00937-15

Wang, W., Xu, C., Jin, Y., Zhang, Z., Yan, R. & Zhu, D. (2020) The accumulation of rare-earth yttrium ions by *Penicillium* sp. ZD28. AMB Expr. 10:25 (11 pp.). https://doi.org/10.1186/s13568-020-0961-8

Wegner, C.-E., Gorniak, L., Riedel, S., Westermann, M. & Küsela, K. (2020) Lanthanide-Dependent Methylotrophs of the Family Beijerinckiaceae: Physiological and Genomic Insights. Appl Environ Microbiol 86:e01830-19. https://doi.org/10.1128/AEM.01830-19.

CHAPTER FOUR

TITANIUM

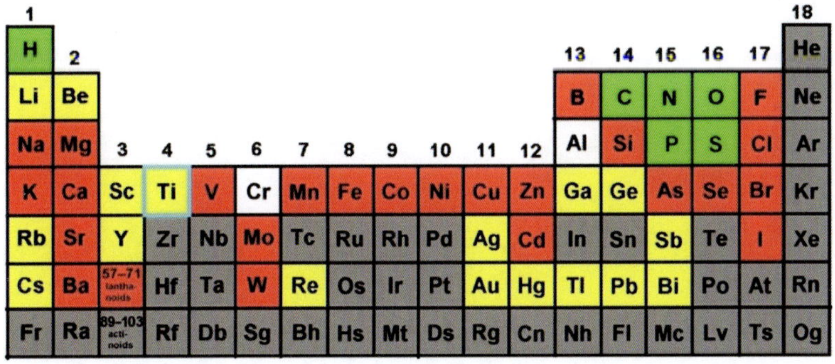

4.1 Titanium

It is usually stated that there is no known biological role for titanium. But there is a detectable amount of titanium in the human body and it has been estimated that we take in about 0.8 mg/day. Most passes through us without being absorbed. It is not a poison metal and the human body can tolerate titanium in large dose.

But I have a special relation to titanium, because my left hip joint is made of it, and I am very pleased with it. Without it I would not be able to walk. Therefore I decided to look a little deeper in the literature for "titanium biology". Expressed in number of atoms, titanium is the ninth most abundant element in the Earth crust (e.g., Zierden & Valentine 2016, https://en.wikipedia.org/wiki/Abundance_of_elements_in_Earth%27s_crust) , and it would be a little surprising if no organism has found any use for it (except for hip joints). Also, it is the only element in that column of the periodic system that could possibly be of biological interest.

Wojcik & Wojcik (2006) found that $TiCl_4$, especially as a solution sprayed on the leaves, increased the growth of young apple plants. The result was interpreted as proceeding via a stimulation of iron (Fe^{2+}) uptake. According to El-Ghamry et al. (2022) TiO_2 sprayed unto the leaves stimulates growth and yield of maize. A concentration of 8 ppm Ti increases the grain yield per plant from 137.2 g dry weight to 168.7 g dry weight. Hussain et al. (2021) found that spraying the leaves of soybean plants with a titanium-containing solution improves the uptake of phosphorus due to changes in root growth.

Yaghoubi et al. (2000) found that Ti^{4+} (5 µg per day) in the feed stimulated the growth of mice (final weight 38 g as compared to 32 g for control). However, no statistics is shown; it seems that the experiment was not repeated, and the result hardly appears trustworthy. But also Schwietert et al. (2001) found that growth of mice was stimulated by titanium as TiO^{2+} (Fig. 4.1). Possibly this is not a direct effect, but depending on the effect of microorganisms in the environment.

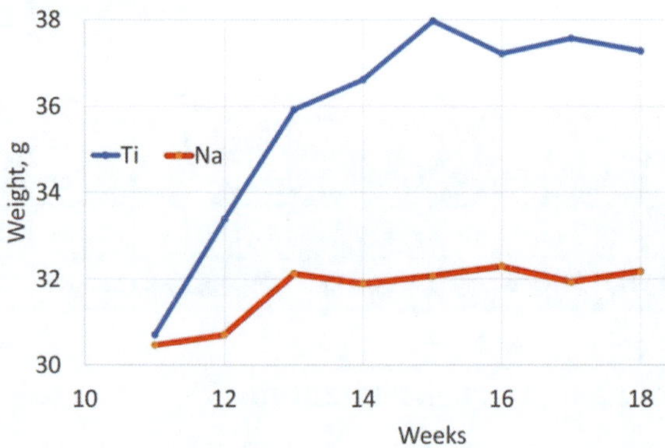

Fig. 4.1. Growth of mice without or with titanium. Group A(–) received 5 µg of Ti (as TiO^{2+} oxalate) daily for 1 week and Group B (–) received comparable amounts of sodium oxalate. At week 12, dietary Ti was increased to 55 µg daily in Group A, and Group B received a comparable increase in sodium oxalate. At week 14, titanium oxalate and sodium oxalate were withheld from both groups for 1 week before the crossover study began. Values from Schwietert et al. (2001).

Titanium exerts various effects on diatoms. Basharina et al. (2012) found that it disrupts normal frustrule formation in *Synedra acus*, but contrary to germanium is not incorporated into the frustrules. This is in contrast to Chauton et al. (2015) and Jeffryes et al. (2008a,b), and Lang et al. (2013), who found incorporation of titanium in the frustrules of the diatom *Pinnularia* sp.

Fistulifera solaris is a particularly titanium-tolerant diatom species (Maeda et al. 2018). With titanium provided as titanium(IV) bis(ammonium lactato) dihydroxide a Ti/Si atomic ratio of 9.7 could be achieved. Brzozowska et al. (2022) found that titanium, provided as titanium(IV) terbutanulate ($C_{16}H_{36}O_4Ti$), increased the diatom biomass up to a Ti concentration of 15 mg/L, and a Ti/Si molar ratio of 1:17.5. Lang et al. (2013)

achieved small modifications of the dimension in the frustrule patterning by feeding *Thalassiosira weissflogii* titanium(IV) bis-(ammonium lactato)-dihydroxide (Fig. 4.2).

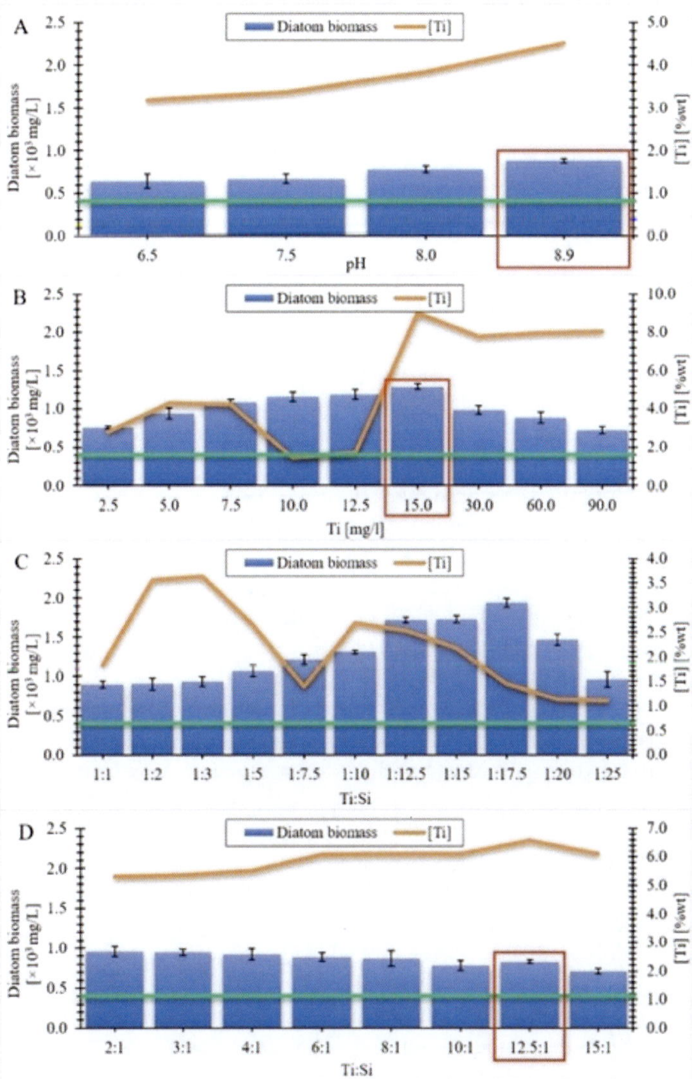

Fig. 4.2. Biomass as a function of (A) pH value; (B) Ti concentration; (C,D) Ti:Si ratio in relation to titanium content in diatom biomass in % weight (orange line).

The undoped cell production value from the diatom species *Pseudostaurosira trainorii* is indicated by the horizontal line. The red rectangle indicates the key samples for a given series of experiments. Reproduced from Brzozowska et al. (2022) under the Creative Commons Attribution (CC BY) license. (https://creativecommons.org/licenses/by/4.0/).

References

Brzozowska, W., Sprynskyy, M,Wojtczak, I., Dąbek, P., Markuszewski, M.J., Witkowski, A. & Buszewski, B. 2022. Metabolically doping of 3D diatomaceous biosilica with titanium. Materials 15: 5210 (26 pp.). https://doi.org/10.3390/ma15155210

Chauton, M.S., Skolem, L.M.B., Olsen, L.M., Vullum, P.E., Walmsley, J. & Vadstein, O. 2015. Titanium uptake and incorporation into silica nanostructures by the diatom *Pinnularia* sp. (Bacillariophyceae). J. Appl. Phycol. 27, 777–786. https://doi.org/10.1007/s10811-014-0373-8

El-Ghamry, A.M., El-Naggar, E.M. & Hassan, H. 2022. An assessment of the influence of foliar spray gradient concentrations of TiO_2 on maize. Middle East J. Agricult. Res. 11, 959–966. https://doi.org/10.36632/mejar/2022.11.3.59

Hussain, S., Shafiq, I., Skalicky, M., Brestic, M., Rastogi, A., Mumtaz, M., Hussain, M., Iqbal, N., Raza, M.A., Manzoor, S., Liu, W. & Yang, W. 2021. Titanium application increases phosphorus uptake through changes in auxin content and root architecture in soybean (Glycine max L.) Front. Plant Sci. 12:743618 (18 pp.). https://doi.org/10.3389/fpls.2021.743618

Jeffryes, C., Gutu, T., Jiao, J. & Rorrer, G.L. 2008a. Metabolic insertion of nanostructured TiO_2 into the patterned biosilica of the diatom *Pinnularia* sp. by a two-stage bioreactor cultivation process. ACS Nano 2, 2104–2112. https://doi.org/10.1021/nn800470x

Jeffryes, C., Gutu, T., Jiao, J. & Rorrer, G.L. 2008b. Peptide-mediated deposition of nanostructured TiO_2 into the periodic structure of diatom biosilica. J. Mater. Res. 23, 3255–3262. https://doi.org/10.1557/JMR.2008.0402

Lang, Y., del Monte, F., Rodriguez, B.J., Dockery, P., Finn, D.P. & Pandit, A. 2013. Integration of TiO_2 into the diatom *Thalassiosira weissflogii* during frustule synthesis. Sci. Rep. 3: 3205 (11 pp.). https://doi.org/10.1038/srep03205

Maeda, Y., Niwa, Y., Tang, H., Kisailus, D., Yoshino, T. & Tanaka, T. 2018. Development of titania-integrated cilica cell walls of the titanium-resistant diatom, *Fistulifera solaris*. ACS Appl. Bio Mater. 1, 2021−2029. https://doi.org/10.1021/acsabm.8b00520

Möckel, H., Giersig, M. & Willig, F. 1999. Formation of uniform size anatase nanocrystals from bis(ammonium lactato)titanium dihydroxide by thermohydrolysis. J. Mater. Chem. 9, 3051–3056. https://pubs.rsc.org/en/content/articlelanding/1999/JM/a905373e

Schwietert, C.W., Yaghoubbi, S., Gerber, C.C., McSharry, J.J. & McCue, J.P. 2001. Dietary titanium and infant growth. Biol. Trace Elem. Res. 83, 149–167.

Wojcik, P. & Wojcik, M. 2006. Growth and nutrition of M.26 Emla apple rootstock as influenced by titanium fertilization. J. Plant Nutrit. 24, 1575–1588.

Yaghoubi, S., Schwietert, C.W. & McCue, J.P. 2000. Biological roles of titanium. Biol. Trace Element Res. 78. 205–216.

Zierden, M.R. & Valentine, A.M. 2016. Contemplating a role for titanium in organisms. Metallomics 8, 9–16. https://doi.org/10.1039/c5mt00231a

CHAPTER FIVE

VANADIUM

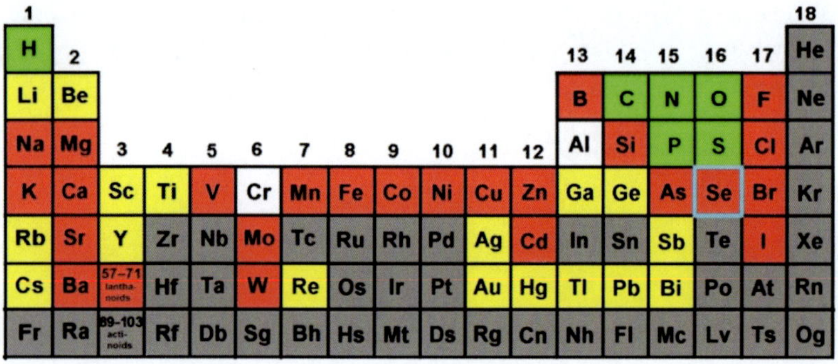

5.1 Vanadium

Vanadium, atomic number 23, a metal, although not well known by people in general, is not rare. The name comes from the vanir, a clan of Nordic gods competing with the æsir. It is the twentieth most abundant element in the Earth's crust. It forms colorful compounds in its many different oxidation states.

Badmaev et al. (1999) write: “The influence of vanadium on biological systems was first demonstrated in 1977 through its effect on ion transport pathways (Cantley et al., 1977)”. However, already 24 years before that Arnon & Wessel (1953) had demonstrated that vanadium greatly stimulates the growth of the green alga *Scenedesmus obliquus*. A very similar experiment was done by Meisch & Bielig (1975) with both *Scenedesmus obliquus* and *Chlorella pyrenoidosa*, which reacted very similarly (the difference between the two species was hardly significant, and only the average between them is shown in Fig. 5.1).

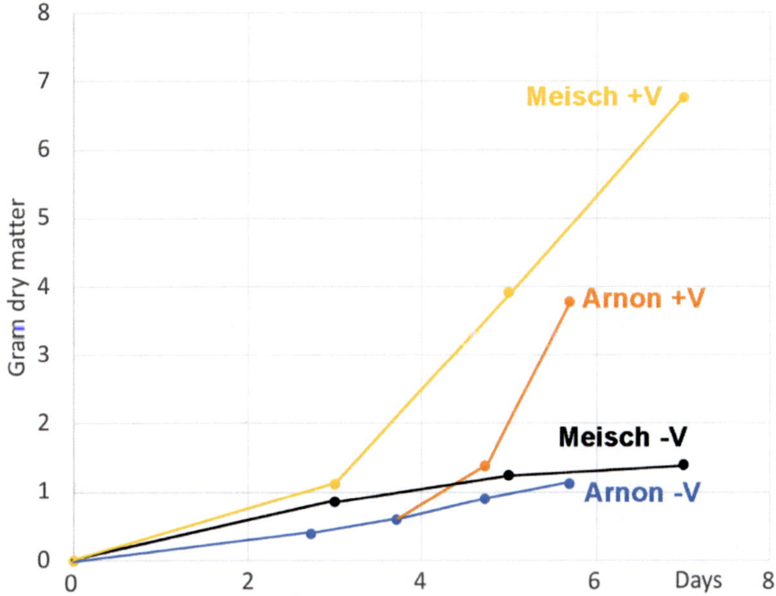

Fig. 5.1. Effect of vanadium (20 mg V/L as ammonium vanadate) on growth of 20 mg V/L as NH_4VO_3. For the experiment by Arnon & Wessel (1953) with *Scenedesmus*, marked "Arnon -V" no vanadate was added from the beginning, but some was probably present in the inoculum, and some dissolved from the glass vessel. For the orange curve Marked "Arnon +V" vanadin was added where it departs from "Arnon -V". For the experiments by Meisch & Bielig (1975) with *Scenedesmus* and *Chlorella*, marked "Meisch" vanadate was added from the beginning.

Also green and brown macroalgae, but not red algae are stimulated by vanadium (Fries 1982, Fig 5.2).

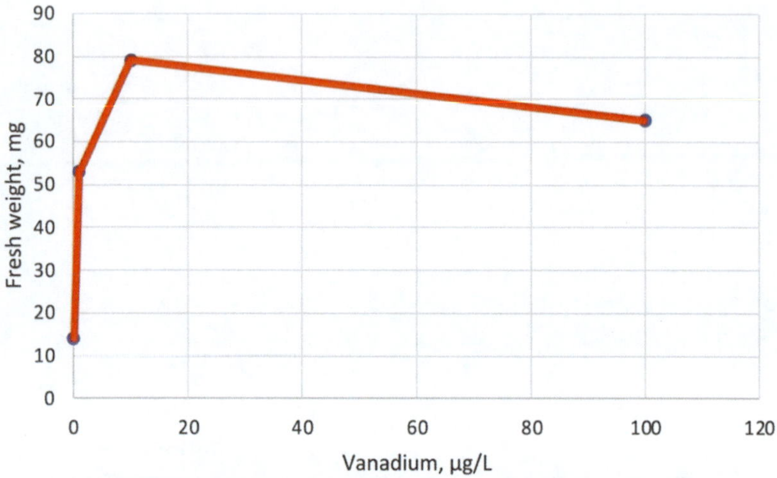

Fig. 5.2. Growth of *Fucus spiralis* with different concentrations of vanadium as ammonium vanadate. Data from Fries (1982).

There is no indication that vanadium is an essential element for any higher plant, but in some cases positive effects have been observed. Aihemaiti et al. (2019) found that vanadium stimulates the uptake of essential elements in *Setaria viridis* (dog tail's grass). According to García-Jiménez et al. (2018) 5 μM (but not higher concentrations) of vanadium as ammonium vanadate increases biomass of *Capsicum annuum* plants grown in nutrient solution (Fig. 5.3).

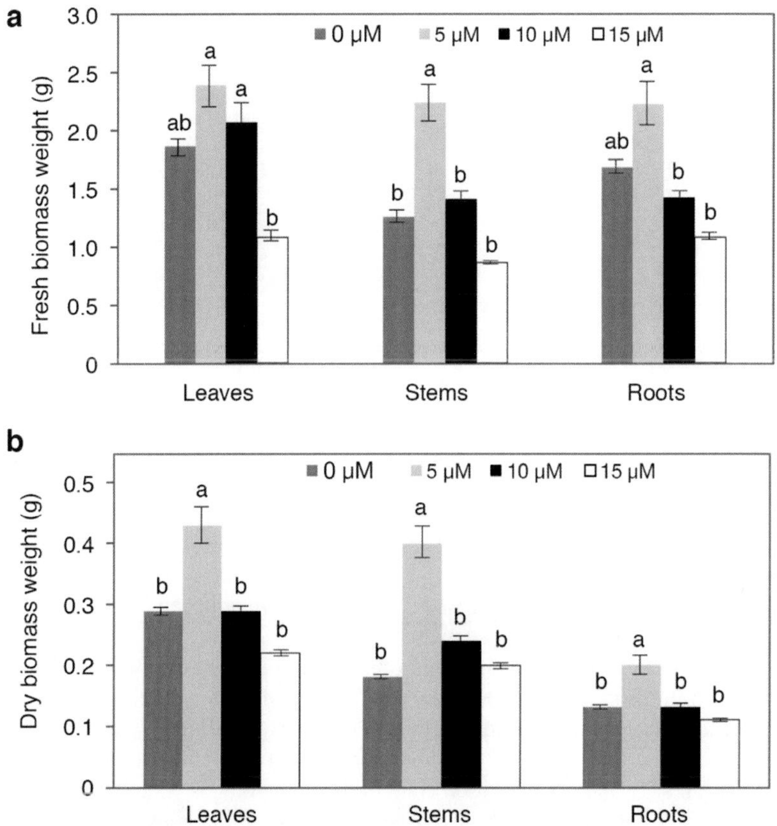

Fig. 5.3. Fresh (top) and dry (bottom) biomass of chili (chili pepper, Cayenne pepper, *Capsicum annuum*) plants grown hydroponically with different concentrations of vanadium as ammonium vanadate. Only the lowest concentration, 5 µ*M*, stimulates biomass accumulation. Reproduced under the Creative Commons Attribution License (CC BY 4.0) (from García-Jiménez et al. 2018).

According to García-Jiménez et al. (2021) vanadium (10 µ*M*) stimulates growth and flower production in tomato (*Solanum lycopersicum* L.) (Fig. 5.4). Many other experiments with various plants show only negative effects of vanadium.

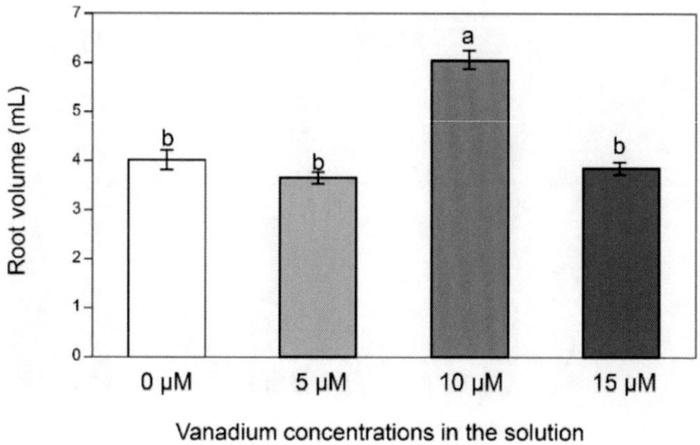

Fig. 5.4. Root volume of tomato plants grown for 28 days in nutrient solution with different concentrations of ammonium vanadate. Reproduced from García-Jiménez et al. (2021) under the Creative Commons Attribution (CC BY 4.0) License.

Since vanadium is toxic to humans and most animals, it is not surprising that others exploit it for their chemical defense. The vanadium concentration in the blood of the ascidian (sea-squirt) *Ascidia gemmate* can reach 347 m*M* which is 107 times that in seawater (Michibata et al. 2002). Some other species in the group Phlebobranchia also have high concentrations, while concentrations in the group Stolidobranchia are only a few micromolar. The highest concentrations reside in the membranes of a kind of blood cells called signet ring cells or vanadocytes. While vanadium is in the +5 oxidation seawater, its oxidation state in the vanadocytes is mostly +3 (and a little as +4), at least in the genus *Ascidia*. An earlier report (Macara & McLeod 1979) states that in the species *Ciona intestinalis* the oxidation state is mostly +5. The pH inside the vanadocytes varies between 1.86 and 4.20 in different *Ascidia*-species, with an inverse relationship between vanadium concentration and pH. From the species *Ascidia sydneiensis samea* Ueki et al. (2003) isolated several vanadium-binding proteins that they termed vanabins. A V-binding protein was also isolated from the coelomic fluid of the sea squirt *Halocynthia roretzi* (Kim & Kim 2019), erroneously called a sea urchin. Vanadium concentrations of fresh blood plasma, intestine, and muscle of *Halocynthia roretzi* were found by Gunasinghe & Kim (2018) to be 6.3, 3.7 and 2.1 mg/kg respectively. They purified two vanadium-binding proteins from vanadium binding proteins from its blood plasma and intestine.

Also some echinoderms accumulate vanadium, although to a lesser extent. Liu et al. (2015) report vanadium contents of several species of sea-cucumbers. The highest concentration, 2.56 μg g^{-1}, of which 85.5% was protein-bound, was found in *Holothuria scabra*.

Vanadium is found also in the "glue" in byssus fibers that anchors mussels such as *Mytilus edulis* to their substrate (usually rock) (Mesko et al. 2021). This "glue" consists of the protein rMfp-1-DOPA (DOPA= 3,4-dihydroxyphenylalanine). This protein can form DOPA−V complexes at the very low pH values relevant to the byssus formation process. Mussels often use vanadium in these complexes to enhance the cohesion, rather than solely iron as often believed. Vanadium provides a nearly twofold increase in cohesion within the protein network compared with iron, relevant for the proposed role of the byssus cuticle as an abrasion-resistant coating. However, also iron seems to play an important role, since the weakest adhesive strengths are produced in iron-deficient waters (Hamada et al. 2021). Mussels store iron and vanadium ions in intracellular metal storage particles (MSPs) complexed with catechol-based biomolecules. During adhesive formation, stockpiled secretory vesicles containing concentrated fluid proteins are mixed with MSPs within a microfluidic-like network of interconnected channels where they coalesce, forming protein-metal bonds within the nascent byssus (Priemel et al. 2021).

The vanadium content of some mushrooms of the genus *Amanita* is also very high. The highest level, 377 mg vanadium per kg dry matter, was found in the stalk of *Amanita velatipes* from Michigan (Koch et al. 1987), but the contents in the hat and in *Amanita regalis* and the common fly agaric, *Amanita muscaria*, from Finland, were also high. The vanadium is present as amavadin (Fig. 5.5), a vanadium (IV)complex with two N-hydroxy-2,2'-imino dipropionate ligands.

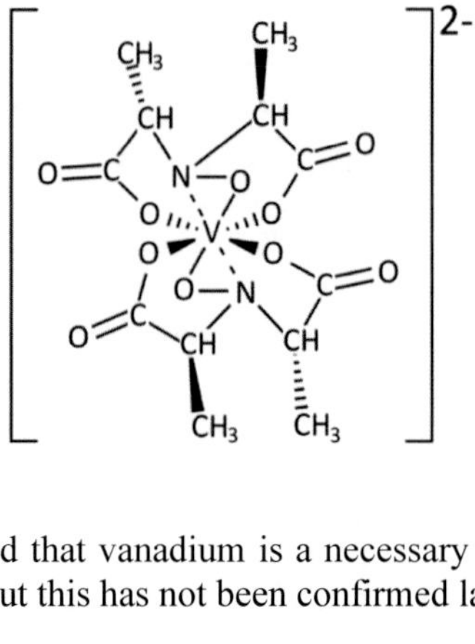

Figure 5.5. Amavadin.

Bertrand (1942) claimed that vanadium is a necessary element for the fungus *Aspergillus niger*, but this has not been confirmed later.

Already by the end of the nineteenth century vanadium was tested as a medicine; see Thompson & Orvig (2006) for a historical overview. More recently there have been many publications regarding the effect of vanadium compounds on diabetes and connection between vanadium status and diabetes incidence. Gestational diabetes is a condition in which a woman without type 1 or 2 diabetes develops high blood sugar levels during pregnancy. Li et al. (2022) found that plasma vanadium levels were significantly lower in the diabetic group than in the control group ($p < 0.001$). Others, e.g., Rezai et al. (2021) have obtained similar results.

The association between plasma vanadium levels and type 2 diabetes was evaluated by Wang et al. (2014). Participants in the highest quartile of plasma vanadium concentration had a notably lower risk of newly diagnosed type 2 diabetes, compared with persons in the lowest quartile. Treviño & Diaz (2020) have published a review with more information on the topic vanadium and insulin. Animal experiments confirm the ability of various vanadium compounds, e.g., a dioxidovanadium complex called cis-[VO_2 (obz) py] (Mbatha et al, 2022) to decrease blood glucose concentrations. Vanadium has an effect also on type 1 diabetes, but it differs from the effect on type 2 diabetes. In type 2 diabetic patients, improved insulin sensitivity, enhancement of nonoxidative glucose disposal rates and higher basal mitogen-activated kinase and ribosomal S6 kinase activity in monocytes were were found (Goldfine et al. 1995, Poucheret et al. 1990). In Type 1 diabetic patients, vanadium lowered insulin requirements without an effect on level of the connecting peptide, suggesting the absence of an influence on insulin release. The connecting peptide, or C-peptide, is a short 31-amino-acid polypeptide that connects the A-chain to the B-chain in the proinsulin molecule. In addition, 2 out of the 5 insulin-dependent subjects showed improved glucose utilization.

Jiang et al. (2021) found that human urinary vanadium concentrations were significantly associated with increased hypertension prevalence and blood pressure levels. Their findings also indicated that vanadium might be a toxic metal for the blood vascular system. On the contrary, Qui et al. (2020) found a negative correlation between vanadium exposure and blood pressure among pregnant women.

The anticancer activity of vanadium complexes has been tested extensively in vitro and in vivo on several types of cancer cell lines (overview by Ferretti & León 2022). Most results concern melanoma (reviewed by Amante et al. 2021). The authors conclude that vanadium compounds and/or vanadium materials show potential anticancer activities that may be used as a useful approach to treat melanoma. Many different vanadium compounds have been tested. In animal experiments sodium metavanadate, $NaVO_3$, proved to be effective against breast cancer (Tian et al. 2021).

Favre et al. (2022) demonstrate the ability of decavanadate, $V_{10}O_{28}^{6-}$, to selectively kill cells infected with SARS-CoV-2 virus.

Schwarz & Milne (1971) found that vanadium is positive for growing rats inside trace-element controlled, all plastic isolators on a highly purified amino acid diet. Vanadium supplementation of the diet increases growth by more than 40%.

Another important biological function of vanadium is in fixation of elementary nitrogen (N_2). There are three classes of nitrogenases, molybdenum, vanadium and iron nitrogenase, but all of them contain iron (Decamps et al. 2022). Vanadium nitrogenases have been found in the free-living bacteria *Azotobacter vinelandii* (Hales et al. 1986) and *Azotobacter croococcum* (Robson et al. 1986), and in the cyanobacterium *Anabaena variabilis* (Attridge & Rowell 2008).

Surprisingly, vanadium (both in the absence and in the presence of molybdenum) inhibits the reduction of acetylene to ethylene (interpreted as nitrogenase activity) in *Anabaena cylindrica* (Fay & de Vasconcelos 1974, Attridge & Rowell 2008). Neither Mo nor V was required for differentiation of the usually nitrogen-assimilating cells (heterocysts) in either *A. cylindrica* or *A. variabilis* (Attridge & Rowell 1997). Vanadium-dependent nitrogenase is present also in other *Anabaena*-species from a rice-field (Boison et al. 2006). There is vanadium-dependent nitrogenase also in the *Nostoc* photobiont in the lichen *Peltigera aphthosa* s.l. Darnajoux et al. (2017). Cyanobacterial nitrogenase is supplied with vanadium by a high-affinity vanadate transport system (Pratte & Thiel 2006).

There exist several CO_2-reduction reaction systems based on V-nitrogenase (Sickerman et al. 2017). They can be ATP-dependent or not, and employ physiological or artificial reductants.

Although some bacteria as well as some archaea are able to reduce the toxic V^V to less toxic V^{IV}, their methods differ (Yan et al. 2022). *Geobacter* and *Anaeromyxobacter* spp. were identified as putative V^V-reducing bacteria. Metagenomic-binning indicated that they may use the two nitrate reductases NarG and NapA for respiratory V^V reduction, as has been demonstrated previously for other species. The archaeon *Methanosarcina* spp. may reduce V^V via anaerobic methane oxidation pathways.

As Wever et al. (2018) write, "Vanadium-dependent haloperoxidases in seaweeds, cyanobacteria, fungi, and possibly phytoplankton play an important role in the release of halogenated volatile compounds in the environment. These halocarbons have effects on atmospheric chemistry since they cause ozone depletion." There are two classes of haloperoxidases: The heme-containing ones and those that contain vanadate as a prosthetic group. Haloperoxidases that can oxidize chloride, bromide, and iodide are named chloroperoxidases, while bromoperoxidases catalyze the oxidation of bromide and iodide and iodoperoxidases oxidize only iodide. Wever et al. (2018) provide a very extensive review of all kinds of haloperoxidases, and Wever & Barnett (2017), despite the title of their article, also review research on both vanadium- and iron-containing chloroperoxidases.

References

Amante, C., De Sousa-Coelho, A.L. & Aurcliano, M. (2021) Vanadium and melanoma: A systematic review. Metals 11: 828 (24 pp.). https://doi.org/10.3390/met11050828

Arnon, D.I. & Wessel, G. (1953) Vanadium is an essential element for green plants. Nature 172, 1039–1040.

Attridge, E.M. & Rowell, P. (1997) Growth, heterocyst differentiation and nitrogenase activity in the cyanobacteria *Anabaena variabilis* and *Anabaena cylindrica* in response to molybdenum and vanadium. New Phytol. 135, 517–526.

Badmaev, V., Prakash, S. & Majeed, M. (1999) Vanadium: A review of its potential role in the fight against diabetes. J. Alternative Complement. Med. 5, 273–291.

Bertrand, D. (1942) Le vanadium comme element oligosynergique pour l'*Aspergillus niger*. Ann. Inst. Pasteur 68, 226–244.

Boison, G., Steingen, C., Stal, L.J. & Bothe, H. (2006) The rice Weld cyanobacteria *Anabaena azotica* and *Anabaena* sp. CH1 express vanadium-dependent nitrogenase. Arch. Microbiol. 186. 367–376. DOI 10.1007/s00203-006-0150-4

Darnajoux, R., Zhang, X., McRose, D.L., Miadlikowska, J., Lutzoni, F., Kraepiel, A.M.L. & Bellenger, J.-P. (2017) Biological nitrogen fixation by alternative nitrogenases in boreal cyanolichens: importance of molybdenum availability and implications for current biological nitrogen fixation estimates. New Phytol. 213: 680–689. doi: 10.1111/nph.14166

Favre, D., Harmon, J.F., Zhang, A., Miller, M.S. & Kaltashov, I.A. (2022) Decavanadate interactions with the elements of the SARS-CoV-2 spike protein highlight the potential role of electrostatics in disrupting the infectivity cycle. J. Inorg. Biochem. 234: 111899 (8 pp.). https://doi.org/10.1016/j.jinorgbio.2022.111899

Fay, P. & de Vasconcelos, L. (1974) Nitrogen metabolism and ultrastructure in *Anabaena cylindrica* II. The effect of molybdenum and vanadium. Arch. Microbiol. 99, 221—230.

Ferretti, V.A. & León, I.E. (2022) An overview of vanadium and cell signaling in potential cancer treatments. Inorganics 10: 47 (11 pp). https://doi.org/10.3390/inorganics10040047

Fries, L. (1982) Vanadium an essential element for some marine macroalgae. Planta 154, 393–396.

García-Jiménez, A., Trejo-Téllez, Peralta Sánchez, M.G., Contreras-Oliva, A. & Gómez-Merino, F.C. (2021) Vanadium stimulates growth and flower production in tomato without affecting seed germination. Notulae Botanicae Horti Agrobotanici Cluj-Napoca 49:12400 (15 pp.). https://doi.org/10.15835/nbha49412400

Gunasinghe, M.A. & Kim, S.M. (2018) Antioxidant and antidiabetic activities of vanadium binding proteins purified from the sea squirt *Halocynthia roretzi*. J. Food. Sci. Technol. 55, 1840–1849. https://doi.org/10.1007/s13197-018-3099-9

Hales, B.J., Case, E.E., Morningstar, J.E., Dzeda, M.F. & Mauterer, L.A. (1986) Isolation of a New Vanadium-Containing Nitrogenase from *Azotobacter vinelandii*. Biochem. 25, 7251–7255.

Hamada, N.A., Gilpin, C. & Wilker, J.J. (2020) Availability of environmental iron influences the performance of biological adhesives produced by blue mussels. Environ. Sci. Technol. 54, 10254−10260. https://dx.doi.org/10.1021/acs.est.0c02392

Hirata, J. & Michibata. H. (1991) Valency of vanadium in the vanadocytes of *Ascidia gemmata* separated by density-gradient centrifugation. J. Exp. Zool. 257, 160–165.

Jiang, S., Zhou, S., Liu, H., Peng, C., Zhang, X., Zhou, H., Wang, Z. & Lu, Q. (2021) Concentrations of vanadium in urine with hypertension prevalence and blood pressure levels. Ecotox. Environm.l Safety 213: 112028 (8 pp.). https://doi.org/10.1016/j.ecoenv.2021.112028

Kim, A.T. & Kim, D.-O. (2019) Anti-inflammatory effects of vanadium-binding protein from *Halocynthia roretzi* in LPS-stimulated RAW264.7 macrophages through NF-κB and MAPK pathways. Internat. J. Biolog. Macromol. 133, 732–738. https://doi.org/10.1016/j.ijbiomac.2019.04.106

Koch, E., Kneifel, H. & Bayer, E. (1987) Das Vorkommen von Amavadin in Pilzen der Gattung *Amanita*/Occurrence of amavadin in mushrooms of the genus *Amanita*. Zetschr.Naturforschg. C 42, 873–878. https://doi-org.ludwig.lub.lu.se/10.1515/znc-1987-7-824

Li, X., Zhu, Y., Yin, J., Li, B., Li, P., Cao, B., Wang, Q., Xu, J. & Liu, L. (2022) Inverse Association of Plasma Vanadium Concentrations with Gestational Diabetes Mellitus. Nutrients 14, 1415. https://doi.org/10.3390/nu14071415

Liu, Y., Zhou, Q., Xu, J., Xue, Y., Liu, X., Wang, J. & Xue, C. (2016) Assessment of total and organic vanadium levels and their bioaccumulation in edible sea cucumbers: tissues distribution, inter-species-specific, locational differences and seasonal variations. Environ. Geochem. Health 38, 111–122. https://doi.org/10.1007/s10653-015-9689-9

Macara, I.G. & McLeod, G.C. (1979) Tunichromes and metal ion accumulation in tunicate blood cells. Comp. Biochem. Physiol. 63B, 299–302.

Mbatha, B., Khathi, A., Sibiya, N., Booysen, I. & Ngubane, P. (2022) A Dioxidovanadium complex cis-[VO_2 (obz) py] attenuates hyperglycemia in Streptozotocin (STZ)-induced diabetic male Sprague-Dawley rats via increased GLUT4 and glycogen synthase expression in the skeletal muscle. Evidence-Based Complementary Alternat. Medicine 2022, Article ID 5372103 (11 pp.). https://doi.org/10.1155/2022/5372103

Meisch, H.-U. & Bielig, H.-J. (1975) Effect of vanadium on growth, chlorophyll formation and iron metabolism in unicellular green algae. Arch. Microbiol. 105, 77–82.

Mesko, M., Xiang, L., Bohle, S., Hwang, D.S., Zeng,H. & Harrington, M.J. (2021) Catechol-vanadium binding enhances cross-linking and mechanics of a mussel byssus coating protein. Chem. Mater. 33, 6530−6540. https://doi.org/10.1021/acs.chemmater.1c02063

Michibata, H., Uyama, T, Ueki, T. & Kanamori, K. (2002) Vanadocytes, cells hold the key to resolving the highly selective accumulation and reduction of vanadium in ascidians. Microscope Res. Techn. 56, 421–434. https://doi.org/10.1002/jemt.10042

Pratte, B.S. & Thiel, T. (2006) High-affinity vanadate transport system in the cyanobacterium *Anabaena variabilis* ATCC 29413. J. Bacteriol. 188, 464–468. https://doi.org/10.1128/JB.188.2.464–468.2006

Priemel, T., Palia, G., Förste, F., Jehle, F., Sviben, S., Mantouvalou, I., Zaslansky, P., Bertinetti, L. & Harrington, M.J. (2021) Microfluidic-like fabrication of metal ion–cured bioadhesives by mussels. Science 374, 206–211. https://www.science.org/doi/10.1126/science.abi9702

Qiu, L., Xu, S., Zhou, A., Liu, W., Liao, J., Cao, Z., Chen, Z., Yao, C., Zhang, Y. & Lia,Y. (2020) Association between changes in gestational blood pressure and vanadium exposure in China. Environm. Toxicol. Pharmacol. 79: 103424 (6 pp.). https://doi.org/10.1016/j.etap.2020.103424

Robson, R.L., Eady, R.R., Richardson, T.H., Miller, R.W., Hawkins, M. & & Postgate, J.R. (1986) The alternative nitrogenase of *Azotobacter chroococcum* is a vanadium enzyme. Nature 322, 388–390.

Schwarz, K. & Milne, D.B (1971) Growth effects of vanadium in the rat. Science 174, 426–428.

Sickerman, N.S., Hu,Y. & Ribbe, M.W. (2017) Activation of CO_2 by vanadium nitrogenase. Chem. Asian J. 12, 1985 – 1996. https://doi.org/10.1002/asia.201700624

Tian, Y., Qi, H., Wang, G., Li, L. & Zhou, D. (2021) Anticancer effect of sodium metavanadate on murine breast cancer both in vitro and in vivo. Biometals 34, 557–571. https://doi.org/10.1007/s10534-021-00295-z

Treviño, S. & Diaz, A. (2020) Vanadium and insulin: Partners in metabolic regulation. J. Inorg. Biochem. 208: 111094 (19 pp.). https://doi.org/10.1016/j.jinorgbio.2020.111094

Ueki, T., Adachi, T., Kawano, S., Aoshima, M., Yamaguchi. N., Kanamori, K. & Michibata, H. (2003) Biochim. Biophys. Acta 1626, 43– 50. https://doi.org/10.1016/S0167-4781(03)00036-8

Wang, X., Sun, T., Liu, J., Shan, Z., Jin, Y., Chen, S., Bao, W., Hu, F.B. & Liu, L. (2014) Inverse association of plasma vanadium levels with newly diagnosed type 2 diabetes in a Chinese population, Am. J. Epidemiol. 180, 378–384. https://doi.org/10.1093/aje/kwu148

Wever, R. & Barnett, P. (2017) Vanadium chloroperoxidases: The missing link in the formation of chlorinated compounds and chloroform in the terrestrial environment? Chem. Asian J. 2017, 12, 1997 – 2007. https://doi.org/10.1002/asia.201700420

Wever, R., Krenn, B.E. & Renirie, R. (2018) Marine vanadium-dependent haloperoxidases, their isolation, characterization, and application. Ch. 6 in Methods in Enzymology 605, 141–201. https://doi.org/10.1016/bs.mie.2018.02.026

Yan, G., Sun, X., Dong, Y., Gao, W., Gao, P., Li, B., Yan, W., Zhang, H., Soleimani, M., Yan, B., Häggblom, M.M. & Sun, W. (2022) Vanadate reducing bacteria and archaea may use different mechanisms to reduce vanadate in vanadium contaminated riverine ecosystems as revealed by the combination of DNA-SIP and metagenomic-binning. Water Res. 226: 119247 (10 pp.). https://doi.org/10.1016/j.watres.2022.119247

CHAPTER SIX

CHROMIUM, MOLYBDENUM, AND TUNGSTEN

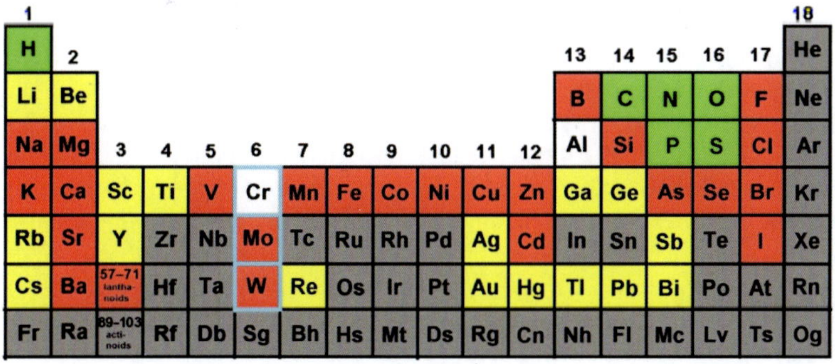

6.1 Chromium

The element chromium was discovered in 1797 by the French chemist Nicolas-Louis Vauquelin, who produced it in metallic form in 1798. Chromium got its name from its many coloured compounds (Greek χρωμα, khroma=colour), which are yellow, orange, green, violet, purple, or blue (Fig. 6.1.1). Emeralds as well as rubies get their colour from their chromium ions. Chromium can exist in all oxidation states from 0 (metal and $Cr(CO)_6$) to +6 (chromate and dichromate); in biology we can restrict ourselves to +3 (III) and +6 (VI). Hexavalent chromium (Cr^{VI}) in our environment is mostly anthropogenic, and there is consensus that it is dangerous for us, causing cancer and mutations.

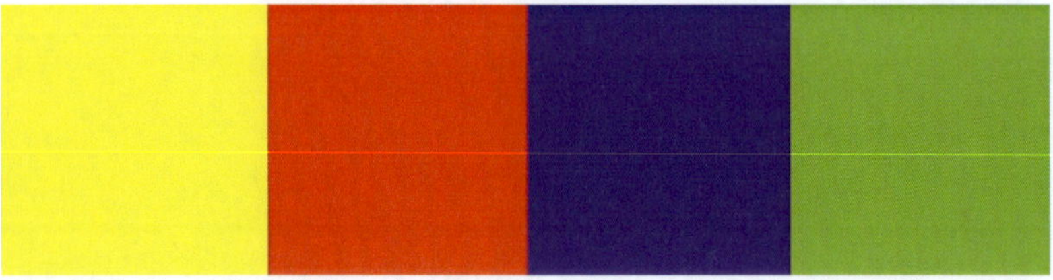

Fig. 6.1.1. Examples of the colours of chromium compounds. From left to right: potassium chromate (K_2CrO_4), ammonium dichromate ($[NH_4]_2Cr_2O_7$), chromium (III) chloride ($CrCl_3$), and chromium oxide (Cr_2O_3).

Trivalent chromium (Cr^{III}), on the other hand, is a bit controversial. Chromium has usually been thought to be an essential trace metal for healthy humans and other mammals (e.g., Kośla et al. 2018, Zoroddu et al. 2019), but everybody does not agree (Kleefstra et al. 2006, Di Bona et al. 2011, Vincent 2017). According to Vincent (2017) chromium has no positive role in healthy humans, although it may for some rodents, and he claims that the ‘‘Glucose Tolerance Factor’’ described in some articles is an artifact. At the end of the section on chromium some more recent literature is cited that voices the opposite opinion. The older literature is discussed in depth by Mertz (1969). As late as in 2021 Genchi et al. write “While Cr(III) is an essential trace element, found in nature in rocks and soil, readily absorbed by plants, hexavalent chromium is mainly an industrial contaminant and is also produced from anthropogenic activities.”

There are many reports claiming that trivalent chromium has beneficial effects on people with certain health problems, especially diabetes. An early indication (Jeejeebhoy et al. 1977) of the importance of chromium was obtained from a woman who had been maintained on total parenteral nutrition (i.e. receiving nutrition directly into the blood through a vein) for more than 5 years. Despite receiving glucose and insulin, she was losing weight. After 250 µg of Cr was added to the daily nutrition, she returned towards normal condition, insulin addition was no longer needed, and glucose infusion had to be reduced to prevent overweight.

Anderson (1997, 2000) stated that “Chromium (Cr) improves the glucose/insulin system in subjects with hypoglycemia, hyperglycemia, diabetes and hyperlipemia”. In other words: Chromium has no function in healthy subjects, but is important for persons with certain health problems, such as diabetes. But later research has shown contradictory results. Thus Pechova & Pavlata (2007) write “Trivalent chromium is essential to normal carbohydrate, lipid and protein metabolism. Chromium is biologically active as part of an oligopeptide, chromodulin, potentiating the effect of insulin by facilitating insulin binding to receptors at the cell surface." We shall return to chromodulin below.

The way chromium is administered is of importance. Feng et al. (2015) found chromium malate to be more effective than the often used chromium picolinate, while Peng et al. (2015) wrote "Chromium phenylalanine sensitizes insulin cell signaling pathway via the activation of phosphorylation of protein kinase B and/or AMP-activated protein kinase". There have been other attempts to pinpoint the enzymatic details regarding how chromium influences glucose metabolism. Hoffman et al. (2014) thought that chromium enhances insulin responsiveness via 5' AMP-activated protein kinase.

Nishimura et al. (2021), working with cultures of erythropoietin-producing so-called HepG2 cells, found that insulin resistance was induced by culturing with insulin resistance induction medium supplemented with palmitic acid for 24 h. When trivalent chromium was added to the medium, the increase in glucose-6-phosphatase and phosphoenolpyruvate carboxykinase 1 mRNA expression levels and the decrease in the ratio of phosphorylated protein kinase B (Akt) to Akt protein were suppressed, and the induction of insulin resistance prevented. Results suggest that the inhibition of insulin resistance induction by chromium in HepG2 cells involves the promotion of erythropoietin production mediated by proliferator-activated receptor γ, in addition to other proliferator-activated receptor γ-mediated activities. Erythropoietin is a glycoprotein cytokine secreted mainly by the kidneys in response to lack of oxygen. It stimulates production of the oxygen-carrying red blood cells.

The presence of chromium increases glucose-6-phosphatase and phosphoenolpyruvate carboxykinase 1 mRNA expression levels and decreases the ratio of phosphorylated to unphosphorylated protein kinase B, and the induction of insulin resistance is prevented. Zhang et al. (2016, 2018) found that even maternal chromium restriction induces insulin resistance in adult mice offspring through miRNA. Asbaghi et al. (2021) cites literature showing that chromium supplementation results in a significant decrease in serum levels of triglyceride and of total cholesterol in persons suffering from type 2 diabetes. Similar results had previously been obtained with mice (Liu et al. 2018). Capone et al. (2018) found that chromium supplementation resulted in better glucose tolerance and calorie delivery during the first week of life, especially in very low birth weight infants. The evidence that there is a human need for chromium nutrition thus seems overwhelming, but Vincent (2017) still maintained a different opinion.

Chromodulin is an oligopeptide that has an essential role for the flawless functioning of insulin (Kircheva et al. 2022). Indispensable for the integrity of the molecule in its active (holo-) form are four chromium cations. Several other cations compete with chromium for the chromodulin, but only the complex with chromium is active for insulin function. LMWCr described by Edwards et al. (2021) appears to be identical to chromodulin.

There are many reports on positive effects of chromium on ruminants. Chromium can increase the growth and milk production of cows (Lashkari et al. 2018, Żarczyńska & Krzebietke 2020) and growth of calves and pigs (for literature, see Lukaski 1999). Chromium was found to improve the feed conversion ratio per dry matter intake and crude protein intake in goats (Lalhriatpuii et al. 2022). In preruminant calves at the transition into ruminant animals the glucose level in blood decreased much faster after injection (or oral administration) of glucose if the calves had been provided with chromium picolinate (Khare et al. 2022, Fig. 6.1.2).

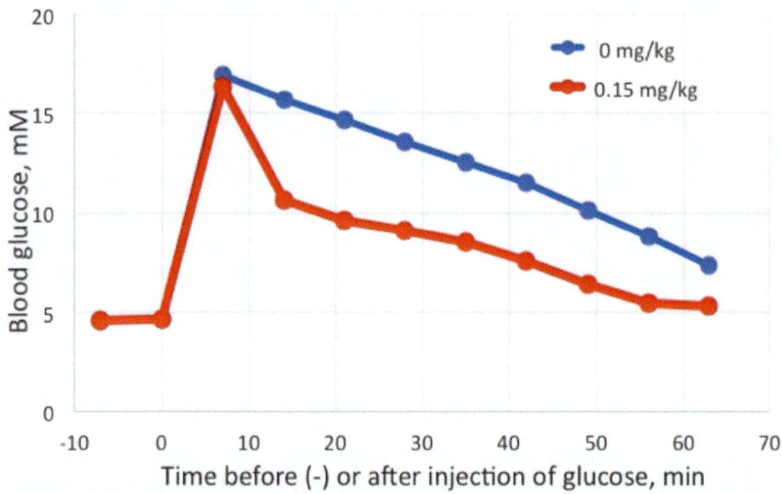

Fig. 6.1.2. Blood concentration (m*M*) in transition calves (age ca 60 days) at different times after intravenous injection of 50% glucose solution (0.50 g/kg of body weight) at 0 minutes. The calves were fed either a basal diet without added chromium (blue) or a basal diet with 0.15 mg/kg chromium as chromium picolinate (red).

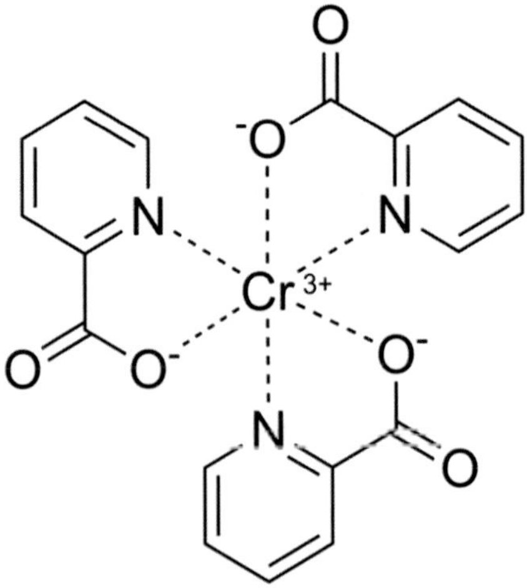

Fig, 6.1.3. Chromium picolinate.

Many papers report alleviation by chromium of heat stress in poultry and other animals (review by Bin-Jumah et al. 2019).

Walker et al. (2020) found that chromium (as Cr_2O_3) had beneficial effects on the metabolic responses of a fish, Nile tilapia (*Oreochormis niloticus*), and Limwachirakhom et al. (2011) describe positive effects on this fish of chromium-L-methionin in combination with a zinc-aminoacid complex. Bagheri et al. (2022), after reviewing the literature, point out that for positive effects on various fish species, the chromium must be given in the food, not dissolved in the water. Sather (1966) studied the chromium content in various organs of the crab *Podophthalmus vigil*, and expressed suspicions that chromium might have a function in this animal. Zhang et al. (2022) write: "Chromium (Cr) is an essential micronutrient for all animals", and more specifically, they claim to show "that the optimal Cr requirement was estimated to be 2.34 mg kg^{-1} for juvenile mud crab".

Authors of literature published after the article by Vincent (2017) appear to agree that trivalent chromium has a positive role to play in the metabolism of humans and some animals. We shall conclude by citing some of the most recent articles (the literature until 2019 was reviewed by Swaroop et al. 2019).

In rats, chromium picolinate appears to have positive effects even on non-diabetic individuals. It was found to lower blood glucose, total cholesterol, and triglyceride, but elevated insulin concentration compared with the control (Pala et al. 2020).

Disturbance of the function of the kidneys is one of the most important complications of diabetes. In rats in which this diabetic disturbance had been induced, administration of chromium picolinate decreased levels of blood glucose, serum insulin, blood urea, serum creatinine, and urinary albumin (Qi et al. 2020).

Chen et al. (2022) found that men with low vs. normal blood chromium levels have increased odds of cardiovascular diseases and diabetes mellitus, and lower odds of depression.

Trivalent chromium is an efficient remedy for rheumatoid arthritis and lacks some unwanted side effects of other treatments (Hassouna et al. 2022).

Chromium has important effects not only in vertebrates, but also in molluscs and crustaceans. As examples we show effects on weight gain and shell size in abalone, *Haliotis discus hannai* (Fu et al. 2022) (Fig. 6.1.4, 6.1.5, and weight gain in the mud crab *Scylla paramamosain* (Zhang et al. 2022) (Fig. 6.1.6).

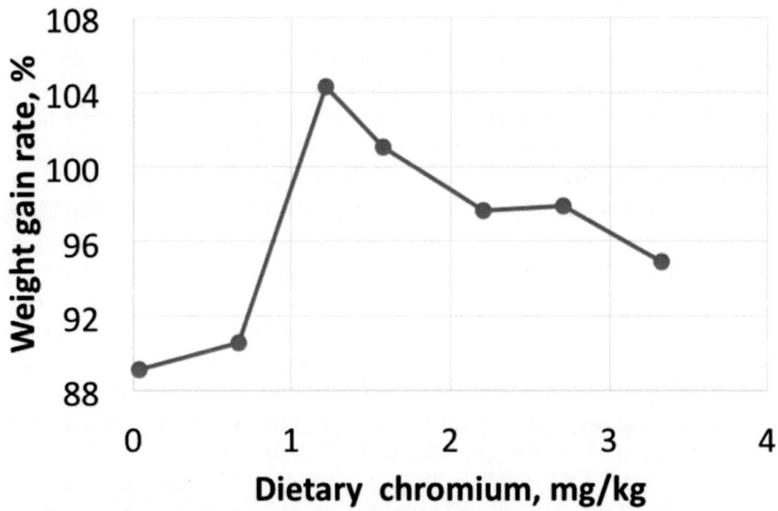

Fig. 6.1.4. The dietary Cr requirement based on weight gain rate of abalone *Haliotis discus hannai*. Redrawn and modified from Fu et al. (2022).

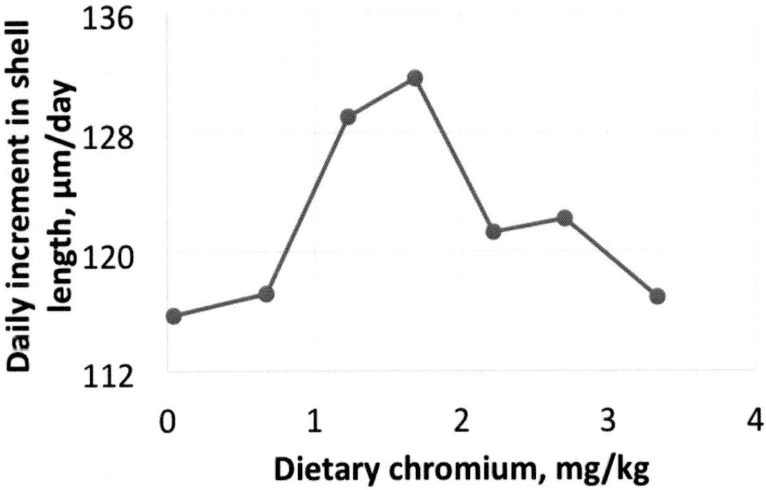

Fig. 6.1.5. The dietary Cr requirement based on daily increment shell length of abalone *Haliotis discus hannai*. Redrawn and modified from Fu et al. (2022).

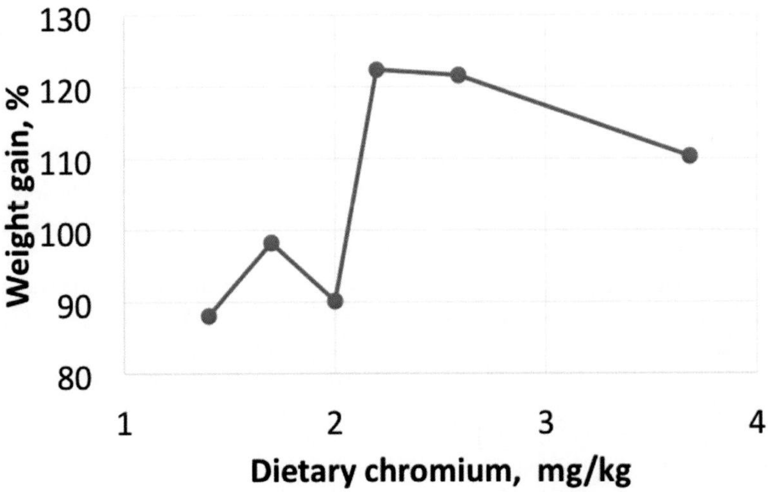

Fig. 6.1.6. Relationship between the percent weight gain and the dietary Cr levels in the mud crab *Scylla paramamosain*. Redrawn and modified from Zhang et al. (2022).

6.2 Molybdenum and tungsten

In column six in the periodic system, as it usually is displayed, we find chromium, molybdenum, tungsten, and some elements having no biochemical role for life on Earth. We discussed chromium separately in the previous section. Molybdenum is the only second-row transition metal that is required by most living organisms, and the few species that do not require molybdenum use tungsten (Hille 2002). Humans have four molybdenum enzymes, and a total of over 50 types of molybdenum enzymes have been characterized while, until recently, tungsten was thought to play no role in human biology, and only a very minor role in other organisms. We shall here put the emphasis on tungsten, because its great biochemical importance has only so recently been appreciated, but also give some information about the roles of molybdenum. For overviews of molybdenum enzymes, see Hille et al. (2014), Leimkühler & Iobbi-Nivol (2016), Mintmier et al. (2020), and Kim et al. (2021).

Tungsten is special in many ways. To start with, in Sweden we use a German name (volfram, from German Wolfram), while English speakers have taken over the original Swedish name, meaning "heavy stone". Pure tungsten has the highest melting point, 3407°C, of all metals. For this reason it is used in the filament of incandescent lamps. Only a couple of other metals come even close to this (the platinum metals rhenium, 3180°C, and osmium, 3045°C). The density, 19.26 g/cm^3 at room temperature, is also among the highest for any element; here tungsten is surpassed only by some platinum and transuranium metals. With the atomic weight, 183.85, corresponding to the weight of the 30 atoms of ten water molecules, tungsten is the "living element" having the highest atomic weight. Second comes barium with 137.33 and third cadmium with 112.41.

Tungsten is an important part of several enzymes in bacteria and archaea. In some cases, but not all, molybdenum can replace tungsten in the same enzyme.

In some respects molybdenum and tungsten are surprisingly similar. The so-called van der Waal radius for molybdenum is 245 pm (picometer, 10^{-12} m), for tungsten 249 pm. It is a measure of how close one atom can get to another atom. Of the electronic orbitals the d orbital for molybdenum is 72.9 pm, for tungsten 77.6 pm. Thus it comes as no surprise that the two elements sometimes can have the same biological function. Perhaps it is more surprising that organisms sometimes can distinguish between them.

Apart from nitrogenase, which is distinct from all other molybdenum-containing enzymes, all molybdenum and tungsten enzymes contain either one or two pyranopterin dithiolene cofactors, known as molybdopterin (Fig. 6.2.1) or tungstopterin (Crawford et al. 2019, Mathew et al. 2022).

Examples of enzymes with tungsten as cofactor are aldehyde reductases and a kind of formyl methanofuran dehydrogenase. The corresponding proteins belong to several different branches on the evolutionary tree (Seelmann et al. 2020). The following enzymes have either tungsten or molybdenum as cofactor: Dimethyl sulfoxide reductase, formate dehydrogenase, thiosulfate reductase, membrane-associated nitrate reductase, periplasmatic nitrate reductase. Seelmann et al. (2020) show the reactions catalyzed by these enzymes.

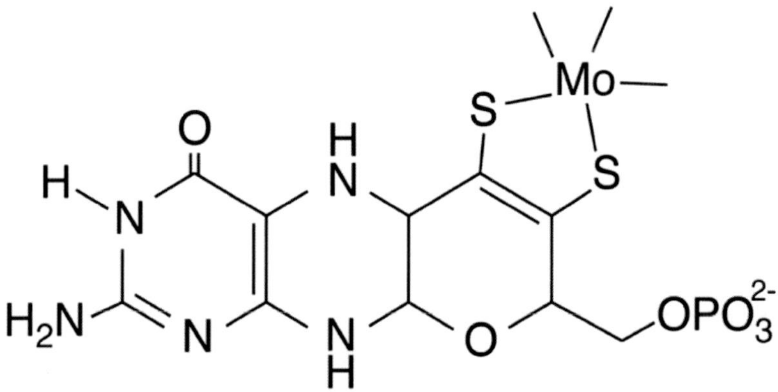

Fig. 6.2.1. Structure of molybdopterin. The "dangling" bonds are usually connected to oxygen.

Those enzymes that have an absolute requirement for molybdenum (and thus cannot function with tungsten) are more numerous, and some of them (e.g., xanthine oxidase), contrary to those mentioned earlier, also occur in eukaryotic organisms. Examples are arsenite oxidase, sulfite oxidase, xanthine oxidase, and carbon monoxide dehydrogenase.

Several strains of *Sulfitobacter dubius* are able to grow in the presence of high tungstate concentrations and their growth is unaffected by 1 m*M* tungstate. It is taken into the bacterial cells in the tungstate ion form (WO_4^{2-}) by several ABC transporters (Coimbra et al. 2017). ABC transporters belong to the ATP-Binding Cassette superfamily, which uses the hydrolysis of ATP to translocate a variety of compounds across biological membranes. References to structures of many molybdate and tungstate transport proteins are given by Hagen (2011).

Not only some bacteria, but also some archaea, in particular methanogens, produce tungsten-containing enzymes. Tungsten stimulation of growth of a methanogen was first reported for *Methanococcus vannielii*

(Jones & Stadtman 1976, 1977), belonging to Euryarchaeota. Its growth was dramatically enhanced by the addition of W, but not by Mo, to the growth medium when formate was used as the carbon and energy source, but not when the organism was grown on H_2 and CO_2.

Another archaeon, *Pyrococcus furiosus*, a strictly anaerobic heterotroph (also belonging to Euryarchaeota) that grows optimally around 100°C produces several tungsten enzymes. Its growth is considerably stimulated by tungstate when grown with maltose, but not when grown with peptides as carbon source (Schicho et al. 1993). Sevsenco et al. (2009) found that it produces the following five tungsten-containing enzymes: formaldehyde ferredoxin oxidoreductase, aldehyde ferredoxin oxidoreductase, glyceraldehyde-3-phosphate ferredoxin oxidoreductase and the tungsten-containing oxidoreductases WOR4 and WOR5 (Fig. 6.2.2).

The oxidoreductase WOR5 is worth further comments. It is increased by "cold shock" (growth at 72°C), and contains a subunit with four [4Fe–4S] clusters. The enzyme was found to be an aliphatic sulfonate ferredoxin oxidoreductase that catalyzes the reaction

$$R(CH_2)_2SO_3^- + 2H_2O \rightarrow SO_3 + R(CH_2)_2O_2^-$$

(Mathew et al. 2022), but can have many different substrates. Perhaps the natural one is taurine, 2-aminoethanesulfonic acid ($^+NH_3CH_2CH_2SO_3^-$). In the ocean huge amounts of taurine are released from planktonic crustaceans (Clifford et al. 2017).

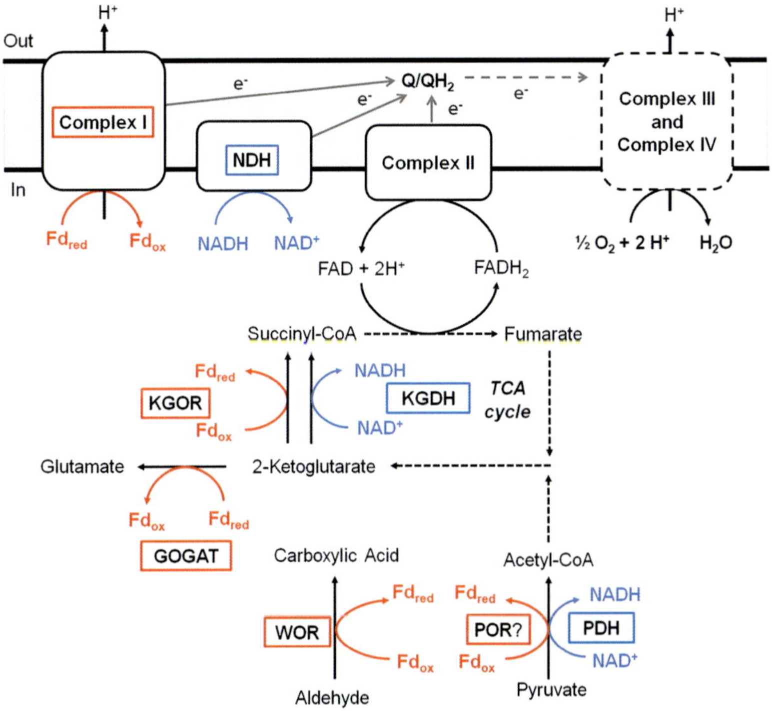

Fig. 6.2.2. Proposed coupling of the tungsten oxidoreductase (WOR) to the respiration system. From Thorgersen et al. (2022). NDH, NADH dehydrogenase; KGOR, 2-ketoglutarate ferredoxin oxidoreductase; KGDH, 2-ketoglutarate dehydrogenase; GOGAT, Fd-dependent glutamate synthase; POR, pyruvate ferredoxin oxidoreductase. Reproduced by Creative Commons Attribution License https://creativecommons.org/ licenses/by/4.0/.

Pyrobaculum aerophilum is a third archaeon equipped with a tungsten enzyme and requires tungstate ions (WO_4^{-2}) for growth. It is the first example of a tungsten-dependent archaeon belonging to the Crenarchaeota. The tungsten enzyme required is probably a aldehyde ferredoxin oxidoreductase (Afshar et al. 1998).

The bacterium *Caldicellulosiruptor bescii* is also thermophilic, with a growth optimum at 78°C. It is equipped with a tungsten-containing oxidoreductase of unknown function that is rather different from the *Pyrococcus furiosus* oxidoreductases. It consists of a large subunit with a single four Mo enzyme site and one iron-sulfur [4Fe-4S] cluster, and a small subunit with four [4Fe-4S] clusters, and uses ferredoxin as electron acceptor (Scott et al. 2019). *Eubacterium acidaminophilum* has a different tungsten-containing oxidoreductase with six iron atoms, and probably a zinc atom per tungsten atom (Rauh et al. 2007). So there exists a great diversity of tungsten-containing oxidoreductases.

Tungsten plays a role even in human biology, although indirectly. Schut et al. (2021) even go so far as to write that tungsten "might be an overlooked micronutrient" for humans. They found that the human gut microbiome produces tungsten-containing oxidoreductase that removes toxic aldehydes generated from cooked foods and microbial metabolism. Thorgersen et al. (2022) identified one of the organisms active in this way as *Brevibacillus massiliensis* strain phR, and purified from it a tungsten-containing oxidoreductase, consisting of a single 65 kDa subunit and containing a single W-pyranopterin (tungstopterin) cofactor and a single [4Fe-4S] cluster. They also found several genes for molybdenum enzymes in it. They also found genes for tungstate (WO_4^{2-}) and molybdate (MoO_4^{2-}) transport proteins. Carbon dioxide is a stable molecule that is difficult to reduce. Its electrochemical reduction usually requires large overpotentials, and thus is energetically wasteful. Reda et al. (2008) describe how formate can be electrochemically produced from carbon dioxide without this wasteful overpotential by using tungsten-containing formate dehydrogenase absorbed on the cathode of an electrochemical cell.

References

Afshar, S., Kim, C., Monbouquette, H.G. & Schröder, I. 1998. Effect of tungstate on nitrate reduction by the hyperthermophilic archaeon *Pyrobaculum aerophilum*. Appl. Environm. Microbiol. 64, 3004–3008.

Anderson, R.A. 1997. Nutritional factors influencing the glucose/insulin system: chromium. J. Am. College Nutrition. 16, 404–410.

Anderson, R.A. 2000. Chromium in the prevention and control of diabetes. Diabetes & Metabolism (Paris) 26, 22–27.

Asbaghi, O., Naeini, F., Ashtary-Larky, D., Moradi, S., Nazanin Zakeri, N., Eslampour, E., Kelishadi, M.R. & Naeini, A.A. 2021. Effects of chromium supplementation on lipid profile in patients with type 2 diabetes: A systematic review and dose-response meta-analysis of randomized controlled trials. Journal of Trace Elements in Medicine and Biology 66: 126741 (12 pp.). https://doi.org/10.1016/j.jtemb.2021.126741

Bagheri, S., Gholamhosseini, A. & Banaee, M. 2022. Investigation of different nutritional effects of dietary chromium in fish: A literature review. Biolog. Trace Elem. Res. (on line 2022, 9 pp.) https://doi.org/10.1007/s12011-022-03326-z

Bin-Jumah, M., El-Hack, M.E.A., Abdelnour, S.A., Hendy, Y.A., Ghanem, H.A., Alsafy, S.A., Khafaga, A.F., Noreldin, A.E., Shaheen, H., Samak, D., Momenah, M.A., Allami, A.A., AlKahtane, A.A., Alkahtani, S., Abdel-Daim, M.M. & Aleya, L. 2020. Potential use of chromium to combat thermal stress in animals: A review. Sci. Total Environm, 707: 135996 (11 pp.). https://doi.org/10.1016/j.scitotenv.2019.135996

Capone, K., Sriram, S., Patton, T., Weinstein, D., Newton, E., Wroblewski, K. & Sentongo, T. 2018. Effects of chromium on glucose tolerance in infants receiving parenteral nutrition therapy. Nutrition in Clinical Practice 33, 426–432.

Chen, J., Kan, M., Ratnasekera, P., Deol, L.K., Thakkar, V. & Davison, K.M. 2022. Blood chromium levels and their association with cardiovascular diseases, diabetes, and depression: National Health and Nutrition Examination Survey (NHANES) 2015–2016. Nutrients 14: 2687 (16 pp.). https://doi.org/10.3390/nu14132687

Clifford, E.L., Hansell, D.A., Varela, M.M., Nieto-Cid, M., Herndl, G.J. & Sintes, E. 2017. Crustacean zooplankton release copious amounts of dissolved organic matter as taurine in the ocean. Limnol. Oceanogr. 62, 2745–2758. https://doi.org/10.1002/lno.10603

Coimbra, C., Farias, P., Branco, R. & Morais, P.V. 2017. Tungsten accumulation by highly tolerant marine hydrothermal *Sulfitobacter dubius* strains carrying a tupBCA cluster. System. Appl. Microbiol. 40, 388–395. http://dx.doi.org/10.1016/j.syapm.2017.06.005

Crawford, A.M., Cotelesage, J.J.H., Prince, R.C. & George, G.N. 2019. The catalytic mechanisms of the molybdenum and tungsten enzymes. Struct. Bond. 179, 63–100. https://doi.org/10.1007/430_2018_30

Di Bona, K.R., Love, S., Rhodes, N.R., McAdory, D.A., Sinha, S.H., Kern, N., Kent, J., Strickland, J., Wilson, A., Beaird, J., Ramage, J., Rasco, J.F. & Vincent, J.B. 2011. Chromium is not an essential trace element for mammals: effects of a ''low-chromium'' diet. J. Biol. Inorg. Chem. 16, 381–390. https://doi.org/10.1007/s00775-010-0734-y

Edwards, K.C., Gannon, M.W., Frantom, P.A. & Vincent, J.B. 2021. Low-molecular-weight chromium-binding substance (LMWCr) may bind and carry Cr(III) from the endosome. J. Inorg. Biochem. 223: 111555 (11 pp.). https://doi.org/10.1016/j.jinorgbio.2021.111555

Feng, W., Mao, G., Li, Q., Wang, W., Chen, Y., Zhao, T., Fang Li, F., Zou, Y., Wu, H., Yang, L. & Wu, X. 2015. Effects of chromium malate on glycometabolism, glycometabolism-related enzyme levels and lipid metabolism in type 2 diabetic rats: A dose–response and curative effects study. J Diabetes Invest 2015; 6: 396–407. https://doi.org/10.1111/jdi.12350

Fu, Y., Guo, J., Wu, Z., Yu, X., Guo, Y., Huang, D., Li, X., Zhang, W. & Mai, K. 2022. Effects of dietary chromium yeast levels on growth performance, anti-oxidative capacity, immune response and flesh quality of abalone *Haliotis discus hannai*. Aquaculture 557: 738291 (11 pp.). https://doi.org/10.1016/j.aquaculture.2022.738291

Genchi, G., Lauria, G., Catalano, A., Carocci, A. & Sinicropi, M.S. 2021. The double face of metals: The intriguing case of chromium. Appl. Sci. 11: 638 (20 pp.). https://doi.org/10.3390/app11020638

Hagen, W.R. 2011. Cellular uptake of molybdenum and tungsten. Coord. Chem-Revs 255, 1117–1128. https://doi.org/10.1016/j.ccr.2011.02.009

Hille, R. 2002. Molybdenum and tungsten in biology. Trends Biochem. Sci. 27, 360–367.

Hassouna, S.S., Sheta, E., Zaki, I., Harby, S.A. & Allam, E.A. 2022. Trivalent chromium supplementation ameliorates adjuvant induced rheumatoid arthritis through up-regulation of FOXP3 and decrease in synovial Cathepsin G expression. Inflammopharmacol. 30, 2181–2195. https://doi.org/10.1007/s10787-022-01025-8

Hille, R., Hall, J. & Basu, P. 2014. The mononuclear molybdenum enzymes. Chem. Rev. 114, 3963−4038. dx.doi.org/10.1021/cr400443z

Jeejeebhoy, K.N., Chu, R.C., Marliss, E.B., Greenberg, G.R. & Bruce-Robertson, A. 1977. Chromium deficiency, glucose intolerance, and neuropathy reversed by chromium supplementation, in a patient receiving long-term total parenteral nutrition. Am. J. Clin. Nutr. 30, 531–538.

Jones, J.B. & Stadtman, T.C. 1976. *Methanococcus canniellii*: growth and metabolism on formate. In: Symposium on the microbial production and utilization of gases (H_2, CH_4, CO) (Schlegel, H.D., Ed.), pp. 197–205. Academie der Wissenschaften zu Göttingen, Göttingen.

Jones, J.B. & Stadtman, T.C. 1977. *Methanococcus cannielii*: culture and effects of selenium and tungsten on growth. J. Bacteriol. 130, 1404–1406.

Khare, S., Kumar, M., Kumar, V., Kushwaha, R., Vaswani, S., Kumar, A., Yadav, R.S., Kumar Singh, S., Singh, Y. & Kumar Shukla, P. 2022. Dietary chromium picolinate supplementation improves glucose utilization in transition calf by ameliorating insulin response. Biol. Trace Elem. Res. (online) https://doi.org/10.1007/s12011-022-03405-1

Kim, D., Lee, J. & Seo, J. 2021. Molybdenum-containing metalloenzymes and synthetic catalysts for conversion of small molecules. Catalysts 11: 217 (34 pp.). https://doi.org/10.3390/catal11020217

Kircheva, N., Toshev, N. & Dudev, T. 2022. Holo-chromodulin: competition between the native Cr^{3+} and other biogenic cations (Fe^{3+}, Fe^{2+}, Mg^{2+}, and Zn^{2+}) for the binding sites. Metallomics 14: mfac082 (9 pp.). https://doi.org/10.1093/mtomcs/mfac082

Kleefstra, N., Houweling, S.T., Jansman. F.G.A., Groenier, K.H., Gans, R.O.B., Meyboom-de Jong, B., Bakker, S.J.L. & Bilo, H.J.G. 2006. Chromium treatment has no effect in patients with poorly controlled, insulin-treated type 2 diabetes in an obese Western population. Diabetes Care 29, 521–525.

Kośla, T., Lasocka, I., Skibniewska, E.M., Kołnierzak, M. & Skibniewski, M. 2018. Chrom trójwartościowy (Cr III) jako pierwiastek śladowy niezbędny dla zwierząt i ludzi (Trivalent chromium (Cr III) as a trace element essential for animals and humans). Med. Weter. 74, 560–567. https://doi.org/dx.doi.org/10.21521/mw.6035

Lashkari, S., Habibian, M. & Krogh Jensen, S. 2018. A Review on the role of chromium supplementation in ruminant nutrition—Effects on productive performance, blood metabolites, antioxidant status, and immunocompetence. Biolog. Trace Elem. Res. 186, 305–321.

Leimkühler, S. & Iobbi-Nivol, C. 2016. Bacterial molybdoenzymes: old enzymes for new purposes. FEMS Microbiol. Revs. 40: fuv043 (18 pp.).

Limwachirakhom, R., Triwutanon, S., Chumkam, S. & Jintasataporn, O. 2022. Effects of chromium-L-methionine in combination with a zinc amino acid complex or selenomethionine on growth performance, intestinal morphology, and antioxidative enzymes in rred *Tilapia oreochromis spp*. Animals 12: 2182 (15 pp.). https://doi.org/10.3390/ani12172182

Liu, L., Zhang, S.-W., Lu, J., Pang, X.-Y. & Lv, J.-P. 2018. Antidiabetic effect of high-chromium yeast against type 2 diabetic KK-Ay mice. J. Food Sci. 83, 1956–1963. https://doi.org/10.1111/1750-3841.14138

Lukaski, H.C. 1999. Chromium as a supplement. Annu. Rev. Nutr. 19, 279–302.

Mathew, L.G., Haja, D.K., Pritchett, C., McCormick, W., Zeineddine, R., Fontenot, L.S., Rivera, M.E., Glushka, J., Adams, M.W.W. & Lanzilotta, W.N. 2022. An unprecedented function for a tungsten-containing oxidoreductase. J. Biol. Inorg. Chem. (Publ. online Oct. 2022). https://doi.org/10.1007/s00775-022-01965-0

Mertz, W. (1989) Chromium occurrence and function in biological systems. Physiol. Revs 49, 163–239.

Mintmier, B., Nassif, S., Stolz3, J.F. & Basu, P. 2020. Functional mononuclear molybdenum enzymes: challenges and triumphs in

molecular cloning, expression, and isolation. J. Biol. Inorg. Chem. 25, 547–569. https://doi.org/10.1007/s00775-020-01787-y
Nishimura, K., Iitaka, S. & Nakagawa, H. 2021. Effect of trivalent chromium on erythropoietin production and the prevention of insulin resistance in HepG2 cells. Arch. Biochem. Biophys. 708: 108960 (8 pp.).
Nolan J. Hoffman, Brent A. Penque, Kirk M. Habegger, Whitney Seallsa, Lixuan Tacketta, Jeffrey S. Elmendorf, J.S. 2014. Chromium enhances insulin responsiveness via AMPK. J. Nutrit. Biochem. 25, 565–572. http://dx.doi.org/10.1016/j.jnutbio.2014.01.007
Pala, R., Sari, M.A., Erten, F., Er, B., Tuzcu, M., Orhan, C., Deeh, P.B.D., Sahin, N., Cinar, V., Komorowski, J.R. & Sahin, K. 2020. The effects of chromium picolinate on glucose and lipid metabolism in running rats. J. Trace Elem. Medicine Biol. 58: 126434 (7 pp.). https://doi.org/10.1016/j.jtemb.2019.126434
Pechova, A. & Pavlata, L. 2007. Chromium as an essential nutrient: a review. Veterinarni Medicina, 52, 1–18.
Peng, M. & Yang, X. 2015. Controlling diabetes by chromium complexes: The role of the ligands. J. Inorg. Biochem. 146, 97–103. http://dx.doi.org/10.1016/j.jinorgbio.2015.01.002
Qi, S.S., Zheng, H.X., Jiang, H., Yuan, L.P. & Dong, L.C. 2020. Protective effects of chromium picolinate against diabetic-induced renal dysfunction and renal fibrosis in streptozotocin-induced diabetic rats. Biomolecules 10: 398 (14 pp.). https://doi.org/10.3390/biom10030398
Rauh, D., Graentzdoerffer, A., Granderath, K., Andreesen, J.R. & Pich, A. 2004. Tungsten-containing aldehyde oxidoreductase of *Eubacterium acidaminophilum*. Isolation, characterization and molecular analysis. Eur. J. Biochem. 271, 212–219. https://doi.org/10.1046/j.1432-1033.2003.03922.x
Reda, T., Plugge, C.M., Abram, N.J. & Hirst, J. 2008. Reversible interconversion of carbon dioxide and formate by an electroactive enzyme. Proc. Natl Acad. Sci. USA 105, 10654–10658. www.pnas.org/cgi/doi/10.1073/pnas.0801290105
Sather. B.T. 1966. Inferences on the physiologic significance of chromium during the molt cycle of the crab, *Podophthalmus vigil*. Comp. Biochem. Physiol. 19, 379– 386.
Schicho, R.N., Snowden, L.J., Mukund, S., Park, J.-B., Adams, M.W.W. & Kelly, R.M. 1993. Influence of tungsten on metabolic patterns in *Pyrococcus furiosus*, a hyperthermophilic archaeon. Arch. Microbiol. 159, 380–385.

Schut, G.J., Thorgersen, M.P., Poole II. F.L., Haja, D.K., Putumbaka, S. & Adams, M.W.W. 2021. Tungsten enzymes play a role in detoxifying food and antimicrobial aldehydes in the human gut microbiome. Proc. Natl Acad. Sci. USA 118: e2109008118 (10 pp.). https://doi.org/10.1073/pnas.2109008118

Scott, I.M., Rubinstein, G.M., Poole II, F.L., Lipscomb. G.L., Schut, G.J., Williams-Rhaesa, A.M., Stevenson, D.M., Amador-Noguez, D., Kelly, R.M. & Adams, M.W.W. 2019. The thermophilic biomass-degrading bacterium *Caldicellulosiruptor bescii* utilizes two enzymes to oxidize glyceraldehyde 3-phosphate during glycolysis. J. Biol. Chem. 294, 9995–10005. https://doi.org/10.1074/jbc.RA118.007120

Seelmann, C.S., Willistein, M., Heider, J. & Boll, M. 2020. Tungstoenzymes: Occurrence, catalytic diversity and cofactor synthesis. Inorganics 8, 44 (23 pp.). https://doi.org/10.3390/inorganics8080044

Sevcenco, A.-M., Pinkse, M.W.H., Bol, E., Krijger, G.C., Wolterbeek, H.T., Verhaert, P.D.E.M., Hagedoorn, P.-L. & Hagen, W.R. 2009. The tungsten metallome of *Pyrococcus furiosus*. Metallomics 1, 395–402. https://doi.org/10.1039/b908175e

Swaroop, A., Bagchi, M., Preuss, H.G., Zafra-Stone, S., Ahmad, T & Bagchi, D. 2019. Benefits of chromium(III) complexes in animal and human health. The Nutritional Biochemistry of Chromium(III), Elsevier, pp. 251–278. https://doi.org/10.1016/B978-0-444-64121-2.00008-8

Thorgersen, M.P., Schut, G.J., Poole II, F.L., Haja, D.K., Putumbaka, S., Mycroft, H.I., de Vries, W.J. & Adams, M.W.W. 2022. Obligately aerobic human gut microbe expresses an oxygen resistant tungsten-containing oxidoreductase for detoxifying gut aldehydes. Front. Microbiol. 13:965625 (14 pp.). https://doi.org/10.3389/fmicb.2022.965625

Vincent, J.B. 2017. New evidence against chromium as an essential trace element. J. Nutrition 2212–2219. https://doi.org/10.3945/jn.117.255901

Walker, A.M., Copatti, C.E., Melo, F.V.S.T. &, Melo, J.F.B. 2020. Metabolic and physiological responses to intraperitoneal injection of chromium oxide in hyperglycaemic Nile tilapia juveniles. Aquaculture 517: 734821 (6 pp.). https://doi.org/10.1016/j.aquaculture.2019.734821

Zhang, Q., Sun, X., Xiao, X., Zheng, J., Li, M., Yu, M., Ping, F., Wang, Z., Qi, C., Wang, T. & Wang, X. 2016. Maternal chromium restriction leads to glucose metabolism imbalance in mice offspring through insulin signaling and Wnt signaling pathways. Int. J. Mol. Sci. 17: 1767 (14 pp.). https://doi.org/10.3390/ijms17101767

Zhang, Q., Sun, X., Xiao, X., Zheng, J., Li, M., Yu, M., Ping, F., Wang, Z., Qi, C., Wang, Y. & Wang, X. 2018. Maternal chromium restriction induces insulin resistance in adult mice offspring through miRNA. Int. J. Molec. Med. 41, 1547–1559.

Zhang., Y., Luo, J., Zhu, T., Zhang, X., Jin, M., Jiao, L., Meng, F., Figueiredo-Silva, C., Hong, Y. & Zhou, Q. 2022. Dietary chromium could improve growth, antioxidant capacity, chromium accumulation in tissues and expression of genes involved into glucose and lipid metabolism in juvenile mud crab *Scylla paramamosain.* Aquaculture Rep. 23: 101088 (12 pp.). https://doi.org/10.1016/j.aqrep.2022.101088'

Żarczyńska, K. & Krzebietke, S.J. 2020. The effect of chromium on ruminant health. J. Elem. 25, 893–903. https://doi.org/10.5601/jelem.2020.25.1.1963

Zoroddu, M.A., Aaseth, J., Guido Crisponi, G., Medici, S., Pean, M. & Nurchid, V.M. 2019. The essential metals for humans: a brief overview. J. Inorg. Biochem. 195, 120–129. https://doi.org/10.1016/j.jinorgbio.2019.03.013

CHAPTER SEVEN

MANGANESE AND RHENIUM

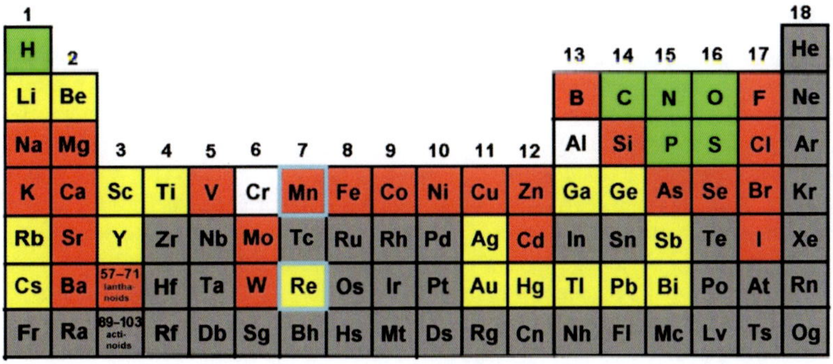

7.1 Manganese

When writing about this metal, it feels necessary to start with a process that makes life as we know it possible on our planet. Four atoms of manganese sit at the heart of the photosynthetic machinery that provides us with life-giving (and, for some, deadly) oxygen (Fig. 7.1.1).

Although one manganese atom does not undergo any change in charge in the cycle, it is necessary for achieving the high positive charge of the complex that is necessary for the oxidation of water (Gao & Fan 2022). Photosystem II can be regarded as a photoenzyme (Björn 2018). But manganese forms part of many other enzymes, too (Crowley et al. 2000). Many of the manganese enzymes have binuclear active sites (Dismukes 1996), i.e., at the active site contain two manganese ions per molecule. Some of them also contain a magnesium ion. Examples of enzymes containing two manganese ions are arginase, catalase (of which there exists also a version with iron instead of manganese), thiosulfate-oxidizing ribonucleotide reductase, xylose isomerase, phosphotriesterase, and enolase.

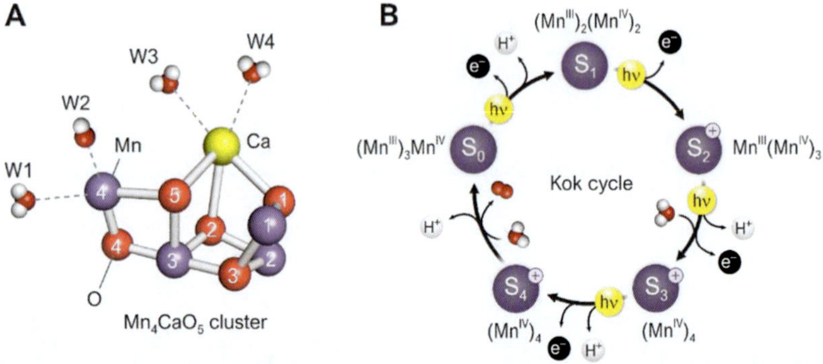

Fig. 7.1.1. A. The "oxygen-evolving cluster" or "oxygen-evolving complex", OEC, in photosystem II of plants and cyanobacteria. Oxygen atoms, red; manganese atoms, violet; hydrogen atoms, grey; calcium atom, yellow; water molecules, W1–4. B. The Kok cycle (Kok-Joliot cycle), driven through five states (S0–S4) by four photons (hn), in which two water molecules are transformed into four electrons, protons, and a molecule of molecular oxygen, while four manganese atoms (ions) change between positively trivalent (III) and tetravalent (IV). Graphics by D. Shevela (SciGraphic, Sweden). Reproduced from Björn (2022). Creative Commons Attribution (CC BY) license (https://creativecommons.org/licenses/by/4.0/).

Other manganese-containing enzymes are carboxy-peptidase, phosphatases (inositol monophosphatase, fructose-1,6 -bisphosphatase), esterases, nucleases (endonucleases and exonucleases), integrases, phosphoenolpyruvate carboxykinase, D-glucarate dehydratase, cyclases, isomerases (keto-aldo isomerases, L-fucose isomerase, cycloisomerases), mandelate racemase, muconate lactonizing enzyme, chloromuconate-lactonizing enzyme, ligases (glutamine synthetase, carbamoyl phosphate synthetase, aminoacyl-tRNA synthetases). In addition there are manganese-containing proteins that are not enzymes. It would carry too far to go into details of this, and we refer to Crowley et al. (2000) for more information.

Also many so-called ribozymes depend on a manganese ion as cofactor (Dange et al. 1990, Kolev et al. 2008, Fig. 7.1.2). Ribozymes are catalytical RNA molecules, which, in many cases, but not always, need a metal ion for activity. For further general information about ribozymes, see Scott (2007) and Maurel et al. (2020). Also ribosomes are regarded as ribozymes by some scientists (Cech 2000), although, contrary to other ribozymes, they contain protein.

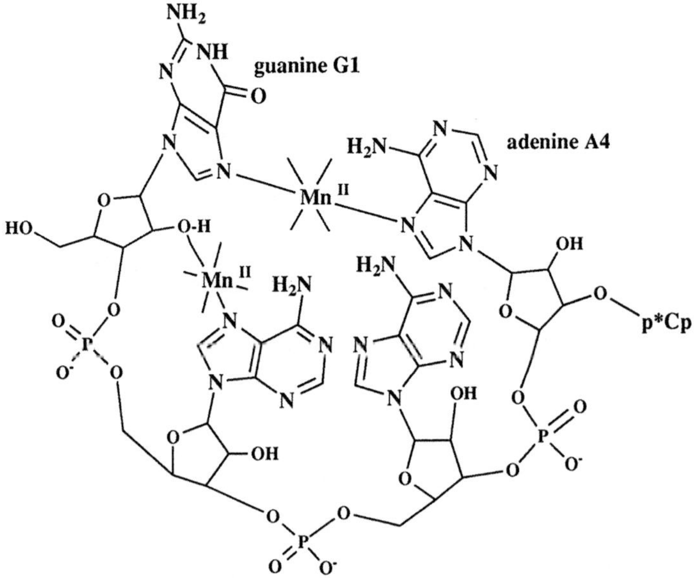

Fig. 7.1.2. Mn^{II} in ribozyme. From Bombard et al. (1998), based on Kazakov & Altman (1992).

A a riboswitch is a highly structured regulatory segment of a messenger RNA molecule that binds a small molecule or ion, resulting in a change in production of the proteins encoded by the mRNA. Manganese ions sometimes act as the regulating element for riboswitches (e.g., Sung & Nesbitt 2019).

Manganese is an essential element for animals, plants, and probably all microorganisms. An adult man requires about 2.3 mg/day. Higher manganese intake has disadvantages: In older persons elevated whole blood manganese is associated with impaired cognition (Larvie et al. 2022). Maintaining serum manganese at a moderate level might be beneficial in preventing diabetes (Chen et al. 2022).

Manganese deficiency is most common on alkaline and poorly drained soils as well as on those high in available iron.

7.2 Rhenium

Rhenium (Re) is one of the rarest metals in the Earth's crust, estimated to come on place 77 in abundance, with a mean concentration of about $7x10^{-8}$ % (Bozhkov & Tzvetkova 2009), and one of the most expensive metals of strategic significance in the world market (ca US $ 85/g). It is present in coal, and if coal fly ash is mixed with soil, plants growing in the mixture accumulate rhenium (He et al. 2018). It is not, as far as is known, an essential element for any organism, yet it is of interest for biologists in various ways.

Rhenium is accumulated by plants. Some plants, as lucerne (*Medicago sativa*), white clover (*Trifolium repens*) and milkvetch (*Astragalus adsurgens*) are even called rhenium hyperaccumulators (Bozhkov & Tzvetkova 2009, He et al. 2018), although this term is not defined for rhenium (Tzvetkova et al. 2021). Following root absorption and translocation, rhenium was stored in the vacuoles of alfalfa leaf cells as water-soluble potassium perrhenate ($KReO_4$). *Brassica juncea* can accumulate up tp 25 g rhenium in the shoot per kg fresh weight (at a soil rhenium, as $KReO_4$, concentration of 80 mg/kg) (Novo et L. 2018). At this soil concentration of rhenium the growth is severely inhibited. Novo et al. (2015) demonstrated how *Equisetum hyemale* (scouring rush) and *Brassica juncea* (Indian mustard) can be used for recovering part of the rhenium present in the substrate in which they are growing.

Yeast (*Saccharomyces cerevisiae*) can be used for collecting rhenium from solution. For good (80%) recovery the pH should be 2 and the temperature 20°C (Zinicovscaia et al. 2021). Bacteria, such as the cyanobacterium *Arthrospira* (*Spirulina*) *platensis* (Zinicovscaia et al. 2022) can also be used for collecting and concentrate rhenium. Other bacteria that have been used for collecting rhenium are *Bacillus* sp. MGG-83 (Mashkani et al. 2007) and *Bacillus megaterium* (Motaghed et al. 2014). *Bacillus megaterium* uses cyanide to make the rhenium accessible, and using it, 98% of the rhenium can be recovered.

The perrhenate ion, ReO^-, containing Re(VII), is the only significant rhenium species found in ocean water (Koide et al. 1986). Macroalgae accumulate rhenium to different extents depending on species. *Fucus vesiculosus* accumulates up to 5.7 µg/g dry matter (Racionero-Gómez et al. 2016).

As gold, various platinum metals can be solubilized by cyanide (see chapter on gold for cyanide-generating bacteria), although the reaction proceeds with different stoichiometry for rhodium as compared to platinum and palladium (Mwase et al. 2012):

$$2Pt + 8NaCN + O_2 + 2H_20 \rightarrow 2Na_2[Pt(CN)_4] + 4NaOH$$
$$2Pd + 8NaCN + O_2 + 2H_20 \rightarrow 2Na_2[Pd(CN)_4] + 4NaOH$$
$$4Rh + 24NaCN + 3O_2 + 6H_20 \rightarrow 4Na_3[Rh(CN)_6] + 12NaOH$$

References

Björn, L.O. 2018. Photoenzymes and related topics: An update. Photochem. Photobiol. 94: 459–465. https://doi.org/10.1111/php.12892

Björn, L.O. 2022. Photosynthetic production of molecular oxygen by water oxidation. Oxygen 2, 337–347. https://doi.org/10.3390/oxygen2030024

Bombard, S., Kozelka, J., Favre, A. & Chottard, J.-C. 1998. Probing the mechanism of an Mn^{2+}-dependent ribozyme by means of platinum complexes. Eur. J. Biochem. 252, 25–35.

Bozhkov, O. & Tzvetkova, C. 2009. Advantages of rhenium phytomining by lucerne and clover from ore dressing soils. Proc. 7 Internat. Conf. Environm. Ecosyst. Developm., p. 127–131.

Cech, T.R. 2000. The ribosome Is a ribozyme. Science 289, 878–879. https://doi.org/10.1126/science.289.5481.878

Chen, H., Cui, Z., Lu, W., Wang, P., Wang, J., Zhou, Z., Zhang, N., Wang, Z., Lin, T., Song, Y., Liu, L., Huang, X., Chen, P., Tang, G., Duan, Y., Wang, B., Zhang, H., Xu, X., Yang, Y., Qin, X. & Fenglin Song, F. 2022. Association between serum manganese levels and diabetes in Chinese adults with hypertension. J. Clin. Hypertens. 24, 918–927. https://doi.org/10.1111/jch.14520

Crowley, J.D., Traynor, D.A. & Weatherburn, D.C. 2000. Enzymes and proteins containing manganese: An overview. *In* Metal ions in biological systems. CRC Press, ISBN 9780429177156.

Dange, V., Van Atta, R.B. & Hecht, S.M. 1990. A Mn^{2+}-dependent ribozyme. Science 248, 585–588.

Dismukes, G.C. 1996. Manganese enzymes with binuclear active sites. Chem. Rev. 1996, 96, 2909–2926.

Gao, X. & Fan, H. 2022. Insight into the huge difference in redox potential between the structural OEC analogues Mn_3CaO_4 and Mn_4CaO_4. Inorganica Chimica Acta 539: 121023. https://doi.org/10.1016/j.ica.2022.121023

He, H., Dong, Z., Pang, J., Wu, G.-L., Zheng, J. & Zhang, X. 2018. Phytoextraction of rhenium by lucerne (*Medicago sativa*) and erect milkvetch (*Astragalus adsurgens*) from alkaline soils amended with coal fly ash. Sci. Total Environm. 630, 570–577. https://doi.org/10.1016/j.scitotenv.2018.02.252

Kazakov, S. & Altman, S. 1992. A trinucleotide can promote metal ion-dependent specific cleavage of RNA. Proc. Natl Acad. Sci. USA 89, 7939–7943.

Koide, M., Hodge, V.F., Yang, J.S., Stallard, M., Goldberg, E.G., Calhoun, J. & Bertine, K.K. 1986. Some comparative marine chemistries of rhenium, gold, silver and molybdenum. Appl. Geochem. 1, 705–714.

Kolev, N.G., Hartland, E.I. & Huber, P.W. (2008) A manganese-dependent ribozyme in the 3'-untranslated region of Xenopus Vg1 mRNA. Nucleic Acids Research 36, 5530–5539. https://doi.org/10.1093/nar/gkn530

Larvie, D.Y., Erikson, K.M. & Armah, S.M. 2022. Elevated whole blood manganese is associated with impaired cognition in older adults, NHANES 2013–2014 cycle. Neurotoxicology 91, 94–99. https://doi.org/10.1016/j.neuro.2022.04.016

Mashkani, S.G., MGhazvini, P.T.M., Ghafourian, H. & Ahmadi, M.A. 2007. Biosorption of rhenium by new strain of *Bacillus* sp. MGG-83. J. Nuclear Sci. Tech. 38, 31–35.

Motaghed, M., Mousavi, S.M., Rastegar, S.O. & Shojaosadati, S.A. 2014. Platinum and rhenium extraction from a spent refinery catalyst using *Bacillus megaterium* as a cyanogenic bacterium: Statistical modeling and process optimization. Bioresource Technol. 171, 401–409. http://dx.doi.org/10.1016/j.biortech.2014.08.032

Mwase, J.M., Petersen, J. & Eksteen, J.J. 2012. A conceptual flowsheet for heap leaching of platinum group metals (PGMs) from a low-grade ore concentrate. Hydrometallurgy 111–112, 129–135. https://doi.org/10.1016/j.hydromet.2011.11.012

Novo, L.A.B.,Mahler, C.F. & González, L. 2015. Plants to harvest rhenium: scientific and economic viability. Environ. Chem. Lett. 13, 439–445. https://doi.org/10.1007/s10311-015-0517-3

Novo, L.A.B., Silva, E.F., Pereira, A., Casanova, A. & González, L. 2018. The effects of rhenium accumulation on Indian mustard. Environm. Sci. Pollution Res. 25, 21243–21250. https://doi.org/10.1007/s11356-018-2547-4

Scott, W.G. 2007. Ribozymes. Curr. Opinion Struct. Biol. 17, 280–286. https://doi.org/10.1016/j.sbi.2007.05.003

Sung, H.-L. & Nesbitt, D.J. 2019. Novel heat-promoted folding dynamics of the yybP-ykoY manganese riboswitch: Kinetic and thermodynamic studies at the single-molecule level. J. Phys. Chem. B 123, 5412−5422. https://doi.org/10.1021/acs.jpcb.9b02852

Racionero-Gómez, B., Sproson, A.D., Selby, D., Gröcke, D.R., Redden, H. & Greenwell, H.C. 2016. Rhenium uptake and distribution in phaeophyceae macroalgae, *Fucus vesiculosus* R. Soc. open sci. 3: 160161 (18 pp.). http://dx.doi.org/10.1098/rsos.160161

Tzvetkova, C., Novo, L.A.B., Atanasova-Vladimirova, S. & Vassilev, T. 2021. On the uptake of rhenium by plants: Accumulation and recovery from plant tissue. J. Cleaner Product. 328: 129534 (7 pp.) https://doi.org/10.1016/j.jclepro.2021.129534

Zinicovscaia, I., Yushin, N., Grozdov, D., Vergel, K., Nekhoroshkov, P. & Rodlovskaya, E. 2021. Treatment of rhenium-containing effluents using environmentally friendly sorbent, *Saccharomyces cerevisiae* biomass. Materials 14: 4763 (15 pp.). https://doi.org/10.3390/ma14164763

Zinicovscaia, I., Cepoi, L., Rudi, L., Chiriac, T., Yushin, N. & Grozdov, D. 2022. *Arthrospira platensis* as bioremediator of rhenium mono- and polymetallic synthetic effluents. Microorganisms 10: 2109 (13 pp.). https://doi.org/10.3390/microorganisms10112109

CHAPTER EIGHT

IRON

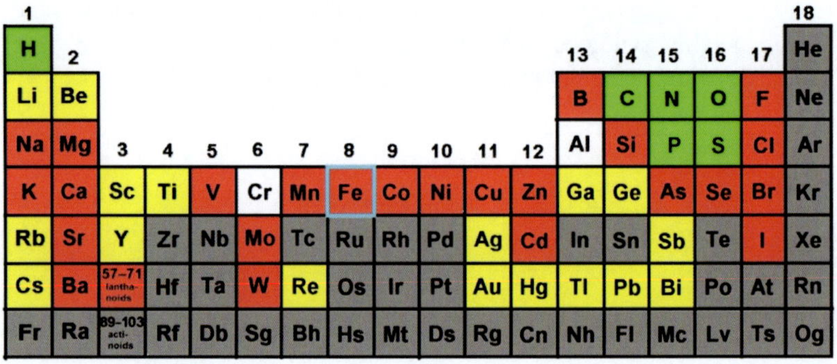

8.1 Iron: Deficiency in the midst of affluence

Synthesis of atomic nuclei by fusion of lighter nuclei in stars cannot proceed any further than iron. The iron nucleus has the highest binding energy of all atomic nuclei. To be more precise, the nucleus with lowest energy is that of the isotope having mass 56 dalton (or more precisely 55.93494 Da).

Our planet consists to almost a third (Allègre et al. 1995, 28.2%; Mc Donough et al. 1995, 29.6–32.7%; Wang et al. 2018, 31.2 ±1.0%) by weight of iron. Much of it is in the core, which is thought to consist to about 80% (Allègre et al. 1995, 79.4%; Mc Donough et al. 1995, 78.0–87.5%; Wang et al. 2018, 82.8 ±2.9%) of iron. Also in the mantle iron is one of the common elements, about 6.6% (Mc Donough et al. 1995) to 6.32 ± 0.06 (Wang et al. 2018) % by weight. The content of the crust is similar to that of the mantle: 6.71% in the continental crust (Rudnick & Gao 2014); about 10% (decreasing to ≈7% with depth) in the oceanic crust (White & Klein 2014).

Iron is thus a quite common element (often the fourth most abundant one) in the environment of organisms. How come then, that iron deficiency is such a common problem, not only for people, but also for crops and other organisms? "Iron deficiency, which affects two billion people, is the primary nutritional disorder in the world" (Cairo et al. 2006). Why have organisms chosen such an element for many different functions, when deficiency so easily develops?

The reason is historical. The iron currently available to organisms is mostly in the ferric, trivalent state (Fe^{3+}), and such compounds are mostly very insoluble. When life started to evolve, the dominant form of iron was ferrous, divalent (Fe^{2+}), and much more soluble. The change started with the emergence of anoxygenic photosynthesis that used divalent iron as an electron source for reduction of carbon dioxide and nitrogen, but the big change came with the emergence of oxygenic photosynthesis and the appearance of molecular oxygen in the environment, "the Great Oxidation Event", GOE, 2400 to 2050 million years ago (Hodgskiss et al. 2019).

When organisms had only trivalent iron available, they started to evolve siderophores, i.e. organic substances that bind iron very tightly. They started to compete for available iron by evolving ever more efficient siderophores. The struggle for iron between parasites and their hosts is described by Golonka et al. (2019), who also show the structures for several siderophores. Calpotectin is an important mammalian metal-binding innate immune protein notable for its ability to limit the availability of multiple transition metal nutrients for parasites.

What properties of iron make it so worth fighting for? The properties that have been particularly exploited by organisms are: (1) the ability to take up and give off electrons, i.e. to become reduced and oxidized; (2) the ability to bind molecular oxygen, and (3) magnetism.

For the well-known Fe oxides occurring in soils, such as goethite, hematite, and ferrihydrite, the solubility constant is usually extremely low: $(Fe^{3+})\,(OH)^3 - 10^{-37} - 10^{-44}\,M^4$ (Schwertmann 1991) Thus $[Fe^{3+}]$ at pH=7 is 10^{-16}–$10^{-23}\,M$, i.e. less than one ion in the volume of one cell.

In some cases plants can use iron that has been sequestered by siderophores (pyoverdines) from bacteria (*Pseudomonas* species). This has been observed for both monocotyledons and dicotyledons (Bar-Ness et al. 1991, Trapet et al. 2016). At the same time the bacterial pyoverdines repress the plant's expression of defense-related genes (Trapet et al. 2016), which possibly may benefit the *Pseudomonas*.

Parasites often employ efficient siderophores to extract iron from their hosts. An example is the fungus *Erwinia chrysanthemi* causing soft rot. Only strains producing efficient siderophores are virulent (Enard et al. 1991).

In the mammal milk the iron (Fe^{3+}) is held very tightly by the multifunctional (Wang et al. 2019) glycoprotein lactoferrin. The dissociation constant is only $10^{-20}\,M$ (Baker & Baker 2004), which protects the milk from many bacteria, since in milk lactoferrin to a large extent is present in the apolactoferrin (iron-free) form, and thus keeps the concentration of iron available to bacteria very low (Steijns & van Hooijdonk 2000). To young mammals the iron is available due to the low pH in the digestive tract (Baker & Baker 2004).

The way mammals handle iron is very sophisticated, but is outsmarted by some bacteria, who have evolved independence of iron, provided they have access to certain organic compounds. Twenty-three strains of lactic acid bacteria belonging to the genera *Lactobacillus, Lactococcus, Leuconostoc, Pediococcus* or *Carnobacterium* were studied for growth and siderophore production under controlled iron-starvation conditions. Nocellular iron incorporation was observed during growth in the presence of radioactive iron (^{59}Fe). It was concluded that those strains do not require iron (Pandey et al. 1994). *Lactobacillus* spp., particularly *Lact. johnsonii*, require iron under particular environmental conditions with limited or specific nucleotide sources (Elli et al. 2000). *Lactobacillus sakei* lives in meat with plenty of iron available, and is able to form iron-containing catalase, but does not require iron for growth (Verplaetse et al. 2020).

Also the Lyme disease pathogen *Borrelia burgdorferi* has bypassed this host defense by eliminating the need for iron. The intracellular concentration of iron in *B. burgdorferi* was estimated to be less than 10 atoms per cell, well below a physiologically relevant concentration (Posey & Gherardini 2000). It is uncertain whether all *Mycoplasma* strains need iron (Perálvarez-Marín et al. 2021). While all eukaryotic organisms require an iron-based ribonucleotide reductase for production of essential parts of DNA, and some prokaryotes use other metals instead of iron, *Mycoplasma* strains have evolved a metal-free enzyme for the purpose (Srinivas et al. 2018).

In complete contrast to what has been described in the preceding paragraphs, Ferrer et al. (2007) found that in the archaeon *Ferroplasma acidophilum*, collected from acidic environments with high iron content, 86% of 189 investigated cellular proteins are iron-metalloproteins.

What do organisms need iron for? One purpose is respiration, another photosynthesis. Several types of cytochrome are active in transferring electrons In the most common form of cell respiration in animals, plants, and bacteria (cytochromes *a*, *b*, and *c*), as well as in photosynthesis of cyanobacteria and plants (cytochromes *f* and b_6). The prosthetic group in the protein that handles the electrons is called heme Fig. 8.1). Another component in the respiratory chain is cytochrome oxidase, which in addition to heme with iron, also contains copper.

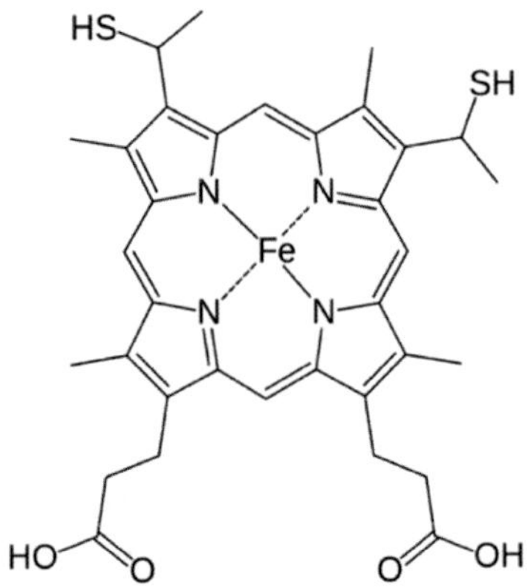

Fig. 8.1. The prosthetic group, heme *c*, of cytochromes, with an iron atom at the centre.

Cytochrome P450 (see Rittle et al. 2010 and Mak & Denisov 2017 for literature) is a heme protein that is not involved in respiration, but is able to insert an atom from molecular oxygen into a variety of organic substrates, often with high degrees of stereoselectivity. It is important in the oxidation of poisonous substances. The oxygen transporting protein hemoglobin in our blood, and the muscle protein myoglobin have a similar prosthetic group, heme *b* (Fig. 8.2).

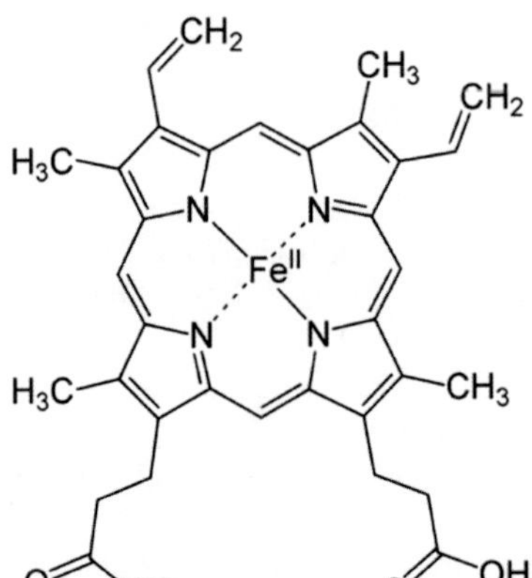

Fig. 8.2. Heme *b*, the prosthetic group of hemoglobin and myoglobin

Cytochromes that have recently attracted considerable interest are the multiheme cytochromes, which may contain from a few to over one hundred heme units in a single protein molecule, e.g., in the archeon *Methanoperedens manganireducens* (Leu et al. 2020).

Other important proteins with similar iron-containing tetrapyrrole groups are catalase, an enzyme that decomposes hydrogen peroxide to water and oxygen, and peroxidase that catalyzes the reduction of organic peroxides to alcohols.

A second group of iron-containing proteins are those that contain iron-sulfur clusters. We can only mention a few in this very large group here, but a more comprehensive review is provided by Beinert (2000). Important representatives among them are the ferredoxins. Figure 8.3 shows the structure of the prosthetic group of a Fe_4S_4 ferredoxin. Such ferredoxins function in photosynthesis of cyanobacteria and plants to transfer electrons, and in this electron transfer also another iron sulfur protein, called the Rieske protein, and the iron-sulfur clusters designated as F_x, F_A, and F_B, take part. There are many other types of iron sulfur proteins, which we cannot enumerate here, and also combinations of iron-sulfur clusters with iron-containing heme, as in sulfite reductase (Gerlach et al. 2013, Khan et al. 2022).

Nitrogenases are enzymes that are very important for all of the biosphere. They make the rather inert molecular nitrogen (N_2) available for the biosphere. Nitrogenases are usually divided into three main groups: Mo-, V-, and Fe-nitrogenases, but also Mo- and V-nitrogenases, in addition to molybdenum or vanadium, contain iron in the form of a Fe_4S_4 cluster (Harris et al. 2019). The Mo-, V-, and Fe-nitrogenases carry out nitrogen reduction to ammonia with different stoichiometries:

Mo-nitrogenase:

$$N_2 + 8H^+ + 16MgATP + 8e^- \rightarrow 2NH_3 + 1H_2 + 16MgADP + 16P_i$$

V-nitrogenase:

$$N_2 + 12H^+ + 24MgATP + 12e^- \rightarrow 2NH_3 + 3H_2 + 24MgADP + 24P_i$$

Fe-nitrogenase:

$$N_2 + 20H^+ + 40MgATP + 20e^- \rightarrow 2NH_3 + 7H_2 + 40MgADP + 40P_i$$

where P_i stands for inorganic orthophosphate.

Besides the reduction of N_2, nitrogenases catalyze a whole range of other reductions, including the reduction of the gaseous substrates carbon monoxide (CO) and carbon dioxide (CO_2) to hydrocarbons (Oehlmann & Rebelein 2022).

There are many other types of iron-dependent enzymes, but we cannot mention them all here.

Bacteria processing iron occurs in many environments, even at depths exceeding one km in the continental subsurface (Dong et al. 2014, Casar et al. 2021, Mateos et al. 2022). Bacteria living in the bottom of the ocean by reducing Fe^{3+} to Fe^{2+} and oxidizing methane are described by Fernandes et al. (2014).

I mentioned in the beginning that one of the properties of iron that has contributed to its use in biology is its magnetism. Many organisms (mammals, birds, fishes, and insects) use the magnetic field of the Earth for navigation (see Gao et al. 2021 for examples). There are two main, not necessary mutually exclusive, theories for how such compasses are constructed (e.g., Warrant 2021). According to one theory the "compass needles" are photoexcited molecules of the flavoprotein cryptochrome forming a radical pair (e.g., Xu et al. 2021), which possess a magnetic moment. According to another theory the "compass needles" consist of iron oxide (Fe_3O_4, with one iron ion in the divalent, and two in the trivalent state, as in the mineral magnetite), or Fe_3S_4, greigite, that is also magnetic (Roberts 1995). There are arguments for both versions of compass (although Nimpf & Keays 2022 have thoroughly criticized all existing explanations of the magnetic sense), but we shall concentrate here on the iron model. Although the existence of the cryptochrome compass has been proven already long ago, it is also certain that another kind of biological compass exists, since some animals can sense the magnetic field in darkness (e.g., Gao et al. 2021). As Qin et al. (2016) express it: "The magnetite-based and radical-pair-reaction-based chemical models both have credible theoretical and experimental foundations, and may not be mutually exclusive; however, evidence from different species tends to favour one hypothesis over the other." According to a recent theory (Qin et al. 2016), cryptochrome functions, together with a protein containing an iron-sulfur cluster, as the compass. For birds also an iron-sulfur protein has been suggested as the "compass needle" (Arai et al. 2021). Sea turtles that undertake long migrations seem to use the magnetic field both for determining their locations by a magnetic map, and for taking out compass directions (Lohmann et al. 2022). Irvin and Lohmann disrupted their navigational ability with a pulsed magnetic field, and interpreted the result as their possession of a magnetite compass.

There are magnetotactic bacteria in several phyla: Many kinds of proteobacteria (from alfa-proteobacteria to zeta-proteobacteria) and also bacteria belonging to other groups (Lin et al. 2020, Uzun et al. 2020) are magnetotactic. There are also examples of horizontal transfer of genes for magnetotaxis between bacterial groups that are widely separated phylogenetically (Uzun et al. 2022). Bacteria use the vertical component of the magnetic field to distinguish up from down. At first thought this may seem to be contradicted by the fact that magnetotactic bacteria occur also at the (geomagnetic) equator, where the magnetic field is horizontal (Frankel 1981, Tan et al. 2021), but the horizontal field makes the bacteria swim in a direction towards a place where the field will be pointing down. At the equator the bacterial population will have "mixed polarity", but at both North and South latitudes they will be of the correct "polarity" for swimming downward. For such small organisms it is not easy to feel gravity. Blakemore (1982) vividly describes his discovery of magnetotaxis of bacteria in the early 1970s.

For bacteria the sensor for magnetic field is always a magnetic iron compound, either magnetite (Fe_3O_4) or greigite (Fe_3S_4) (see Amor et al. 2020 for a thorough review); as far as we know never cryptochrome, even though bacterial cryptochromes exist. Some magnetotactic bacteria deposit either mineral depending on conditions (Lefèvre et al. 2011), or even within the same cell, others contain only magnetite (Frankel et al. 1983, Frankel & Blakemore 1984) or only greigite (Lefèvre et al. 2011). In general, magnetite-producing magnetotactic bacteria are found at or very close to the oxic-anoxic interface, while greigite producers are present in reducing biotopes, below the oxic-anoxic interface, in the sulfidic anoxic zone.The magnetic minerals are enclosed in membrane-bounded organelles called magnetosomes (Fig. 8.3).

Magnetotactic bacteria (Fig. 8.3) are present in very different habitats, from volcanic lakes (Liu et al. 2022) to the sea, and as fossils in the bottom sediment of the sea (Petersen et al. 1986). Bacterial magnetite is mostly formed under aerobic conditions, but can be formed also anaerobically (Bazylinski et al. 1988).

The number of publications about magnetotactic bacteria are now about 120 per year. Only a small selection of them can be mentioned here.

Magnetofossils produced by magnetotactic bacteria are sensitive to redox conditions in sediments and water columns, and have been used to reconstruct paleoredox conditions over a range of geological settings (Xue et al. 2022).

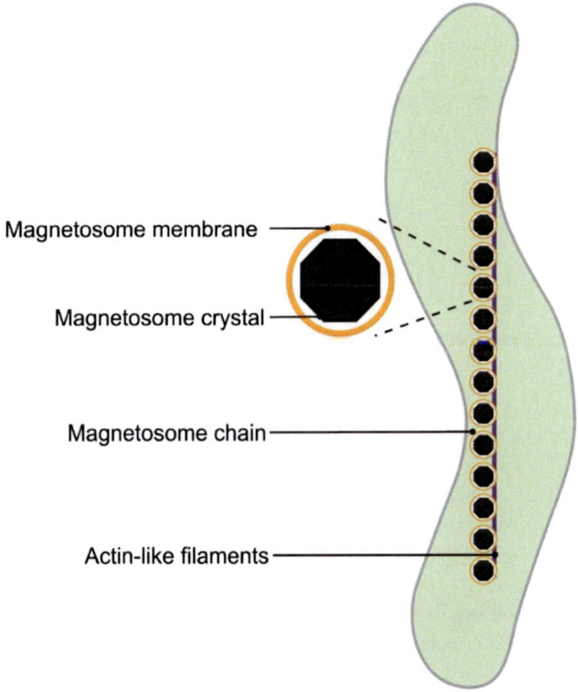

Fig. 8.3. Magnetotactic bacterium, schematic. From Lin et al. (2020). Reproduced under a Creative Commons Attribution 4.0 International License (http://creativecommons.org/licenses/by/4.0/).

Several authors have described a bacterial iron protein called MagR that may have magnetic properties, possibly together with a cryptochrome, but the results seem disparate, e.g., Guo et al. (2021), Pekarsky et al. (2021).

Leão et al. (2020) describe a protist exhibiting magnetotaxis, in which the "compass" consists of an iron compound, probably magnetite. Foraminifera in the Mariana Trench containing magnetite that are able to orient in a magnetic field are described by Yang et al. (2022). Euglenoids of the genus *Anisonema* containing magnetite are able to orient in a magnetic field (de Araujo et al. 1986).

Magnetic navigation in fishes is probably not dependent on iron.

Termites navigate by the magnetic field of the earth, and it is believed that they use a magnetite compass in the dark, and a cryptochrome compass in the open during daytime (Gao et al. (2021).

Much of the iron in the human body (Fig. 8.4) is in hemoglobin in red blood cells and their precursors in red bone marrow, and in the muscle protein myoglobin. A vital muscle is the heart, and a common cause of heart failure is iron deficiency. Iron deficiency is a common comorbidity in patients with heart failure, with a prevalence reaching up to 59% even in non-anaemic patients. It is a serious condition, with a mortality rate reaching up to 75% at 5 years (Shah et al. 2017).

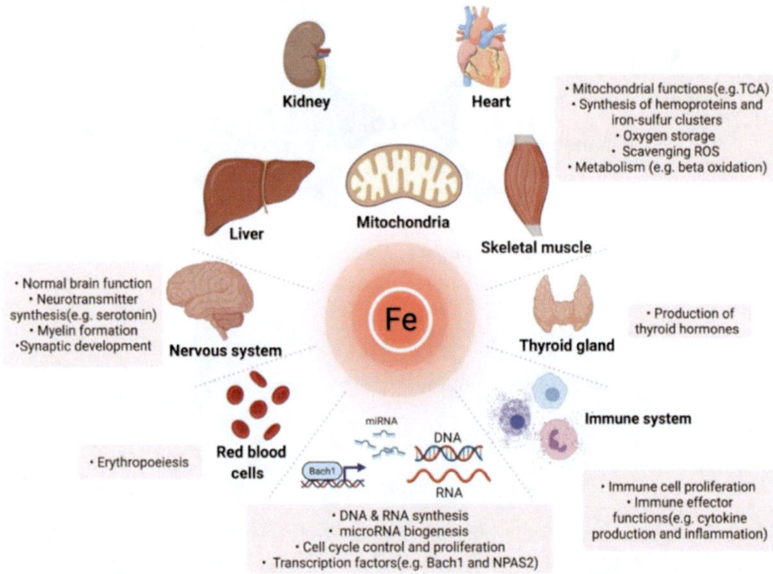

Fig. 8.4. Overview of the multifaceted roles of iron in diverse organs and molecular processes. TCA: tricarboxylic acid cycle; miRNA: microRNA, ROS: reactive oxygen species. Reproduced from Alnuwaysir et al. (2022) in accordance with Creative Commons Attribution (CC BY) license (https://creativecommons.org/licenses/by/4.0/).

Before leaving iron, we must mention one very important iron atom, on which much of present life on Earth depends: The one that mediates electron transfer between the plastoquinones in photosystem II of the photosynthetic machinery of cyanobacteria and plants.

References

Allègre, C.J., Poirier, J.-P., Humler, E. & Albrecht W. Hofmann, A.W. (1995) The chemical composition of the Earth. Earth Planet. Sci. Lett. 134, 515–526.

Alnuwaysir, R.I.S., Hoes, M.F., van Veldhuisen, D.J., van der Meer, P. & Beverborg, N.G. (2022) Iron deficiency in heart failure: Mechanisms and pathophysiology. J. Clin. Med. 11, 125 (30 pp.). https://doi.org/10.3390/jcm11010125

Amor, M., Mathon, F.P., Monteil, C.L., Busigny, V. & Lefevre, C.T. (2020) Iron-biomineralizing organelle in magnetotactic bacteria: function, synthesis and preservation in ancient rock samples. Environm. Microbiol. 22, 3611–3632. https://doi.org/10.1111/1462-2920.15098

Arai, S., Shimizu, R., Adachi, M. & Hirai, M. (2021) Magnetic field effects on the structure and molecular behavior of pigeon iron–sulfur protein. Protein Science 31:e4313 (13 pp.). https://doi.org/10.1002/pro.4313

Baker, H.E. & Baker, E.N. (2004) Lactoferrin and Iron: structural and dynamic aspects of binding and release. BioMetals 17, 209–216.

Bar-Ness, E., Chen, Y., Hadar, Y, Marschner, H. & Romheld, V. (1991) Siderophores of *Pseudomonas putida* as an iron source for dicot and monocot plants. Plant Soil 130, 231–241.

Bazylinski, D., Frankel, R.B. & Jannasch, H.W. (1988) Anaerobic magnetite production by a marine, magnetotactic bacterium. Nature 334, 518–519. https://doi.org/10.1038/334518a0

Beinert, H. (2000) Iron-sulfur proteins: ancient structures, still full of surprises. J. Biol. Inorg. Chem. 5, 2–15.

Cairo, G., Bernuzzi, F. & Recalcati, S. (2006) A precious metal: iron, and essential nutrient for all cells. Genes & Nutrit. 1, 25–40.

Casar, C.P., Momper, L.M., Kruger, B.R. & Osburn, M.R. (2021) Iron-fueled life in the continental subsurface: Deep mine microbial observatory, South Dakota, USA. Appl. Environm. Microbiol. 87: e00832-21 (14 pp.).

Dong, Y., Kumar, C.G., Chia, N. et al. (2014) *Halomonas sulfidaeris*-dominated microbial community inhabits a 1.8 km-deep subsurface Cambrian Sandstone reservoir. Environm. Microbiol. 16, 1695–1708. https://doi.org/10.1111/1462-2920.12325

Elli, M., Zink, R., Rytz, A., Reniero, R. & Morelli, L. (2000) Iron requirement of *Lactobacillus* spp. in completely chemically defined growth media. Journal of Applied Microbiology 88, 695–703.

Enard, C., Franza, T., Neema, C., Gill, P.R., Persmark, M., Neilands, J.B. & Expert, D. (1991) The requirement of chrysobactin dependent iron transport for virulence incited by *Erwinia chrysanthemi* on *Saintpaulia ionantha*. Plant Soil 130, 263–271.

Fernandes, C.E.G., Gonsalves, M.J.B.D., Nazareth, D.R., Nagarchi, L. & Kamaleso, S.A. (2015) Microbial iron reduction and methane oxidation in subsurface sediments of the Arabian Sea. Marine Petroleum Geology 67, 327–335. http://dx.doi.org/10.1016/j.marpetgeo.2015.05.002

Ferrer, M., Golyshina, O.V., Beloqui, A., Golyshin, P.N. & Timmis, K.N. (2007) The cellular machinery of *Ferroplasma acidiphilum* is iron-protein-dominated. Nature 445, 91–94. https://doi.org/10.1038/nature05362

Frankel, R.B. & Blakemore, R.P. (1984) Precipitation of Fe_3O_4 in magnetotactic bacteria. Phil. Transact. Roy. Soc. B. 304, 567–574. https://doi.org/10.1098/rstb.1984.0047

Frankel, R.B., Blakemore, R.P., Torres De Araujo, F.F., Esquivel, D.M.S. & Danon, J. (1981) Magnetotactic bacteria at the geomagnetic equator. Science 212, 1269–1270.

Frankel, R.B., Papaefthymiou, G.C., Blakemore, R.P. & Obrien, W. (1983) Fe_3O_4 precipitation in magnetotactic bacteria. Biochim. Biophys. Acta 763, 147–159.

Gao, Y., Wen, P., Cardé, R.T., Xu, H. & Huang, Q. (2021) In addition to cryptochrome 2, magnetic particles with olfactory co-receptor are important for magnetic orientation in termites. Communic. Biol. 4:1121 (12 pp.). https://doi.org/10.1038/s42003-021-02661-6

Gerlach, D.L., Coucouvanis, D. & Lehnert, N. (2013) Eur. J. Inorg. Chem. 2013, 3883–3890. https://doi.org/10.1002/ejic.201300308

Golonka, R., Yeoh, B.S. & Vijay-Kumar, M. (2019) The Iron tug-of-war between bacterial siderophores and innate immunity. J. Innate Immun. 11, 249–262. https://doi.org/10.1159/000494627

Guo, Z., Xu, S., Chen, X., Wang, C., Yang, P., Qin, S., Zhao, C., Fei, F., Zhao, X., Tan, P.-H., Wang, J. & Xie, C. (2021) Modulation of MagR magnetic properties via iron–sulfur cluster binding. Sci. Rep. 11:23941 (12 pp.). https://doi.org/10.1038/s41598-021-03344-2

Harris, D.F., Lukoyanov, D.A., Kallas, H., Trncik, C., Yang, Z.-Y., Compton, P., Kelleher, N., Einsle, O., Dean, D.R., Hoffman, B.M. & Seefeldt, L.C. (2019) Mo-, V-, and Fe-nitrogenases use a universal eight-electron reductive-elimination mechanism to achieve N_2 reduction. Biochemistry 58, 3293−3301. https://doi.org/10.1021/acs.biochem.9b00468

Hodgskiss, M.S.W., Crockford, P.W., Peng, Y., Wing, B.W. & Horner, T.J. (2019) A productivity collapse to end Earth's Great Oxidation. Proc. Natl Acad. Sci. USA 116, 17207–17212. www.pnas.org/cgi/doi/10.1073/pnas.1900325116

Irwin, W.P. & Lohmann, K.J. (2005) Disruption of magnetic orientation in hatchling loggerhead sea turtles by pulsed magnetic fields. J. Comp. Physiol. A) 191, 475–480. J Comp Physiol A (2005) 191: 475–480

https://doi.org/10.1007/s00359-005-0609-9
Khan, S.N., Griffith, A., De Proft, F., Miliordos, E., Havenith, R.W.A., Bykov, D. & Cunha, A.V. (2022) [Fe4S4] cubane in sulfite reductases: new insights into bonding properties and reactivity. Phys. Chem. Chem. Phys. 24, 18543–18551. https://doi.org/10.1039/d2cp02124b
Leão, P., Le Nagard, L., Yuan, H., Cypriano, J., Da Silva-Neto, I., Bazylinski, D.A., Acosta-Avalos, D., de Barros, H.L., Hitchcock, A.P., Lins, U. & Abreu, F. (2020) Magnetosome magnetite biominerallization in a flagellated protist: evidence for an early evolutionary origin for magnetoreception in eukaryotes. Environm. Microbiol. 22, 1495–1506. https://doi.org/10.1111/1462-2920.14711
Lefèvre, C.T., Menguy, N., Abreu, F., Lins, U., Pósfai, M., Prozorov, T., Pignol, D., Frankel, R.B. & Bazylinski, D.A. (2011) A cultured greigite-producing magnetotactic bacterium in a novel group of sulfate-reducing bacteria. Science 334, 1720–1723. https://doi.org/10.1126/science.1212596
Leu, A.O., Cai, C., McIlroy, S.J., Southam, G., Orphan, V.J., Yuan, Z., Hu, S. & Tyson & G.W. (2020) Anaerobic methane oxidation coupled to manganese reduction by members of the Methanoperedenaceae. ISME J. 14, 1030–1041. https://doi.org/10.1038/s41396-020-0590-x
Lin, W., Zhang, W., Paterson, G.A., Zhu, Q., Zhao, X., Knight, R., Bazylinski, D.A., Roberts, A.P. & Pan, Y. (2020) Expanding magnetic organelle biogenesis in the domain Bacteria. Microbiome 8:152 (13 pp.). https://doi.org/10.1186/s40168-020-00931-9
Liu, T., Da, H., Zhang, S., Wang, W., Pan, H. & Yan, L. (2022) Magnetotactic bacteria in vertical sediments of volcanic lakes in NE China appear Alphaproteobacteria dominated distribution regardless of waterbody types. World J. Microbiol. Biotechnol. 38:76 (14 pp.). https://doi.org/10.1007/s11274-022-03262-z
Lohmann, K.J., Goforth, K.M., Mackiewicz, A.G., Lim, D.S. & Lohmann, C.M.F. (2022) Magnetic maps in animal navigation. J. Comp. Physiol. A, 208, 41–67. https://doi.org/10.1007/s00359-021-01529-8
Mak, P.J. & Denisov, I.G. (2018) Spectroscopic studies of the cytochrome P450 reaction mechanisms. BBA - Proteins and Proteomics 1866, 178–204. http://dx.doi.org/10.1016/j.bbapap.2017.06.021
Mateos, G.; Bonilla, A.M.; de Francisco de Polanco, S.; Martínez, J.M.; Escudero, C.; Rodríguez, N.; Sánchez-Andrea, I.; Amils, R. (2022) *Shewanella* sp. T2.3D-1.1 a Novel Microorganism Sustaining the Iron Cycle in the Deep Subsurface of the Iberian Pyrite Belt. Microorganisms 2022, 10, 1585. https://doi.org/10.3390/microorganisms10081585

McDonough, W.F. & Sun, S.-S. (1995) The composition of the Earth. Chem. Geol. 120, 223–253.

Nimpf, S. & Keays, D.A. (2022) Myths in magnetosensation. iScience 104454 (10 pp.). https://doi.org/10.1016/j.isci.2022.104454

Oehlmann, N.N. & Rebelein, J.G. (2022) The conversion of carbon monoxide and carbon dioxide by nitrogenases. ChemBioChem 23, e202100453 (13 pp.). https://doi.org/10.1002/cbic.202100453

Pandey, A., Bringel, F. & Meyer, J.-M. (1994) Iron requirement and search for siderophores in lactic acid bacteria. Appl. Microbiol. Biotechnol. 40, 735–739.

Pekarsky, A., Michor, H. & Spadiut, O. (2021) Revisiting the potential functionality of the MagR protein. Magnetochemistry 7: 147 (8 pp.). https://doi.org/10.3390/magnetochemistry7110147

Perálvarez-Marín, A., Baranowski, E., Bierge, P., Pich, O.Q. & Lebrette, H. (2021) Metal utilization in genome-reduced bacteria: Do human mycoplasmas rely on iron? Comput. Struct. Biotechnol. J. 19, 5752–5761. https://doi.org/10.1016/j.csbj.2021.10.022

Petersen, N., von Dobeneck, T. & Vali, H. (1986) Fossil bacterial magnetite in deep-sea sediments from the South Atlantic Ocean. Nature 320. 611–615.

Posey, J.E. & Gherardini, F.C. (2000) Lack of a role for iron in the Lyme disease pathogen. Science 288, 1651–1653. https://www.science.org/doi/10.1126/science.288.5471.1651

Qin, S., Hang Yin, H., Yang, C. et al. (2016) A magnetic protein biocompass. Nature Materials 15, 217–229. https://doi.org/10.1038/NMAT4484

Rittle, J., Younker, J.M. & Green, M.T. (2010) Cytochrome P450: The active oxidant and its spectrum. Inorg. Chem. 49, 3610–3617. https://doi.org/10.1021/ic902062d

Roberts, A.P. (1995) Magnetic properties of sedimentary greigite (Fe_3S_4). Earth Planetary Sci. Lett. 134, 227–236.

Rudnick, R.L. & Gao, S. (2014) Composition of the continental crust. Ch. 4.1, pp. 1–51 in Treatise on Geochemistry, 2nd Edition (). http://dx.doi.org/10.1016/B978-0-08-095975-7.00301-6

Schwertmann, U. (1991) Solubility and dissolution of iron oxides. Plant Soil 130, 1–25.

Shah, K.S., Xu, H., Matsouaka, R.A., Bhatt, D.L., Heidenreich, P.A., Hernandez, A.F., Devore, A.D., Yancy, C.W. & Fonarow, Y.G. (2017) Heart failure with preserved, borderline, and reduced ejection fraction. 5-Year Outcomes. J. Am. College Cardiol. 7 0, 2476–2486.

Srinivas, V., Lebrette, H., Lundin, D., Kutin, Y., Sahlin, M., Lerche, M., Eirich, J., Branca, R.M.M., Cox, N., Sjöberg, B.-M. & Högbom, M. (2018) Metal-free ribonucleotide reduction powered by a DOPA radical in *Mycoplasma* pathogens. Nature 563, 416–420. https://doi.org/10.1038/s41586-018-0653-6

Steijns, J. M. & Van Hooijdonk, A.C.M. (2000) Occurrence, structure, biochemical properties and technological characteristics of lactoferrin. British J. Nutrition 84, 11–17. https://doi.org/10.1017/S0007114500002191

Tan, S.M., Ismail, M.H. & Cao, B. (2021) Biodiversity of magnetotactic bacteria in the tropical marine environment of Singapore revealed by metagenomic analysis. Environm. Res. 194: 110714 (14 pp.).

Torres de Araujo, F.F., Pires, M.A., Frankel, R.B. & Bicudo, C.E.M. (1986) Magnetite and magnetotaxis in algae. Biophys. J. 50, 375–378.

Trapet, P., Avoscan, L., Klinguer, A., Pateyron, S., Citerne, S., Chervin, C., Mazurier, S., Lemanceau, P., Wendehenne, D. & Besson-Bard, A. (2016) The *Pseudomonas fluorescens* siderophore pyoverdine weakens *Arabidopsis thaliana* defense in favor of growth in iron-deficient conditions. Plant Physiol. 171, 675–693.

Uzun, M., Alekseeva, L., Krutkina, M., Koziaeva, V. & Grouzdev, D. (2020) Unravelling the diversity of magnetotactic bacteria through analysis of open genomic databases. Scientif. Data 7:252 (13 pp.). https://doi.org/10.1038/s41597-020-00593-0

Uzun, M., Koziaeva, V., Dziuba, M., Leão, P., Krutkina, M. & Grouzdev, D. (2022) Detection of interphylum transfers of the magnetosome gene cluster in magnetotactic bacteria. Front. Microbiol. 13:945734 (15 pp.). https://doi.org/10.3389/fmicb.2022.945734

Verplaetse, E., André-Leroux, G., Duhutrel, P., Coeuret, G., Chaillou, S., Nielsen-Leroux, C. & Champomier-Vergès, M.-C. (2020) Heme uptake in *Lactobacillus sakei* evidenced by a new energy coupling factor (ECF)-like transport system. Appl. Environ. Microbiol. 86:e02847-19 (14 pp.). https://doi.org/10.1128/AEM.02847-19.

Wang, B., Timilsena, Y.P., Blanch, E. & Adhikari, B. (2019) Lactoferrin: Structure, function, denaturation and digestion. Crit. Revs Food Sci. Nutr. 59, 580–596. https://doi.org/10.1080/10408398.2017.1381583

Wang, H.S., Lineweaver, C.H. & Ireland, T.R. (2018) The elemental abundances (with uncertainties) of the most Earth-like planet. Icarus 299, 460–474. http://dx.doi.org/10.1016/j.icarus.2017.08.024

Warrant, E.J. (2021) Unravelling the enigma of bird magnetoreception. Nature 594, 497–498.

White, W.M. & Klein, E.M. (2014) Composition of the oceanic crust. Ch. 4.13, pp.457–496, pp. 1–51 in Treatise on Geochemistry, 2nd Edition http://dx.doi.org/10.1016/B978-0-08-095975-7.00315-6

Xu, J., Jarocha, L.E., Zollitsch, T. et al. (2021) Magnetic sensitivity of cryptochrome 4 from a migratory songbird. Nature 594, 535–540. https://doi.org/10.1038/s41586-021-03618-9

Yang, H., Peng, X., Gooday, A.J., Jones, B., Li, J., Liu, S., Huang, W., Z. Sun, Z., Chen, S., Dasgupta, S., Xu, H., Liu, S., Xu, W. & Ta, K. (2022) Magnetic foraminifera thrive in the Mariana Trench. Geochem. Persp. Lett. 21, 23–27. https://doi.org/10.7185/geochemlet.2212

CHAPTER NINE

COBALT

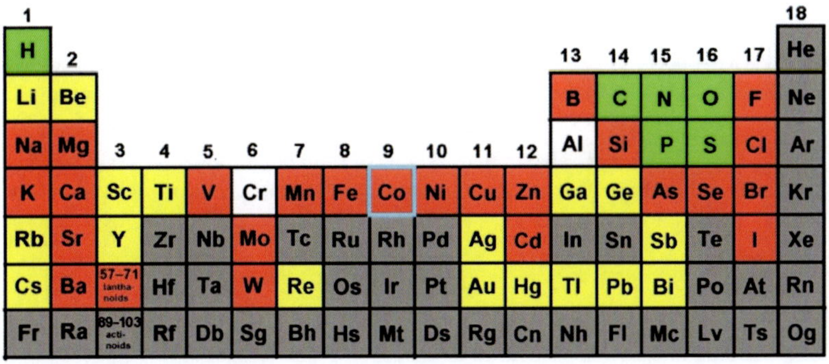

9.1 Cobalt

In the coccolithophore *Emiliania huxleyi,* and in several diatoms, cobalt and zinc can partly replace each other (Sunda & Huntsman 1995, Kellogg et al. 2022). In the bacterium *Salmonella typhimorum* cobalt can fulfill the functions of zinc in zinc deficient cells (Ammendola et al. 2020).

Cyanobacteria need cobalt, but not much. Saito et al. (2002) found that *Prochlorococcus* strain MED4-Ax has an absolute cobalt requirement and that zinc cannot substitute for cobalt. Hawco et al. (2020) estimate that *Prochlorococcus* strain MIT9215 can sustain growth with less than 50 cobalt atoms per cell, and maximum growth rate is obtained with 200 cobalt atoms per cell. Also for for *Microcystis aeruginosa* cobalt is necessary for sustained growth (Facey et al. 2022).

Cobalamin (Fig. 9.1) is a very important compound both in eukaryotes and in many prokaryotes.

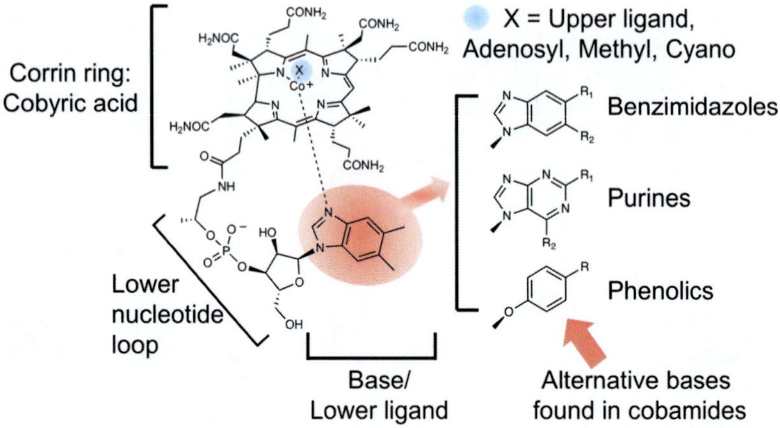

Fig. 9.1. The structure of cobalamin (vitamin B_{12}) and related substances. In vitamin B_{12} the ligand marked X is a cyano group (CN). Reproduced from Osman et al. (2021) under the CC BY License (http://creativecommons.org/licenses/by/4.0/.)

Vitamin B_{12} and the related compounds shown in Fig. 9.1 act as coenzyme in various isomerases, methyltransferases and dehalogenases. The dehalogenases occur only in some species of anaerobic bacteria, while the isomerases and methyltransferases occur in eukaryotes and various prokaryotes, but are synthesized only by certain prokaryotes. The biosynthesis is described by Fang et al. (2017). In addition to acting as cofactor in these enzymes, vitamin B_{12} (or, more correctly, a vitamin B_{12}-like compound with adenosyl as X; see Fig. 1) is also part of a photoreceptor protein, a bacterial transcriptional regulator, CarH, that controls the biosynthesis of carotenoids in response to light (Kutta et al. 2015). Also another photoreceptor protein, AerR, uses a vitamin B_{12}-like chromophore (Cheng et al. 2014, 2016).

In addition to the proteins containing vitamin B_{12}-related functional groups, there are some other cobalt-containing enzymes. At least two types of eukaryotic methionyl aminopeptidases contain cobalt (Arfin et al. 1995). Another enzyme type is nitrile hydratase (Pogorelova et al. 1966. Fig. 9.2), which has proved useful in industry for production of acrylamide and nicotinamide. They exist with either iron or cobalt as active metal (Lan et al. 2017). A thiocyanate hydrolase from *Thiobacillus thioparus* THI115 that composes thiocyanate to carbonyl sulfide and ammonia contains cobalt in a distorted square-pyramidal geometry that is not known from any other protein (Arakawa et al. 2007).

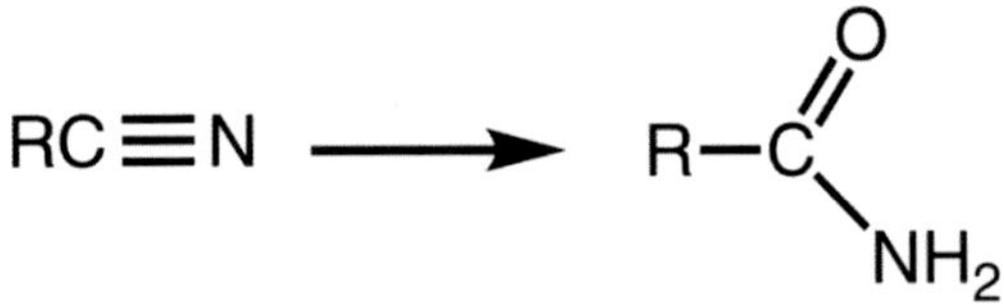

Fig. 9.2. Nitrile hydratase converts a nitrile to an amide by the addition of water.

Cobalt is not an absolute requirement for higher plants in general, but since some plants depend on cobalt-requiring nitrogen fixation, it is important for many plants (Hu et al. 2021). Also in some plants that do not bind N_2 internally, cobalt can have very positive effects. An example is sugarcane (*Saccharum* spp), in which application of cobalt fertilizer increased the number of open stomata, the number of leaves and tillers and sugar yield (Silva et al. 2022). Some of the effects are probably mediated by microorganisms.

High concentrations of cobalt ions are toxic to most organisms, but some are able to live in environments with high concentrations. One species, *Maribacter cobaltidurans*, has even go its name for this property. It can tolerate a record concentration of 10 m*M* Co^{2+} (Wu et al. 2018).

Duckweed, *Lemna minor*, is a well-known plant that can accumulate cobalt to very high concentration, more than 5,000 μg g^{-1} dry matter. Some plants, as *Rinorea* cf. *bengalensis*

(Violaceae) on Borneo, also accumulate cobalt, but mainly nickel to very high concentrations (e.g., Paul et al. 2020), probably as a defense against herbivores.

References

Ammendola, S., Ciavardelli, D., Consalvocd, A. & Battistoni, A. (2020) Cobalt can fully recover the phenotypes related to zinc deficiency in *Salmonella typhimurium*. Metallomics 12, 2021 –2031. https://doi.org/10.1039/d0mt00145g

Arakawa, T., Kawano, Y., Kataoka, S., Katayama, Y., Kamiya, N., Yohda, M. & Odaka, M. (2007) Structure of thiocyanate hydrolase: A new nitrile hydratase family protein with a novel five-coordinate cobalt(III) center. J. Mol. Biol. 366, 1497–1509. https://doi.org/10.1016/j.jmb.2006.12.011

Arfin, S.M., Kendall, R.L., Hall, L., Weaver, L.H., Stewart, A.E., Matthews, B.W. & Bradshaw, R.A. (1995) Eukaryotic methionyl aminopeptidases: Two classes of cobalt-dependent enzymes. Proc. Natl Acad. Sci. USA 92, 7714–7718.

Cheng, Z., Li, K., Hammad, L.A., Karty, J.A. & Bauer, C.E. (2014) Vitamin B12 regulates photosystem gene expression via the CrtJ antirepressor AerR in *Rhodobacter capsulatus*. Mol Microbiol 91, 649–664. https://doi.org/10.1111/mmi.12491

Cheng, Z., Yamamoto, H. & Bauer, C.E. (2016) Cobalamin's (vitamin B12) surprising function as a photoreceptor. Trends Biochem. Sci. 41:647–650. https://doi.org/10.1016/j.tibs.2016.05.002

Facey, J.A., King, J.J., Apte, S.C. & Mitrovic, S.M. (2022) Assessing the importance of cobalt as a micronutrient for freshwater cyanobacteria. J. Phycol. 58, 71–79. https://doi.org/10.1111/jpy.13216

Fang, H., Jie Kang, J. & Zhang, D. (2017) Microbial production of vitamin B_{12}: a review and future perspectives. Microb. Cell Fact. 16:15 (14 pp.). https://doi.org/10.1186/s12934-017-0631-y

Hawco, N.J., McIlvin, M.M., Bundy, R.M., Tagliabue, A., Goepfert, T.J., Moran, D.M., Valentin-Alvarado, L., DiTullio, G.R. & Saito, M.A. (2020) Proc. Natl Acad. Sci. USA 117, 15740–15747.

Hu, D., Cheng, M., Hu, K., Zhang, W., Yang, Y. & Xu, Q. (2019) Evaluation of cobalt hyperaccumulation and tolerance potential of the duckweed (*Lemna minor* L.). Ecotoxicol. Environm. Safety 179, 79–87. https://doi.org/10.1016/j.ecoenv.2019.04.058

Hu, X., Wei, X., Ling. J. & Chen, J. (2021) Cobalt: An essential micronutrient for plant growth? Front. Plant Sci. 12:768523 (24 pp.). https://doi.org/10.3389/fpls.2021.768523

Kellogg, R.M., McIlvin, M.R., Vedamati, J., Twining, B.S., Moffett, J.W., Marchetti, A., Moran, D.M. & Saito, M.A. (2020) Efficient zinc/cobalt inter-replacement in northeast Pacific diatoms and relationship to high surface dissolved Co: Zn ratios. Limnol. Oceanogr. 65, 2557–2582. https://doi.org/10.1002/lno.11471

Kutta, R.J., Hardman, S.J.O., Johannissen, L.O., Bellina, B., Messiha, H.L., Ortiz-Guerrero, J.M., Elías-Arnanz, M., Padmanabhan, S., Barran, P., Scrutton, N.S. & Jones, A.R. (2015) The photochemical mechanism of a B_{12}-dependent photoreceptor protein. Nature Comm. 6:7907 (11 pp.) https://doi.org/10.1038/ncomms8907

Lan, Y., Zhang, X., Liu, Z., Zhou, L., Shen, R., Zhong, X., Cui, W. & Zhou, Z. (2017) Overexpression and characterization of two types of nitrile hydratases from *Rhodococcus rhodochrous*. PLoS ONE 12: e0179833 (14 pp.). https://doi.org/10.1371/journal.pone.0179833

Osman, D., Cooke, A., Young, T.R., Deery, E., Robinson, N.J. & Warren, M.J. (2021) The requirement for cobalt in vitamin B_{12}: A paradigm for protein metalation. BBA - Molecular Cell Res. 1868: 118896 (25 pp.). https://doi.org/10.1016/j.bbamcr.2020.118896

Paul, A.L.D., Nkrumah, P.N., Echevarria, G., Erskine, P.D., Chaney, R.L., M. Spiers, M., Sumai, S. & van der Ent, A. (2020) Cobalt hyperaccumulation in *Rinorea cf. bengalensis* (Violaceae) from Sabah: accumulation potential and tissue and cellular-level distribution of cobalt. Plant Soil 455, 289–303. https://doi.org/10.1007/s11104-020-04629-7

Pogorelova, T.E., Ryabchenko, L.E., Sunzov, N.I. & Yanenko, A.S. (1996) Cobalt-dependent transcription of the nitrile hydratase gene in *Rhodococcus rhodochrous* M8. FEMS Microbiol. Lett. 144, 191–195.

Saito, M.A., Moffett, J.W., Chisholm, S.W. & Waterbury, J.B. (2002) Cobalt limitation and uptake in *Prochlorococcus*. Limnol. Oceanogr., 47, 1629–1636.

Silva, D.P., Johnson, R.M. & Crusciol, C.A.C. (2022) The effects of cobalt on sugarcane growth and development in plant cane and two ratoon crops. Sugar Tech., published online. https://doi.org/10.1007/s12355-022-01108-4

Sunda, W.G. & Huntsman, S.A. (1995) Cobalt and zinc interreplacement in marine phytoplankton: Biological and geochemical implications. Limnol. Oceanogr., 40, 1404-1417.

Wu, Y.-H., Fang, C., Zhou, P., Wang, C.-S. & Xu, X.-W. (2018) Complete genome sequence of a heavy metal resistant bacterium *Maribacter cobaltidurans* B1T, isolated from the deep-sea sediment of the South Atlantic Ocean. Marine Genomics 39, 19–21. https://doi.org/10.1016/j.margen.2017.12.005

CHAPTER TEN

NICKEL

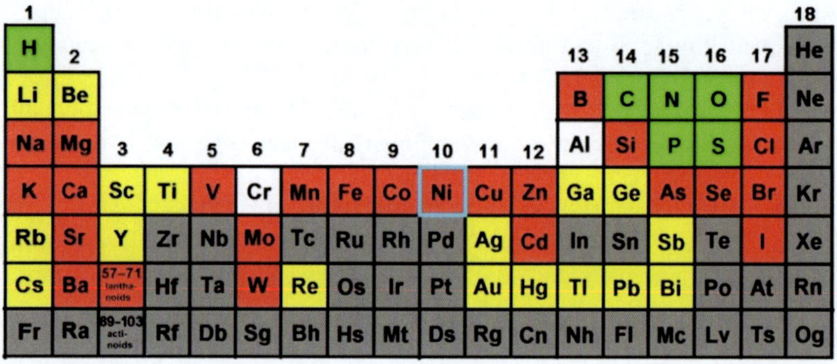

10.1 Nickel-dependent enzymes

Nickel was more easily available before the "Great Oxygenation" about 2.5 billion years ago. This is to some extent reflected in the biological use of nickel as part of enzymes. There are nine types of nickel enzymes, and nickel enzymes are present in archaea, bacteria, plants, fungi, and various protists, but not in animals (Boer et al. 2014, Alfano & Cavazza 2020). The relative abundance of nickel enzymes in anaerobic microbial metabolisms involving reduced gases (hydrogen, carbon monoxide, and methane), and its limited use by eukaryotes, suggest that the biological functions of nickel are, to a large extent, evolutionary relicts of the anoxic Archean (Glass & Dupont 2017).

Table 10.1 Overview of nickel enzymes (from Alfano & Cavazza 2020)

Redox enzymes	**Occurrence**	**Enzyme reaction**
[NiFe]-hydrogenase	Bacteria, Archaea	H_2 < == >$2H^+$ + $2e^-$
CO-dehydrogenase	Bacteria, Archaea	$CO+H_2O$<=>$CO_2+2H^++2e^-$
Acetyl-SCoA synthase	Bacteria, Archaea	CO + CoA-$S^-+CH_3CO_3$+FeSP < == > $CH_3C(O)$-S-CoA+CO^+FeSP
Methyl-SCoM reductase	Archaea	CH_3-SCoM+CoBSH <=>CH_4+CoBS-SCoM
Superoxide dismutase	Bacteria, Marine Eukary.	$2O_2^-+2H^+$=> $H_2O_2+O_2^-$
Non-redox enzymes	**Occurrence**	**Enzyme reaction**
Urease	Bacteria, Archaea, plants, fungi, algae	$(NH_2)CO_2+H_2O$=> CO_2+2NH_3
Lactate racemase	Bacteria, Archaea?	L-lactate < == > D-lactate
Glyoxalase	Bacteria, protists, plants	CH_3-CO-C(OH)-SG => CH_3-CH(OH)-CO-SG
Acireductone dioxygenase	Bacteria, Eukaryota?	1,2-dihydroxy-3-keto-5-methylhtiopentane + O_2 < == > 3-methyl thiopropanoate +formate + CO

In lactate racemase and related enzymes Ni(II) is bound to the protein via a so-called Ni-pincer nucleotide (NPN), covalently bound to the protein by forming a thioamide with lysine residue 184 (Fig. 10.1).

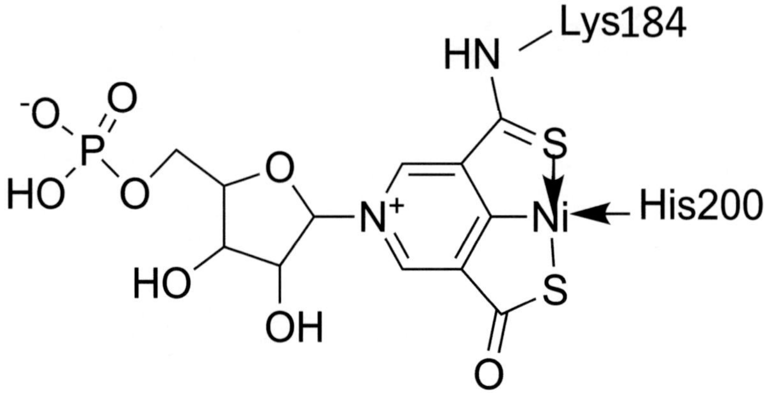

Fig. 10.1. Structure of the (SCS)Ni(II) pincer cofactor covalently bound to *Lactobacillus plantarum* lactate racemase (LarA). Redrawn from Desguin et al. (2016).

Desguin et al. (2020) have shown that the lactate racemase listed above is just one of several related enzymes having five more racemization or epimerization activities involving α-hydroxycarboxylic acids. One enzyme in this group is malate racemase, for which Gatreddi et al. 2022 have determined the detailed structure.

Gafurov et al. (2021) have reviewed the synthesis of artificial nickel pincer complexes and their use in artificial catalysts.

10.2 *Helicobacter pylori* and nickel

A special section is devoted to the pathogenic bacterium *Helicobacter pylori*, both because it is interesting and because it is very important. *Helicobacter pylori* is a Gram-negative helical bacterium and has been estimated to infect at least half of the world's human population (Ansari & Yamaoka 2017). It was identified rather recently (Warren & Marshall 1983, Marshall & Warren 1984, the latter publication cited 3943 times). It is specialized for living in our our stomachs (corpus and antrum), where it causes ulcer and often gastric cancer, killing almost a million people per year (The EUROGAST Study Group 1993).

Our stomachs are inhospitable environments for most invaders. The pH (1.5–2.0) of the human gastric juice is lower than the gastric juice in many other mammals (Fujimori 2020). It has evolved for two benefits: aiding the breakdown of our food, and killing invaders of our interior. *Helicobacter pylori* can live in this hostile environment because it produces urea and has an enzyme, urease, that produces pH-increasing ammonia from it. The gastric urea concentration (≈ 1 mM is not sufficient, and the bacterium probably produces more of it. For details about *H. pylori* urease, see Scott et al. (1998). Ammonia is also produced by the enzyme aspartase, but urease action on urea appears more important. Maximum activity of urease occurs at low pH (Miederer & Grübel 1996).

Helicobacter pylori is strictly dependent on nickel for several reasons. An obvious one is that urease contains nickel, and it needs urease for improving its environment. Nickel is also used for hydrogenase. A general opinion is that when *H. pylori* has infected a person, it is difficult to eradicate it, but Campanale et al. (2011, 2014) claim that 30 days of nickel free diet make it easier.

Experiments by Saylor et al. (2018) indicate that *H. pylori* possesses a large number of nickel proteins in addition to those known. Two proteins, Hpn and Hpnl, were previously known as nickel storage and detoxification proteins (Seshadri et al. 2007) but have also been found to be mediators of urease activity, Hpn acting as a nickel-sequestering protein (Vinella et al. 2015). Saylor et al. (2018) suspect that in addition Hpn and Hpnl interact with proteins across other cellular functions.

10.3 Nickel contact allergy (nickel contact dermatitis)

Other metals, e.g., cobalt (Silverberg et al. 2022) may also cause allergic reactions, but nickel is the most common cause. Problems are particularly common for women, who often wear jewelry or clothing details in contact with their skin. According to Liu et al. (2022) 17% of women and 3% of men are allergic to nickel.

Hypersensity is classified into four main types (Marwa & Kondmudi 2022):

- Type I: reaction mediated by IgE antibodies
- Type II: cytotoxic reaction mediated by IgG or IgM antibodies
- Type III: reaction mediated by immune complexes
- Type IV: delayed reaction mediated by cellular response

Nickel contact allergy is a cell-mediated type IV hypersensitivity reaction (a delayed-type hypersensitivity). This means that the reaction takes several days to develop.

Exposure to nickel causes changes in leukocyte composition including influx and activation of natural killer cells, macrophage polarization, and T-cell immunity (Wisgrill et al. 2020). An upregulation takes place of the enzymes matrix metallopeptidase 12 (MMP12) and suppressor of cytokine signaling 3 (SOCS3). A few days after the first contact over a thousand genes are upregulated, and a similar number downregulated. One of the upregulated proteins is semaphorin 3A (Liu et a. 2022). Semaphorins are both secreted and membrane proteins. Semaphorin-3A protein is vital for normal neuronal pattern development and it downregulates T cell proliferation via the neuropilin-1 (Nrp1)-plexin-A4 receptor complex (Takamutsu et al. 2010) and increases cytokine production in response to toll-like receptor agonists or bacterial sepsis in mice (Yamashita et al. 2015).

10.4 Nickel hyperaccumulation

Many elements, even those considered poisonous, are accumulated by plants to much higher concentrations than needed for the physiological functions of the plant. In some cases this will benefit the plant by deterring grazers or competitors. Examples of such hyperaccumulated elements or ions are aluminum (Jansen et al. 2002), cadmium (Pielichowska & Wierzbicka 2004, Dai et al. 2020, Borovička et al. 2022, Yan et al. 2022), copper (Corzo Remigio et al. 2021), gadolinium (Good et al. 2022), various lanthanides (Ashraf et al. 2022), manganese (Bihanic et al. 2021), and selenium (Valdez Barillas et al. 2011, Statwick et al. 2016). A general review on hyperaccumulating plants has been published by Reeves et al. 2018.

Cobalt and nickel are also sometimes hyperaccumulated. They are closely related metals, neighbors in the periodic table of elements: Cobalt with atomic number 27 and atomic mass 58.933, nickel with atomic number 28, and atomic mass 58.693. Really like twins, but still different in a biological context. Actually they were neighbors even before they appeared in the periodic table, being names of goblins in mines and other scary dark places. The words cobalt and kobolt and kobold, in fact, are just other forms of the word goblin.

The word nickel is derived from the Swedish word koppamickel (Kupfernickel in German), a mineral containing copper and nickel, from which mineralogist Axel Fredrik Cronstedt in 1751 purified metallic nickel and named the metal in a publication (Cronstedt 1751—1754). But the word nickel is older, referring to an elf or goblin.

Nickel, as ion, is part of several important enzymes and other proteins. We shall deal with this separately, and here focus on the fact that some plants hyperaccumulate nickel (and some hyperaccumulate cobalt, or both cobalt and nickel. The most extreme nickel hyperaccumulator is a New Caledonian tree, *Sebertia acuminata* (Jaffré et al. 1976, Boyd & Jaffré 2001,

Perrier et al. 2004, Grohmann et al. 2008, Callahan et al. 2008), also called *Pycnandra acuminata* (Sagner et al. 1998, Schaumlöffel et al. 2002). It absorbs divalent nickel, and stores it as nickel citrate in the latex, which holds an almost unbelievable concentration of nickel, 11.2 percent by weight on a wet weight basis, 25.7% by weight on a dry weight basis (Jaffré et al. 1976). This means that 80% of the dry matter of the latex is nickel citrate (trinickel dicitrate). However, Schaumlöffel et al. (2002) regard nicotianamine to be an important ligand of nickel in the plant, while Callahan et al. (2008) regard aldaric acid (2,4,5-trihydroxy-3-methoxy-1,6-hexan-dioic acid) to be an important ligand, second to the dominating citric acid.

The color of the latex in *Pycnandra acuminata* is blue-green from divalent nickel, as it is in two related species, *Pycnandra caeruleilatex* and *Pycnandra kouakouensis*, also growing in New Caledonia (Isnard et al, 2020). The total nickel content of a single mature tree was estimated to be 37 kg (Callahan et al, 2008). A repellent effect of the plant sap was observed on the fruit fly *Drosophila melanogaster* indicating that in hyperaccumulating plants nickel functions as an agent to prevent predation (Sagner et al. 1998). Old leaves falling on the ground under the trees still contain a high amount of nickel, which makes the resulting soil nickel-rich (Boyd & Jaffré 2001).

There are many other nickel hyperaccumulating plants, in New Caledonia and elsewhere. Baker & Brooks (1989) have compiled a list of 145 species and subspecies, and note that they are mostly occurring in 7 geographical regions: New Caledonia, Western Australia, Southern Europe and Asia Minor, The Malay Archipelago, Cuba, Western United States, and Zimbabwe. They also note that these areas have that in common that they have not been ice-covered during the last glaciation, and that the time elapsed since it ended is probably too short for this trait to evolve. The reasoning does not appear convincing to me, since the plants now living in previously ice-covered regions have mostly migrated from other regions (some species have survived on nunataks).

10.5 Cable bacteria: Nickel electricity conductors

Oxygen consumption in marine sediments is often coupled to the oxidation of sulphide generated by degradation of organic matter in deeper, oxygen-free layers. This coupling can be mediated by electric currents carried by filamentous bacteria, "cable bacteria" over centimeter distances (Pfeffer et al. 2012). They were first described in the beginning of this century, and since then the literature about them has grown rapidly. Tentatively they are divided into a marine genus, *Electrothrix*, and a freshwater genus, *Electronema*

(Trojan et al. 2016, Kjeldsen et al. 2019), both belonging to the family family Desulfobulbaceae in the phylum Thermodesulfobacteriota, a kind of delta-Proteobacteria.

Recently Boschker et al. (2021), using an unnamed species of cable bacterium came to the conclusion that electricity is conducted through the bacterial filaments in periplasmic wires consisting of a conductive nickel protein core surrounded by an insulating protein shell layer. The presence of nickel in the conductor was confirmed by Eachambadi et al. 2021), who also confirmed that the conducting protein layer is surrounded by an insulating layer. The conductor is not metallic nickel, since the electric properties point to organic conductors (Bonné et al. (2020). They conduct electrons over centimeter distances. Bonné et al. (2020) found that the bacteria show n-type transport, suggesting that electrons are the charge carriers.

The cable bacteria referred to above are Gram-negative. Recently (Yang et al. 2021) long-distance electron transfer was also discovered in Gram-positive bacteria (*Lysinibacillus varians* GY32).

References

Alfano, M. & Cavazza, C. (2020) Structure, function, and biosynthesis of nickel-dependent enzymes. Protein Sci. 29, 1071–1089. https://doi.org/10.1002/pro.3836

Ansari, S. & Yamaoka, Y. (2017) Survival of *Helicobacter pylori* in gastric acidic territory. Helicobacter 22:e12386 (13 pp). https://doi.org/10.1111/hel.12386

Ashraf, N., Rodrigues, E.S., de Almeida, E., Montanha, G.S., Abreu-Junior, C.H., Vitová, M., Lazzari Garcia, R.H., Küpper, H. & Pereira de Carvalho, H.W. (2022) Identification of potential plant species hyperaccumulating light rare earth elements (LREE) in a mining area in Minas Gerais, Brazil. Environm. Sci., Pollut. Res. https://doi.org/10.1007/s11356-022-22009-9

Baker, A.J.M. & Brooks, R.R. (1989) Terrestrial higher plants which hyperaccumulate metallic elements - A review of their distribution, ecology and phytochemistry. Biorecovery, 1989, 81–126.

Bihanic, C., Petit, E., Perrot, R., Cases, L., Garcia, A., Pelissier, F., Poullain, C., Rivard, C., Hossaert-McKey, M., McKey, D. & Grison, C. (2021) Manganese distribution in the Mn-hyperaccumulator *Grevillea meisneri* from New Caledonia. Sci. Reps 11:23780 (14 pp.) https://doi.org/10.1038/s41598-021-03151-9

Boer, J.L., Mulrooney, S.B. & Hausinger, R.P. (2014) Nickel-dependent metalloenzymes. Arch Biochem Biophys. 544, 142–152.

Bonné, R., Hou, J.-L., Hustings, J., Wouters, K., Mathijs Meert, M., Hidalgo-Martinez, S., Cornelissen, R., Morini, F., Thijs, S., Vangronsveld, J., Valcke, R., Cleuren, B., Meysman, F.J.R. & Manca, J.V. (2020) Intrinsic electrical properties of cable bacteria reveal an Arrhenius temperature dependence. Sci. Rep. 10:19798 (8 pp.). https://doi.org/10.1038/s41598-020-76671-5

Borovička, J., Braeuer, S., Walenta, M., Hršelová, H., Leonhardt, T., Sácký, J., Kaňa, A. & Goessler, W. (2022) A new mushroom hyperaccumulator: Cadmium and arsenic in the ectomycorrhizal basidiomycete *Thelephora penicillate*. Sci. Tot. Environm. 826: 154227 (10 pp.).

Boschker, H.T.S., Cook, P.L.M., Polerecky, L., Eachambadi, R.T., Lozano, H., Silvia Hidalgo-Martinez, S., Khalenkow, D., Spampinato, V., Claes, C., Kundu, P., Wang, D., Bals, S., Sand, K.K., Cavezza, F., Hauffman, T., Bjerg, J.T., Skirtach, A.G., Kochan, K., McKee, M., Wood, B., Bedolla. D., Gianoncelli, A., Geerlings, N.M.J., Van Gerven, N., Remaut, H., Geelhoed, J.S., Millan-Solsona, R., Fumagalli, L., Nielsen, L.P., Franquet, A., Manca, J.V., Gomila, G. & Filip J. R. Meysman, F.R.J. (2021) Efficient long-range conduction in cable bacteria through nickel protein wires. Nature Commun. 12:3996 (12 pp.) https://doi.org/10.1038/s41467-021-24312-4

Boyd, R.S. & Jaffré, T. (2001) Phytoenrichment of soil Ni content by *Sebertia acuminata* in New Caledonia and the concept of elemental allelopathy. South African J. Sci. 97, 535–538.

Campanale, M., Nucera, E., Cesario, V., Tortora, A., F. Purchiaroni, F., Bertucci, F., Di Rienzo, T.A., De Pasquale, T., Gasbarrini, G., Ojetti, V., Schiavino, D. & Gasbarrini, A. (2011) Nickel free-diet enhances *Helicobacter pylori* eradication rate. Abstr. XVII Nat. Congr. Digestive Dis./Dig. Liver Dis. OC.04.1.

Campanale, M., Nucera, E., Ojetti, V., Cesario, V., Di Rienzo, T.A., D'Angelo, G., Pecere, S., Barbaro, F., Gigante, G., De Pasquale, T., Rizzi, A., Cammarota, G., Schiavino, D., Fraceschi, F. & Gasbarrini, A. (2014) Nickel free-diet enhances the *Helicobacter pylori* eradication rate: A pilot study. Dig. Dis. Sci. 59, 1851–1855. https://doi.org/10.1007/s10620-014-3060-3

Corzo Remigio, A., & Edraki, M., & Baker, A.J.M. & Antony van der Ent, A. (2021) Is the aquatic macrophyte *Crassula helmsii* a genuine copper hyperaccumulator? Plant Soil (2021) 464, 359–374. https://doi.org/10.1007/s11104-021-04955-4

Cronstedt, A.F. (1774) Rön och försök, gjorda med en malmart från Los kobolt grufvor, Vetenskapsakademins handlingar 12, 287–292, 1751—54. (Experiences and experiments with an ore from the cobolt mines at Los, Proceedings of the Swedish Academy of Sciences 12, 287–292, 1751—1754.

Dai, Z.-C., Cai, H.-H., Qi, S.-S., Li, J., Zhai, D.-L., Wan, J.S.H. & Du, D.-L- (2020) Cadmium hyperaccumulation as an inexpensive metal armor against disease in Crofton weed. Environm,Pollut. 267: 115649 (9 pp.). https://doi.org/10.1016/j.envpol.2020.115649

Eachambadi, R.T., Boschker, H.T.S., Franquet, A., Spampinato, V., Hidalgo-Martinez, S., Valcke, R., Meysman, F.J.R. & Manca, J.V. (2021) Enhanced laterally resolved ToF-SIMS and AFM imaging of the electrically conductive structures in cable bacteria. Anal. Chem. 93, 7226−7234. https://doi.org/10.1021/acs.analchem.1c00298

Desguin, B., Soumillion, P., Hols, P. & Hausinger, R.P. (2016) Nickel-pincer cofactor biosynthesis involves LarB catalyzed pyridinium carboxylation and LarEdependent sacrificial sulfur insertion. Proc. Natl Acad. Sci. USA 223, 5598–5603.

Desguin, B., Urdiain-Arraiza, J., Da Costa, M., Fellner, M., Hu, J., Hausinger, R.P., Desmet, T., Hols, P. & Soumillion, P. (2020) Uncovering a superfamily of nickel-dependent hydroxyacid racemases and epimerases. Scientific Reports 10:18123 (11 pp.) https://doi.org/10.1038/s41598-020-74802-6

Fujimori, S. (2020) Gastric acid level of humans must decrease in the future. World J. Gastroenterol. 26, 6706–6709. https://doi.org/10.3748/wjg.v26.i43.6706

Gafurov, Z.N., Kantyukov, A.O., Kagilev, A.A., Kagileva, A.A., Sakhapov, I.F., Mikhailov, I.K. & Yakhvarov, D.G. (2021) Recent advances in chemistry of unsymmetrical phosphorus-based pincer nickel complexes: From design to catalytic applications. Molecules 2021, 26, 4063. https://doi.org/10.3390/molecules26134063

Glass, J.B. & Dupont, C.L. (2017) Oceanic nickel biogeochemistry and the evolution of nickel use. Ch. 2 in Metallobiology Series No. 10, The biological chemistry of nickel (Zamble, D., Rowińska-Żyrek, M., and Kozłowski, H., Eds).

Good, N.M., Lee, H.D., Hawker, E.R., Su, M.Z., Gilad, A.A. & Martinez-Gomez, N.C. (2022) Hyperaccumulation of gadolinium by *Methylorubrum extorquens* AM1 reveals impacts of lanthanides on cellular processes beyond methylotrophy. Front. Microbiol. 13:820327 (25 pp.). https://doi.org/10.3389/fmicb.2022.820327

Grohmann, A., Lerch, M., Ressler, T. & Krumeich, F. (2008) Nickel-Nanopartikel aus dem Baum *Sebertia acuminata*. Z. Anorg. Allg. Chem. 2081–2081. https://doi.org/10.1002/zaac.200870145

Gatreddi. S., Urdiain-Arraiza, J., Desguin, B., Hausinger, R.P. & Hu, J. (2022) Structural and mutational characterization of a malate racemase from the LarA superfamily. Biometals, published online 2022. https://doi.org/10.1007/s10534-022-00372-x

Jaffré, T., Brooks, R.R., Lee, J. & Reeves, R.D. (1976) *Sebertia acuminata*: A hyperaccumulator of nickel from New Caledonia. Science 193, 579–580. https://www.jstor.org/stable/1742764

Jansen, S., Broadley, M.R., Robbrecht, E. & Smets, E. (2002) Aluminum hyperaccumulation in angiosperms: A review of its phylogenetic significance. Bot. Rev. 68, 235–269. http://www.bioone.org/doi/full/10.1663/0006-8101%282002%29068%5B0235%3AAHIAAR%5D2.0.CO%3B2

Kjeldsen, K.U., Schreiber, L., Thorup, C.A., Boesen, T., Bjerg, J.T., Yang, T., Dueholm, M.S., Larsen, S., Risgaard-Petersen, N., Nierychlo, M., Schmid, M., Bøggild, A., van de Vossenberg, J., Geelhoed, J.S., Meysman, F.J.R., Wagner, M., Nielsen, P.H., Nielsen, P. & Schramm, A. (2019) On the evolution and physiology of cable bacteria. Proc. Natl Acad. Sci. USA 16, 19116–19125.

Konhauser, K.O., Robbins, L.J., Pecoits, E., Peacock, C., Kappler, A. & Lalonde, S.V. (2015) The archean nickel famine revisited. Astrobiology 15, 804–815.

Liu, L., Watanabe, M., Minami, N., Yunizar, M.F. & Ichikawa, T. (2022) Semaphorin 3A: A potential target for prevention and treatment of nickel allergy. Comm. Biol. 5:671 (13 pp.) https://doi.org/10.1038/s42003-022-03641-0

Marshall, B. & Warren, J.R. (1984) Unidentified curved bacilli on gastric epithelium in active chronic gastritis and peptic ulceration. Lancet 1984, 1311–1315.

Marwa, K. & Kondamudi, N.P. (2022) Type IV hypersensitivity reaction. National Library of Medicine. https://www.ncbi.nlm.nih.gov/books/NBK562228/

Perrier, N., Colin, F., Jaffré, T., Ambrosi, J.-P., Rose, J. & Bottero, J.-Y. (2004) Nickel speciation in *Sebertia acuminata*, a plant growing on a lateritic soil of New Caledonia. C. R. Geosci. 336, 567–577. https://doi.org/10.1016/j.crte.2003.12.014

Pfeffer, C., Larsen, S., Song, J., Dong, M., Besenbacher, F., Meyer, R.L., Kjeldsen, K.U., Schreiber, L., Gorby, Y.U., El-Naggar, M.Y., Leung, K.M., Schramm, A., Risgaard-Petersen, N. & & Nielsen, L.P. (2012)

Filamentous bacteria transport electrons over centimetre distances. Nature 491, 2018–221. https://doi.org/10.1038/nature11586
Pielichowska, M. & Wierzbicka, M. (2004) Uptake and localization of cadmium. Acta Niologica Cracoviensia Series Botanica 46, 57–63.
Reeves, R.D., van der Ent, A. & Baker, A.J.M. (2018) Global distribution and ecology of hyperaccumulator plants. In A. van der Ent et al. (eds.), Agromining: Farming for Metals, Mineral Resource Revs, pp. 75–92. https://doi.org/10.1007/978-3-319-61899-95
Sagner, S., Kneer, R., Wanner, G., Cosson, J.-P. Deus-Neumann, B. & Zenk, M.H. (1998) Hyperaccumulation, complexation and distribution of nickel in *Sebertia acuminata*. Phytochemistry 41, 339–347.
Scott, D.R., Weeks, D., Hong, C., Postius, S., Melchers, K. & Sachs, G. (1998) The role of Internal urease in acid Resistance of *Helicobacter pylori*. Gastroenterology 114, 58–70.
Seshadri, S., Benoit, S.L. & Maier, R.J. (2007) Roles of His-rich hpn and hpn-like proteins in Helicobacter pylori nickel physiology. J. Bacteriol. 189, 4120–4126.
Silverberg, J.I., Patel, N., Warshaw, E.M., Maibach, H.I., Belsito, D.V., DeKoven, J.G., A. Zug, K.A., Taylor, J.S., Sasseville, D., DeLeo, V.A., Pratt, M.D., Reeder, M.J., Atwater, A.R., Fowler Jr, J.R. & Houle, M.-C. Patch testing with cobalt in children and adolescents: North American contact dermatitis group experience, 2001–2018. Contact Dermatitis. 2022;1–10. https://doi.org/10.1111/cod.14185
Statwick, J., Majestic, B.J. & Sher, A.A. (2016) Characterization and benefits of selenium uptake by an Astragalus hyperaccumulator and a non-accumulator. Plant Soil 404, 345–359. https://doi.org/10.1007/s11104-016-2842-0
Takamatsu, H. et al. (2010) Semaphorins guide the entry of dendritic cells into the lymphatics by activating myosin II. Nat. Immunol. 11, 594–600.
The EUROGAST Study Group (1993) Epidemiology of, and risk factors for, Helicobacter pylori infection among 3194 asymptomatic subjects in 17 populations. Gut 34, 1672–1676.
Trojan, D., Schreiber, L., Bjerg, J.T., Bøggild, A., Yang, T., Kjeldsen, K.U. & Schramm, A. (2016) A taxonomic framework for cable bacteria and proposal of the candidate genera Electrothrix and Electronema. Systematic and Appl. Microbiol. 39, 297–306. http://dx.doi.org/10.1016/j.syapm.2016.05.006
Valdez Barillas, J.R., Quinn, C.F. & Pilon-Smits, E.A.H. (2011) Selenium accumulation in plants – phytotechnological appliations and ecological implications. Int. J. Phytoremed. 13(S1), 166–178. https://doi.org/10.1080/15226514.2011.568542

Vinella, D., Fischer, F., Vorontsov, E., Gallaud, J., Malosse, C., Michel, V., Cavazza, C., Robbe-Saule, M., Richaud, P., Chamot-Rooke, J., Brochier-Armanet, C. & De Reuse, H., (2015) Evolution of *Helicobacter*: acquisition by gastric species of two histidine-rich proteins essential for colonization. PLoS Pathog. 11:e1005312 (32 pp.). https://doi.org/10.1371/journal.ppat.1005312

Warren, J.R. (1983) Unidentified curved bacilli on gastric epithelium in active chronic gastritis. Lancet 1, 1273–1273.

Wisgrill, L., Werner, P., Jalonen, E., Berger, A., Lauerma, A., Alenius, H. & Fyhrquist, N. (2020) Integrative transcriptome analysis deciphers mechanisms of nickel contact dermatitis. Allergy. 2021;76:804–815. https://doi.org/10.1111/all.14519

Yan, J., Tang, Z., Fischel, M., Wang, P., Siebecker, M.G., Aarts, M.G.M., Sparks, D.L. & Zhao, F.-J. (2022) Variation in cadmium accumulation and speciation within the same population of the hyperaccumulator *Noccaea caerulescens* grown in a moderately contaminated Soil. Plant Soil 475, 379–394.
https://doi.org/10.1007/s11104-022-05373-w

Yamashita, N. et al. (2015) Anti-Semaphorin 3A neutralization monoclonal antibody prevents sepsis development in lipopolysaccharide-treated mice. Int. Immunol. 27, 459–466.

Yang, Y., Wang, Z., Gan, C., Hyldgaard Klausen, L., Bonné, R., Kong, G., Luo, D., Meert, M., Zhu, C., Sun, G., Guo, J., Ma, Y., Bjerg, J.T., Manca, J., Xu, M., Nielsen, L.P. & Dong, M. (2021) Long-distance electron transfer in a filamentous Gram-positive bacterium. Nature Comm. (8 pp.). https://doi.org/10.1038/s41467-021-21709-z

CHAPTER ELEVEN

COPPER, SILVER, AND GOLD

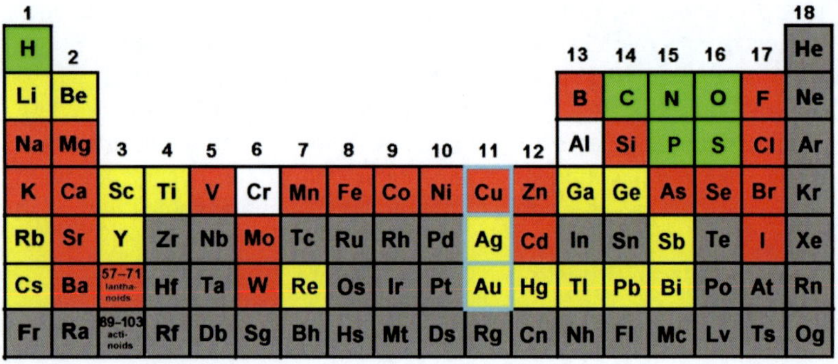

11.1 Copper

There are similarities and differences between copper and iron in relation to life. Both easily change between two oxidation states; between Cu^{+I} and Cu^{+II}, and between Fe^{+II} and Fe^{+III}, respectively. Iron is much more common in the Earth's crust (5.6% by weight) than is copper (0.006% by weight), and for early life iron was the natural choice, also because any copper in the environment, mostly bound as sulfide, was difficult to assimilate. A change came with "the Great Oxidation Event", caused by the rise of oxygenic photosynthesis, about 2500 million years ago. Sulfide was oxidized, while iron was precipitated in trivalent form. But the dependence of life on iron remains from the early days, and lactic acid bacteria, that do not need iron, are an exception.

Thus copper biochemistry is mostly of younger date than iron biochemistry. Most contemporary organisms need copper, but life, in its richness, provides us with exceptions also in this respect. The common bacterium *Pseudomonas aeruginosa* uses copper when it is available, but can also live without it (Frangipani et al. 2008), using a cytochrome *bd*-type cyanide-insensitive oxidase in place of copper-containing respiratory enzymes. This is not the only case in which iron- and copper-containing proteins can substitute for one another. In the photosynthetic electron transport chain copper-containing plastocyanin serves the same function as iron-containing cytochrome c_6. Some organisms, e.g., the diatoms *Phaeodactylum tricornutum* and *Thalassiosira weissflogii* use only cytochrome c_6 (Bernal-Bayard et al. 2015, Inda et al. 1999), while another diatom, *Thalassiosira oceanica*, is thought to use only plastocyanin (Peers & Price 2006). On the other hand, the green alga *Scenedesmus vacuolatus* codes for both, and if copper is available, this suppresses cytochrome c_6 synthesis (Miramar et al. 2003). To serve the same purpose, cytochrome c_6 and plastocyanin have evolved to have many common properties: the same redox potential (ca. 350 mV at pH 7), similar molecular mass, similar isoelectric point, and certain structural similarities (De la Rosa et al. 2002).

Most copper proteins are oxidoreductases. Doerrer (2018) describes more than twenty kinds of them and the reactions they catalyze. Copper proteins can be broadly classified into types 1–4.

Type 1: These proteins contain one copper atom, Cu^{+II}, per molecule, and are blue-coloured, and give a well-defined electron spin resonance (ESR) signal. A well-known example is plastocyanin, a protein that transfers electrons in photosynthesis. Type 1 copper proteins occur also in Archaea. An example is halocyanin from the haloalkaliphilic *Natronobacterium pharaonic* (Scharf & Engelhard 1993). It is a peripheral membrane protein which serves as a mobile electron carrier.

Type 2: These proteins are not blue, but also give an ESR signal. Examples are nitrite reductase, that catalyzes the reaction $NO_2^- + 2H^+ + e^- \longrightarrow NO + H_2O$, and CuZn superoxide dismuttase, that catalyzes $2O_2^- + 2H^+ \longrightarrow O_2 + H_2O_2$. Further examples are described by MacPherson & Murphy (2007).

Type 3: These proteins (Fig. 1.1.1) contain two copper atoms per molecule, and their electron spins cancel each other. Examples are the enzymes laccase, tyrosinase (Fig. 1), and the invertebrate oxygen-transporting hemocyanins; see also Pretzler & Rompel (2018). Arthropod and mollusc hemocyanins bind oxygen in the same manner, but are distinct in their molecular structures. Other examples of type 3 copper proteins are given by Kaintz et al. (2014). More details about arthropod hemocyanins are found in the review by Burmester (2002), and all types of hemocyanins were thoroughly reviewed by van Holde & Miller (1995). More recently Costa-Paiva et al. (2022) provided a dated tree of hemocyanin evolution.

Type 4: In this category we lump together all copper proteins that do not fit into any of the other categories.

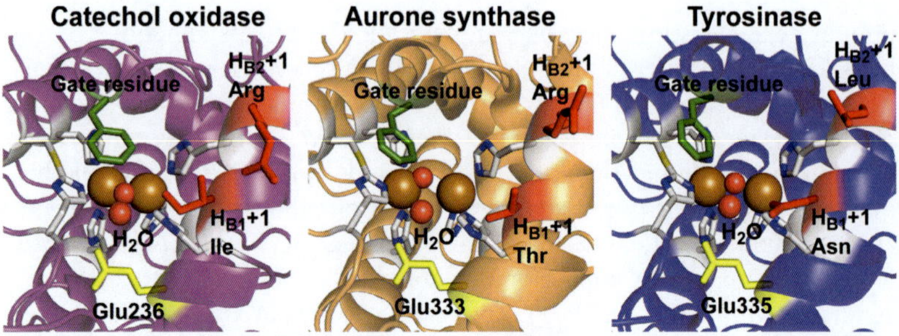

Fig. 11.1.1. Structures of the active centers of three type 3 copper proteins from plants. Large brown spheres are copper ions (left: CuA, right CuB), smaller red spheres oxygen atoms. Reproduced from Pretzler & Rompel (2018) under the CC BY license (http://creativecommons.org/licenses/by/4.0/).

From an anthropocentric view, the most important copper-containing protein is cytochrome *c* oxidase. It is a copper-iron enzyme, and its origin has been traced back to soil dwelling, probably iron-oxidizing bacteria, likely ancestors of extant acidophilic iron-oxidizers such as *Acidithiobacillus spp*., which belong to basal lineages of the phylum Proteobacteria, living on emerged land over two billion years ago (Esposti 2020). Already Castresana et al. (1994) had a similar view of a very ancient origin of cytochrome oxidase, long before the oxygenation of the biosphere by oxygenic photosynthesis. These authors even think that some subunits of cytochrome oxidase can be traced back to the common origin of bacteria and archaea.

A very important type 3 copper protein is tyrosinase. This enzyme causes the breakdown of "protecting" phenolic substances in soil, and thereby speeds up the breakdown and release of climate-important carbon dioxide (Panis & Rompel 2022).

Methane is another gas of great importance for the climate; per molecule it is a much more potent greenhouse gas than carbon dioxide is. One of the main enzymes involved in converting methane to the less potent carbon dioxide is the copper-containing methane monooxygenase present in methanotrophic microbes, and copper is often limiting for their growth and activity. They can secrete so-called chalkophores, copper-binding proteins, to acquire copper from the environment. A well-characterized chalkophore is methanobactin, exuded by the methanotroph *Methylosinus trichosporium* OB3b (Rushworth et al. 2021). This member of Aphaproteobacteria is able to obtain copper from solid copper sulfide minerals.

Ceruloplasmin, which we can categorize as a type 4 copper protein (although it is as blue as a type 1 protein, because three of its copper sites are of type 1, while one is of type 2, and two of type 3; Hellman & Gitlin

2002), more precisely as belonging to the multicopper ferroxidase family, is present in mammal blood serum, and also in some other fluids and tissues. It was discovered by Holmberg and Laurell (1948). Although it has often been stated that it contains about 95% of the copper in blood serum, the percentage is probably lower, since about half of ceruloplasmin is present in a copper-free (apo) form (Linder 2021). When fully loaded it carries six copper ions (Fig. 11.1.2).

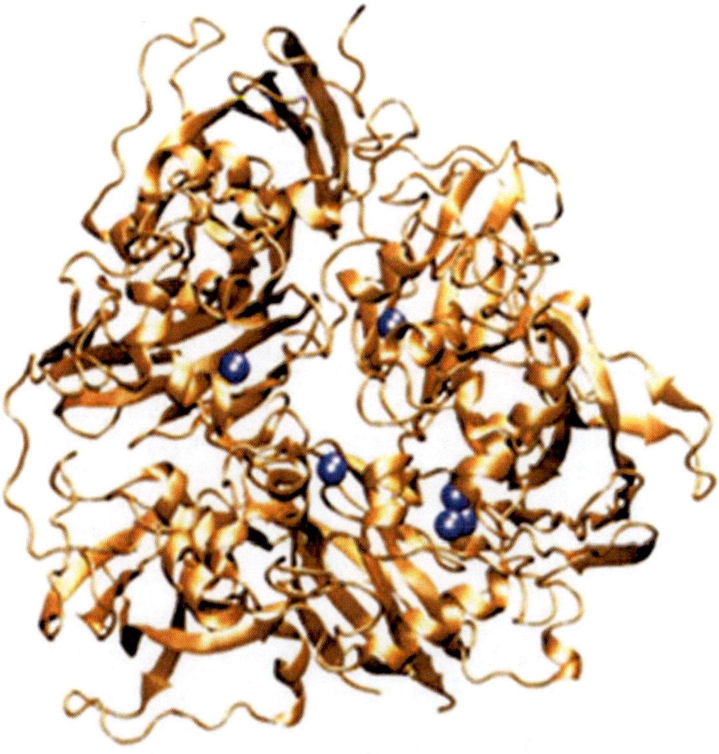

Fig. 11.1.2. The ceruloplasmin molecule carries six copper ions when fully loaded. From Linder (2021), reproduced under the Creative Commons Attribution (CC BY) license (https://creativecommons.org/licenses/by/4.0/). Based on research by Zaitseva et al. (1996).

A very high copper intake is poisonous, although the tolerance varies very much between organisms. For humans 10 mg per day is regarded as tolerable, while less than 0.9 mg per day leads to deficiency, and 2.6 mg per day is optimal (Chambers et al. 2010).

Copper plays several important roles in the battles between bacterial parasites and their hosts (Festa & Thiele 2012, Besold et al. 2016). Hosts can fight their invaders by using copper to disrupt active sites of their metalloproteins, such as Fe-S enzymes. In particular, blood cells called macrophages can attack invading microbes with high copper and this metal is elevated, e.g., at sites of lung infection. Microbial intruders, on the other hand, are equipped with various defenses against copper, such as copper-exporting ATPases and copper binding metallothioneins. But since copper is also an essential nutrient for the invaders, another way of fighting them is to bind copper so tightly that it becomes unavailable for the attacker.

11.2 Silver

Silver is not an essential element for any organism. On the contrary, silver ions (Ag^+) are quite toxic for many organisms, and the old Latin name for silver nitrate is lapis infernalis, stone of hell. "Lapis" is still used as a name for silver nitrate.

The medicinal and preservative properties of silver have been known for more than two thousand years. Because of its antimicrobial effect, silver nitrate, silver acetate (Mannix-Fisher & McLean 2021), and sometimes other silver compounds or metallic silver as small particles (nanoparticles), has been used to treat wounds and burns to combat infection (Moir & Serra 2012). The antimicrobial effects are partly due to binding of Ag^+ to DNA, especially to the purine bases adenine and guanine (Arakawa et al. 2001). Another explanation for the toxicity of silver ions, Ag^+, is that they displace nickel (Ni^{2+}) in nickel enzymes (Dobrev et al. 2022). Metallic silver nanoparticles execute their antimicrobial effect by being oxidatively converted to silver ions:

$4Ag + O_2 \longrightarrow 2Ag_2O$ followed by $2Ag_2O + 4H^+ \longrightarrow 4Ag^+ + 2H_2O$
(Khina & Krutyakov 2021).

Colloidal silver is effective against both Gram-positive and Gram-negative bacteria (Domínguez et al. 2020).

Bacteria may develop resistance to silver ions. One resistance mechanism is using an Ag^+ efflux pump.

Silver ions have been used also in dentistry, against caries. Silver nitrate has been used, but permanently stains caries lesions black (Gao et al. 2018). Another way of delivering the silver ions is application of silver diammine fluoride (often erroneously spelt silver diamine fluoride), $AgF(NH_3)_2$, (Bridge et al. 2021, Srisomboon et al. 2022). The the healing effect of silver ions is due to formation of a precipitate of silver phosphate, Ag_3PO_4, which acts as a barrier against bacterial attack.

Rarely can medical use of silver cause argyria, black spots in the skin, probably consisting mostly of silver sulfide. More common is that argyria is caused by ingestion of silver ions, e.g., from mouthwash containing silver ions, or from silver jewelry.

11.3 Gold

Although gold is relatively inert and has a reputation as a "noble metal", it cycles between metallic and ionic forms. Nuggets and grains of gold in nature are usually covered by a layer of bacteria (Rea et al. 2016). Au^{3+} ions are toxic to about the same extent as mercury (Hg^{2+}) and silver (Ag^{+}) ions, and microorganisms handle them in various ways to avoid poisoning. One important way is to reduce them to gold metal. Examples of bacteria that do this are *Desulfovibrio* sp. (Thermodesulfobacteriota) (Lengke & Southam 2007), *Cupriavidus metallidurans* (β-Proteobacteria) (Reith et al. 2009), *Shewanella oneidensis* (γ-Proteobacteria) (De Corte et al. 2011), *Rhodobacter sphaeroides*, a phototrophic α-proteobacterium (Italiano et al. 2018), *Bacillus licheniformis* FZUL-63 (Bacillota) (Cheng et al. 2019), and *Bacillus subtilis* (Johnson & Fein 2019) (Bacillota). Wang et al. (2022) describe a gold reductase (GolR) from an *Erwinia* species (γ-Proteobacteria) that uses NADH to reduce Au(III) to metallic gold. *Cupriavidus metallidurans* often dominates biofilm communities on natural gold grains

Dyer et al. (1994) found that the filamentous cyanobacterium *Plectonema terebrans* would accumulate gold from a solution of $AuCl_3 \cdot H_2O$ in its sheath (glycocalyx). But even non-living organisms may cause formation of metallic gold. Thus Lengke et al. (2007) write: "Addition of [≈2 m*M*] $AuCl^{4-}$ to cyanobacteria killed the cultures instantly, and Au was precipitated throughout the cells as nanoparticles." Parial et al. (2016) used lower concentrations, 5 μM –0.5 mM Au^{3+}, and several species of cyanobacteria. They found that 75 μ*M* Au^{3+} was the best concentration to maintain formation of metallic gold particles.

There exists a considerable literature on the relations of fungi to gold (Fig. 11.3.1). The older part has been reviewed by Kitching et al. (2014). Many of the publications, as is the case with bacteria, deal with conversion of gold ions to metallic gold. Bindschedler et al. (2017), citing Bouquet (2013) describe how fungi such as *Aureobasidium pullulans* can produce cracks in sheets of gold metal and accumulate gold. Xu et al. (2019) describe how the filamentous fungus *Phomopsis* sp. XP-8 can be used to recover gold from low concentration in electronic wastewater. The fungus accumulates gold together with other metals. The recovery rate from dilute Au^{3+} was, under optimum conditions, more than 80%. The formation of immobilized nanoparticles of metallic gold in the fungal mycelium was indicated by a change from white to purple within one hour. Borovicka et al. (2005) examined 89 species of macrofungi growing in unpolluted areas far from known any gold deposit. The species had very different gold contents. The highest, expressed as ng gold per g dry matter, were in *Amanita strobiliformis* (136), *Russula claroflava* (148), *Cantharellus lutescens* (156 – 210), *Boletus edulis* (235), *Langermannia gigantea* (160), and *Morchella esculenta* (189). Obviously the fungi concentrated gold from the environment.

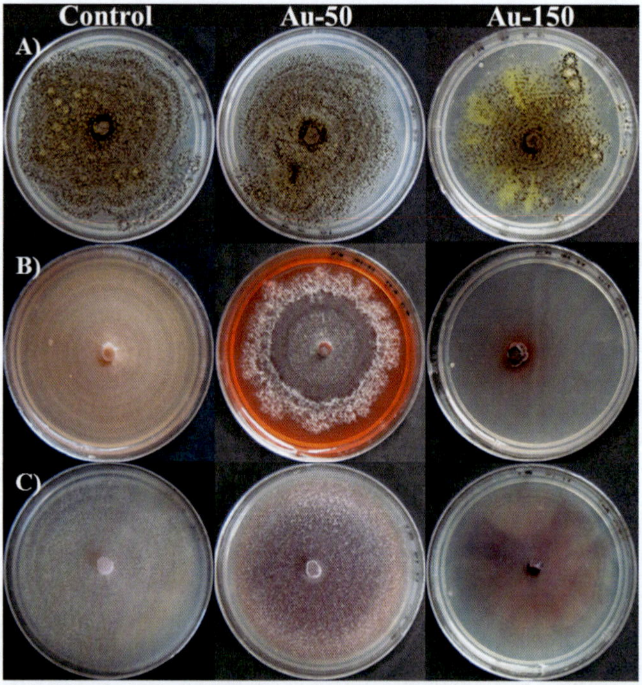

Fig. 11.3.1. Morphological changes and pigment production by three strains of filamentous fungi exposed to three doses of gold for 20 days. (A) *Aspergillus niger* MX5, (B) *Fusarium solani* MXPE15 and (C) *Fusarium oxysporum* MX17 treated with 0, 50, and 150 mg/L Au^{3+} as trichloride. From Argumedo-Delira et al. (2020). Reproduced under the terms and conditions of the Creative Commons Attribution (CC BY) license (http://creativecommons.org/licenses/by/4.0/).

But gold can also change from metallic to ionic state. One important way seems to be by being complexed with cyanide ions secreted by bacteria.

Knowles (1976) gives a list of bacteria that produce cyanide. Campbell et al. (2001) as well as Fairbrother et al. (2009) directly showed how gold was dissolved by cyanide secreted by *Chromobacterium violaceum*:

$$4\ Au + 8\ CN^{-} + 2\ H_2O + O_2 \rightarrow 4\ Au(CN)_2^{-} + 4\ OH^{-}.$$

Also some other bacteria, such as *Pseudomonas fluorescens* and *P. plecoglossicida*, algae (e.g. *Chlorella vulgaris*) and fungi (e.g. *Marasmius oreades*, *Stemphylium loti* and *Gloeocercospora sorghi*), have been shown to solubilize metallic gold by releasing cyanide (Brandl et al. 2008; Fairbrother et al. 2009).

Gold ions do not appear "nude"; they are complexed, if not with cyanide, with chloride as $AuCl_2^{-}$ or $AuCl_4^{-}$, with thiosulphate as $Au(S_2O_3)_2^{3-}$, or with amino acids.

The ways in which various bacteria (*Chromobacterium violaceum*, *Pseudomonas aeruginosa*, *Bacillus megaterium*, *Micrococcus* sp., and *Roseovarius tolerans*) can solubilize gold are reviewed by Wang et al. (2021) and Jorjani & Sabzkoohi (2022). In addition to cyanide and thiosulphate they mention dissolution by triiodide, I_3^{-}. Bacteria such as *Roseovarius tolerans* and *Rhodothalassium salexigens* can oxidize iodide (I^{-}) to elemental iodine (I_2), which with remaining iodide forms triiodide. This can react with metallic gold: $2Au + I_3^{-} + I^{-} \rightarrow 2\ AuI_2^{-}$ and $2Au + 3I_3^{-} \rightarrow 2\ AuI_4^{-} + I^{-}$. Other methods are the thiourea method: $Au + 2CS(NH_2)_2 + Fe^{3+} \rightarrow [Au(CS(NH_2)_2)_2]^{+} + Fe^{2+}$; the ammonia method: $Au + 2NH_3 \rightarrow Au(NH_3)_2^{+} + e^{-}$; the glycine method: $4Au + 8NH_2CH_2COOH + 4OH^{-} + O_2 \rightarrow 4Au(NH_2CH_2COO)_2 + 6H_2O$, or, using hydrogen peroxide as oxidant, $2Au + 4NH_2CH_2COOH + 2OH^{-} + H_2O_2 \rightarrow 2Au(NH_2CH_2COO)_2 + 4H_2O$; the bromide method: $Au + 4Br \rightarrow AuBr_4 + 3e^{-}$; the chloride method: $Au + 4Cl^{-} + Cu^{2+} \rightarrow AuCl_4^{-} + Cu^{+} + 2e^{-}$; also other methods have been proposed (Wang et al. 2021).

At elevated temperature (150–200°C) metallic gold is soluble in crude petroleum, and, depending on the kind of oil, may stay in solution after cooling to room temperature (Williams-Jones & Midgisov 2007, Emsbo et al. 2009). It is even soluble in pure *n*-dodecane (Crede et al. 2018) and in 1-dodecanethiol (Crede et al. 2019). It is suspected that gold that is found in organic matter may be remains of a gold solution in petroleum (Crede et al. 2020).

An Au(III)-reductase from the *Erwinia sp*. IMH that reduces trivalent gold to metallic gold has been characterized by Wang et al. (2022). It is localized to the cytoplasm of the bacterium.

Approximate gold content in various organisms is shown in Table 11.3.1.

Table 11.3.1.

Organism	Au, mg/ton	Reference
Algae (average)	150	Vaganov et al. (1978) Vaganov & Shtangeeva (1978)
Sea urchins	100	Korobushkina et al. (1983)
Starfish	150	Korobushkina et al. (1983)
Crabs	250	Korobushkina et al. (1983)
Bryozoa	750	Korobushkina et al. (1983)
Microorganims	32–85	Korobushkina et al. (1983)

References

Arakawa, H., Neault, J.F. & Tajmir-Riahi, H.A. 2001. Silver(I) complexes with DNA and RNA studied by fourier transform infrared spectroscopy and capillary electrophoresis. Biophys. J. 81, 1580–1587.

Argumedo-Delira, R., Gómez-Martínez, M.J. & Uribe-Kaffure, R. 2020. Fungal Tolerance: An alternative for the selection of fungi with potential for the biological recovery of precious metals. Appl. Sci. 10, 8096 (12 pp.). https://doi.org/10.3390/app10228096

Bernal-Bayard, P., Pallara, C., Castell, M.C., Molina-Heredia, F.P., Fernández-Recio, J., Hervás, M. & Navarro, J.A. 2015. Interaction of photosystem I from *Phaeodactylum tricornutum* with plastocyanins as compared with its native cytochrome c_6: Reunion with a lost donor. Biochim. Biophys. Acta 1847, 1549–1559. http://dx.doi.org/10.1016/j.bbabio.2015.09.006

Besold, A.N., Culbertson, E.M. & Culotta, V.C. 2016. The Yin and Yang of copper during infection. J. Biol. Inorg. Chem. 21, 137–144. https://doi.org/10.1007/s00775-016-1335-1

Borovicka, J., Randa, Z. & Jelínek 2005. Gold content of ectomycorrhizal and saprobic macrofungi from non-auriferous and unpolluted areas. Mycol. Res. 109, 951–955. https://doi.org/10.1017/S095375620500328X

Brandl, H., Lehmann, S., Faramarzi, M.A. et al. 2008. Biomobilization of silver, gold, and platinum from solid waste materials by HCN-forming microorganisms. Hydrometallurgy 94, 14–17.

Bridge, G., Martel, A.-S. & Lomazzi, M. 2021. Silver diamine fluoride: Transforming community dental caries program. Internat. Dental J. 71, 458 – 461. https://doi.org/10.1016/j.identj.2020.12.017

Burmester, T. 2002. Origin and evolution of arthropod hemocyanins and related proteins. J. Comp. Physiol. B 172, 95–107. https://doi.org/10.1007/s00360-001-0247-7

Campbell, S.C., Olson, G.J., Clark, T.R. & McFeters, G. 2001. Biogenic production of cyanide and its application to gold recovery. J. Industr. Microbiol. Biotechnol. 26, 134–139.

Castresana, J., Lübben, M., Saraste, M. & Higgins, D.G. 1994. Evolution of cytochrome oxidase, an enzyme older than atmospheric oxygen. EMBO J. 13, 2516–2525.

Chambers, A., Krewski, D., Birkett, N., Plunkett, L., Hertzberg, R., Danzeisen, R., Aggett, P.J., B. Starr, T.B., Baker, S., Dourson, M., Jones, P., Keen, C.L., Meek, B., Schoeny, R. & Slob, W. 2010. An exposure-response curve for copper excess and deficiency. J. Toxicol. Environm. Health, Part B, 13, 546–578. https://doi.org/10.1080/10937404.2010.538657

Cheng, Y., Ke, Z., Bian, X., Zhang, J., Huang, Z., Lv, Y. & Liu, M. 2019. Selective mineralization and recovery of Au(III) from multi-Ionic aqueous systems by *Bacillus licheniformis* FZUL-63. Minerals 9, 392 (10 pp.). https://doi.org/10.3390/min9070392.

Costa-Paiva, E.M., Mello, B., Bezerra, B.S., Coates, C.J., Halanych, K.M., Brown, F., de Moraes Leme, J. & Trindade, R.I.F. 2022. Molecular dating of the blood pigment hemocyanin provides new insight into the origin of animals. Geobiology 20, 333–345. https://doi.org/10.1111/gbi.12481

Crede, L.S., Rempel, K.U., Hu, S.Y. & Evans, K.A. 2018. An experimental method for gold partitioning between two immiscible fluids: Brine and n-dodecane. Chemical Geology 501, 35–50. https://doi.org/10.1016/j.chemgeo.2018.09.034

Crede, L.S., Evans, K.A., Rempel, K.U., Grice, K. & Sugiyama, I. 2019. Gold partitioning between 1-dodecanethiol and brine at elevated temperatures: Implications of Au transport in hydrocarbons for oil-brine ore systems. Chem. Geol. 504, 28–37. https://doi.org/10.1016/j.chemgeo.2018.11.011

Crede, L.S., Evans, K.A., Rempel, K.U., Brugger, J., Etschmann, B., Bourdet, J. & Reith, F. 2020. Revisiting hydrocarbon phase mobilization of Au in the Au–Hg McLaughlin Mine, Geysers/Clear Lake area, California. Ore Geology Revs 117: 103218. https://doi.org/10.1016/j.oregeorev.2019.103218

De Corte, S., Hennebel, T., Verschuere, S., Cuvelier, C., Verstraete, W. & Boon, N. (2011) Gold nanoparticle formation using *Shewanella oneidensis*: a fast biosorption and slow reduction process. J. Chem. Technol. Biotechnol. 86, 547–553.

De la Rosa, M.A., Navarro, J.A., Díaz-Quintana, A., De la Cerda, B., Molina-Heredia, F.P., Balme, A., del S. Murdoch, P., Díaz-Moreno, I., Durán, R.V. & Hervás, M. (2002) An evolutionary analysis of the reaction mechanisms of photosystem I reduction by cytochrome c_6 and plastocyanin. Bioelectrochemistry 55, 41– 45.

Dobrev, S., Kircheva, N., Nikolova, V., Angelova, S. & Dudev, T. (2022) Competition between Ag+ and Ni^{2+} in nickel enzymes: Implications for the Ag^+ antibacterial activity. Comput. Biol. Chem. 101: 107785 (9 pp.). https://doi.org/10.1016/j.compbiolchem.2022.107785

Doerrer, L.H. (2018) Cu in biology: Unleashed by O_2 and now irreplaceable. Inorganica Chimica Acta 481, 4–24. https://doi.org/10.1016/j.ica.2017.11.051

Domínguez, A.V., Algaba, R.A., Canturri A,M., Rodríguez Villodres, Á. & Smani, Y. (2020) Antibacterial activity of colloidal silver against Gram-negative and Gram-positive bacteria. Antibiotics 9: 36 (10 pp.). https://doi.org/10.3390/antibiotics9010036

Dyer, B.D., Krumbein, W.R. & Mossman, D.J. (1994) Accumulation of gold in the sheath of *Plectonema terebrans* (filamentous marine cyanobacteria). Geomicrobiol. J. 12, 91–98. https://doi.org/10.1080/01490459409377975

Emsbo, P., William.Jones, A.E., Koenig, A.E. & Wilson, S.A. (2009) Petroleum as an agent of metal fransport: metallogenic and exploration implications. Proc. Tenth Biennial SGA Meet., Townsville, 2009, 99–101.

Fairbrother, L., Shapter, J., Brugger, J., Southam, G., Pring, A. & Reith, F. (2009) Effect of the cyanide-producing bacterium *Chromobacterium violaceum* on ultraflat Au surfaces. Chemical Geol. 265, 313–320. https://doi.org/10.1016/j.chemgeo.2009.04.010

Festa, R.A. & Thiele, D.J. (2012) Copper at the front line of the host-pathogen battle. PLoS Pathog 8: e1002887 (4 pp.). https://doi.org/10.1371/journal.ppat.1002887

Frangipani, E., Slaveykova, V.I., Reimmann, C. & Haas, D. (2008) Adaptation of aerobically growing *Pseudomonas aeruginosa* to copper starvation. J. Bacteriol. 190, 6706–6717. https://doi.org/10.1128/JB.00450-08

Gao, S.S., Zhao, I.S., Duffin, S., Duangthip, D., Lo, E.C.M. & Chu, C.H. (2018) Revitalising silver nitrate for caries management. Int. J. Environ. Res. Public Health 15: 80 (9 pp.). https://doi.org/10.3390/ijerph15010080

Hellman, N.E & Gitlin, J.D. (2002) Ceruloplasmin metabolism and function. Annu. Rev. Nutr. 22, 439–58. https://doi.org/10.1146/annurev.nutr.22.012502.114457

Inda, L.A., Erdner, D.L., Peleato, M.L. & Anderson, D.M. (1999) Cytochrome c_6 isolated from the marine diatom *Thalassiosira weissflogii*. Phytochemistry 51, 1-–4.

Italiano, F., Agostiano, A., Belviso, B.D., Caliandro, R., Carrozzini, B., Comparelli, R., Melillo, M.T., Mesto, E., Tempesta, G. & Trotta. M. (2018) Interaction between the photosynthetic anoxygenic microorganism *Rhodobacter sphaeroides* and soluble gold compounds. From toxicity to gold nanoparticle synthesis. Colloids Surfaces B: Biointerfaces 172, 362–371. https://doi.org/10.1016/j.colsurfb.2018.06.010

Johnson, C.R. & Fein, J.B. (2019) A mechanistic study of Au(III) removal from solution by *Bacillus subtilis*. Geomicrobiol. J. 36, 506–514. https://doi.org/10.1080/01490451.2019.1573279

Jorjani, E. & Sabzkoohi, H.A. (2022) Gold leaching from ores using biogenic lixiviants – A review. Curr. Res. Biotechnol. 4, 10–20. https://doi.org/10.1016/j.crbiot.2021.12.003

Kaintz, C., Mauracher, S.G. & Rompel, A. (2014) Type-3 Copper proteins: Recent advances on polyphenol oxidases. Ch. 1 in Adv. Protein Chemistry and Structural Biol. 97, 1–35. http://dx.doi.org/10.1016/bs.apcsb.2014.07.001

Khina, A.G. & Krutyakov, Yu.A. (2021) Similarities and differences in the mechanism of antibacterial action of silver ions and nanoparticles. Appl. Biochem. Microbiol. 57, 683–693Knowles, C.J. (1976) Microorganisms and cyanide. Bacteriolog. Revs 40, 652–680.

Korobushkina, E.D., Karavaiko, G.I. & Korobushkin, I.M. (1983) Biochemistry of gold. Ecol. Bull. 35, 325–333.

Krauskopf KB (1951) The solubility of gold. Economic Geology 46, 858–870.

Lengke, M. & Southam, G. (2007) The deposition of elemental gold from gold(I)-thiosulfate complexes mediated by sulfate-reducing bacterial conditions. Economic Geology 102, 109–126.

Lengke, M.F., Fleet, M.E. & Southam, G. (2006) Bioaccumulation of gold by filamentous cyanobacteria between 25 and 200°C. Geomicrobiology Journal, 23, 591–597. https://doi.org/10.1080/01490450600964326

Linder, M.C. (2021) Apoceruloplasmin: Abundance, detection, formation, and metabolism. Biomedicines 9, 233 (15 pp.). https://doi.org/10.3390/biomedicines9030233

MacPherson, I.S. & Murphy, M.E.P. (2007) Type-2 copper-containing enzymes. Cell. Mol. Life Sci. 64, 2887 – 2899. https://doi.org/10.1007/s00018-007-7310-9

Mannix-Fisher, E. & McLean, S. (2021) The antimicrobial activity of silver acetate against *Acinetobacter baumannii* in a *Galleria mellonella* infection model. PeerJ 9:e11196 (19 pp.). https://doi.org/10.7717/peerj.11196

Miramar, M.D., Inda, L.A., Saraiva, L.M. & Peleato, M.L. (2003) Plastocyanin/cytochrome c_6 interchange in *Scenedesmus vacuolatus*. J. Plant Physiol. 160, 1483–1486. http://www.urbanfischer.de/journals/jpp

Panis, F. & Rompel, A. (2022) The novel role of tyrosinase enzymes in the storage of globally significant amounts of carbon in wetland ecosystems. Environ. Sci. Technol. 56, 11952−11968. https://doi.org/10.1021/acs.est.2c03770

Parial, D., Gopal, P.K., Santanu Paul, S., & Pal, R. (2016) Gold (III) bioreduction by cyanobacteria with special reference to in vitro biosafety assay of gold nanoparticles. J. Appl. Phycol. 28, 3395–3406. https://doi.org/10.1007/s10811-016-0880-x

Peers, G. & Price, N.M. (2006) Copper-containing plastocyanin used for electron transport by an oceanic diatom. Nature 441, 341–344. https://doi.org/10.1038/nature04630

Pretzler, M. & Rompel, A. (2018) What causes the different functionality in type-III-copper enzymes? A state of the art perspective. Inorganica Chimica Acta 481, 25–31. http://dx.doi.org/10.1016/j.ica.2017.04.041

Rea, M.A., Zammit, C.M. & Reith, F. (2016) Bacterial biofilms on gold grains—implications for geomicrobial transformations of gold. FEMS Microbiol, Ecol. 92: fiw082 (12 pp.). https://doi.org/10.1093/femsec/fiw082

Reith, F., Etschmann, B., Grosse, C., Moors, H., Benotmane, M.A., Monsieurs, P., Grass, G., Doonan, C., Vogt, S., Lai, B., Martinez-Criado, G., George, G.N., Nies, D.H., Mergeay, M., Pring, A., Southam, G. & Brugger, J. (2009) Mechanisms of gold biomineralization in the bacterium *Cupriavidus metallidurans*. Proc. Natl Acad. Sci. USA 106, 17757–17762. https://doi.org/10.1073.pnas.0904583106

Rushworth, D.D., Christl, I., Kumar, N., Hoffmann, K., Kretzschmar, R., Lehmann, M.F., Schenkeveld, W.D.C. & Kraemer, S.M. (2022) Copper mobilisation from Cu sulphide minerals by methanobactin: Effect of pH, oxygen and natural organic matter. Geobiol. 20, 690–706. https://doi.org/10.1111/gbi.12505

Scharf, B. & Engelhard, M. (1993) Halocyanin, an Archaebacterial Blue Copper Protein (Type I) from *Natronobacterium pharaonic*. Biochemistry 32, 12894–12900.

Srisomboon, S., Kettratad, M., Stray, A., Pakawanit, P., Rojviriya, C., Patntirapong, S. & Panpisut, P. (2022) Effects of Silver Diamine Nitrate and Silver Diamine Fluoride on Dentin Remineralization and Cytotoxicity to Dental Pulp Cells: An In Vitro Study. J. Funct. Biomater. 13: 16 (12 pp.). https://doi.org/10.3390/jfb13010016

Vaganov, P.A. & Shtangeeva, I.V. 1978. On the concentration of gold and its geochemical satellites by marine organisms. - Uch. Zap. LGU 392, 168-173. (in Russian).

Vaganov, P.A., Kulikov, R.D. & Shtangeeva, I.V. 1978. Biogeochemical characteristics of algae of the Barents Sea (based on the results of the neutron-activation analysis). - Geochimiya 11:1740-1745. (in Russian).

van Holde, K.E. & Miller, K.I. (1995) Hemocyanins. Adv. Protein Chem. 47, 1–81.

Wang, J., Faraji, F., Ramsay, J. & Ghahreman, A. (2021) A review of biocyanidation as a sustainable route for gold recovery from primary and secondary low-grade resources. J. Cleaner Prod. 296: 126457 (18 pp.). https://doi.org/10.1016/j.jclepro.2021.126457

Wang, L., Yan, L., Ye, L., Chen, J., Li, Y., Zhang, Q. & Jing, C. (2022) Identification and characterization of a Au(III) reductase from *Erwinia* sp. IMH. JACS Au 2, 1435−1442. https://doi.org/10.1021/jacsau.2c00170

Williams-Jones & A.E. & Midgisov, A.A. (2007) An experimental study of the solubility of gold in crude oil: Implications for ore genesis. Goldschmidt Conf. Abstr. 2006, A703. https://doi.org/10.1016/j.gca.2006.06.1526

Xu, X., Yang, Y., Zhao, X., Zhao, H., Lua, Y., Jiang, C., Shaoa, D. & Shi, J. (2019) Recovery of gold from electronic wastewater by *Phomopsis* sp. XP-8 and its potential application in the degradation of toxic dyes. Bioresource Technology 288: 121610 (8 pp.). https://doi.org/10.1016/j.biortech.2019.121610

Zaitseva, I., Zaitsev, V., Moshkov, G.C.K., Bax, B., Ralph, A. & Lindley, P. (1996) The X-ray structure of human serum ceruloplasmin at 3.1 Å: nature of the copper centres. J. Biol. Inorg. Chem. 1, 15–23.

CHAPTER TWELVE

ZINC, CADMIUM, AND MERCURY

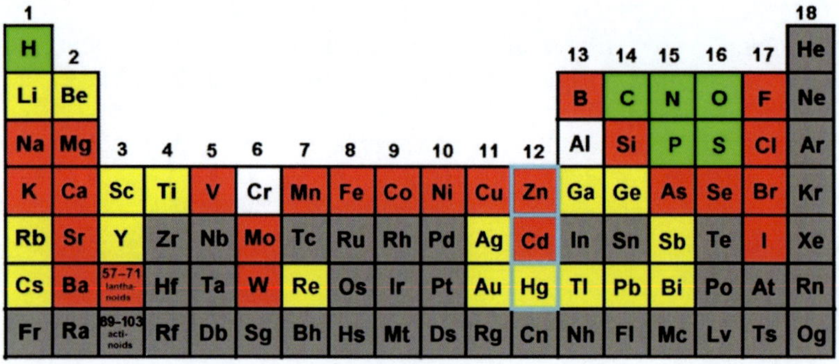

12.1 Zinc

12.1.1 The metal that has a finger everywhere in the machinery of life

Zinc (as Zn^{2+}) is essential for all animals, plants, and microorganisms, required for numerous enzymes, transcription factors, and signaling molecules. It differs from many other metals in that it, in addition to the metal state, appears only as divalent ions. One important function of zinc in proteins is in so-called zinc fingers, peptide loops held together by zinc ions. Andreini et al. (2006) found 1684 zinc proteins in the human proteome, identified by three different methods. Several hundred more were suspected based on one or two methods. Single methods give somewhat uncertain results, partly because Fe-S clusters may be masquerading as zinc in finger proteins (Pritts & Michel 2022). Silver ions may replace zinc in fingers and thereby disrupt protein function, thus contributing to silver toxicity (Kluska et al. 2022).

Human and plant zinc deficiency is quite common (Stanton et al. 2022, Fig. 12.1.1, Fig. 12.1.2).

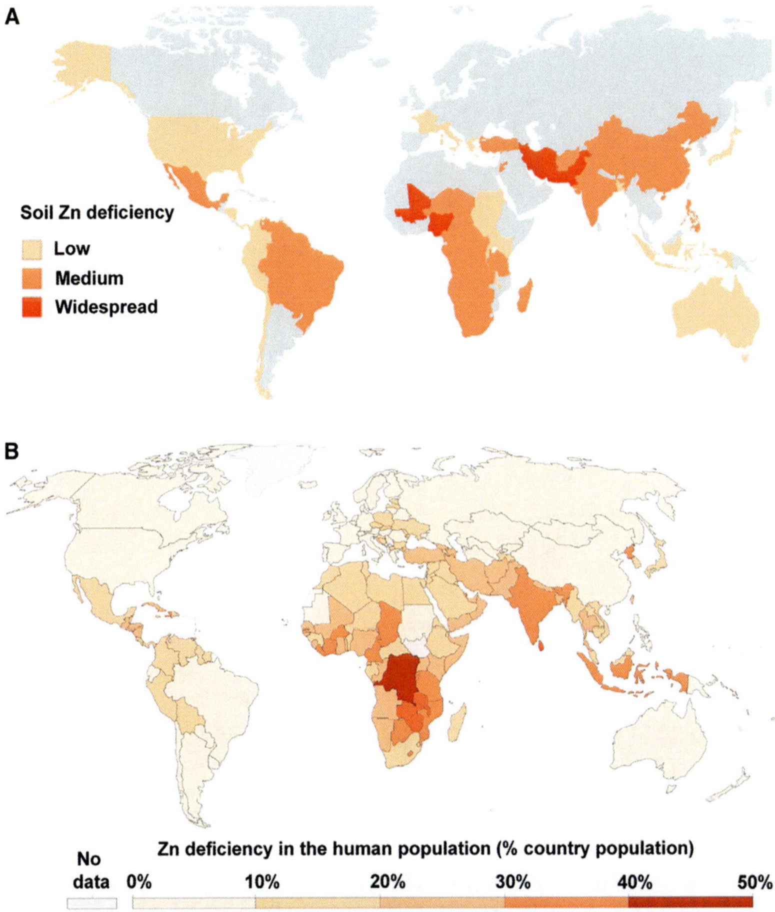

Fig. 12.1.1. The global prevalence of Zn deficiency in agricultural soils and the human population. (A) Countries and areas affected by Zn deficiency in agricultural soils. (B) Global map showing the prevalence of Zn deficiency as the percentage of the population with intakes below physiological requirements for each country. From Stanton et al. (2022) under the CC BY-NC-ND license (http://creativecommons.org/licenses/by-nc-nd/4.0/).

Zinc deficiency affects approximately a third of the world population, principally in the developing countries (Restrepo Caro et al. (2016), and is estimated to reduce the GDP of some countries by as much as 11% (Khan et al. 2022).

Insulin forms a stable hexamer with four zinc ions, resulting in a slow-acting form of insulin (Smith et al. 1984). Mutation in Zn transporter 8 – a key protein in insulin secretion – has been associated with Type 2 diabetes (Poudel et al. 2017). Zinc also modulates the insulin signaling pathway (Norouzi et al. 2018).

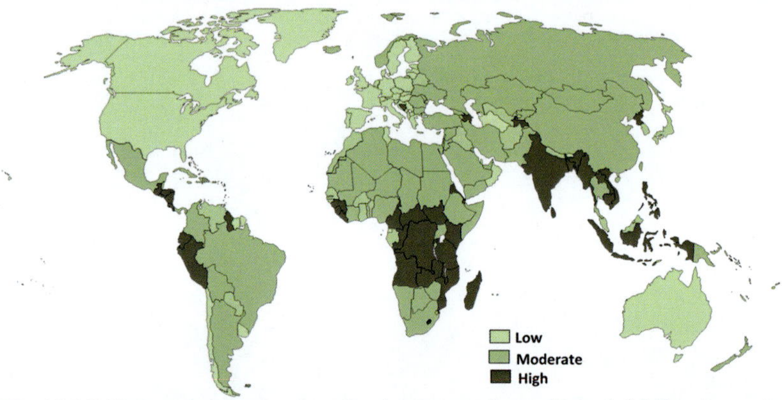

Fig. 11.1.2. Nationwise Zn status based on inadequate dietary Zn and childhood growth stunting. From Khan et al. (2022).CC BY license (http://creativecommons.org/licenses/by/4.0/).

When a eukaryote host is attacked by a bacterium, it tries to make zinc unavailable to the attacker by binding it tightly (Murdoch & Skaar 2022); the corresponding strategy is used also for some other metal ions.

The concentration of free Zn^{2+} in a cell is in the picomolar (10^{-12} molar) range, much lower than the total concentration of the element (typically in the micromolar (10^{-6} molar) range. Thus free Zn^{2+} ions constitute only one millionth of the total zinc content (Castillo-González et al. 2018, Clemens 2022). Various metabolites and peptides, including nicotianamine, glutathione, and phytochelatins, serve as Zn^{2+} buffers. Most of the protein-bound Zn in eukaryotic cells is found in the cytosol.

Nearly 10% of all plant proteins belong to the zinc proteome. They require zinc either for catalysis or as a structural element.

More than one-third of the world population is suffering zinc deficiency, which results in various types of diseases or nutritional deficiencies (Duan et al. 2021). Human Zn deficiency accounts for over 400 000 deaths annually (Stanton et al. 2022).

Some plant species, for example *Arabidopsis halleri* and *Noccaea* (*Thlaspi*) *caerulescens*, are able to hyperaccumulate Zn^{2+} (Verbruggen et al. 2009, Krämer 2010, Balafrej et al. 2020). Twenty-eight Zn-hyperaccumulating plant species have been described so far, most of them belonging to Brassicaceae, but other families are also represented, such as Caryophyllaceae and Dichapetalaceae. High Zn^{2+} content can protect plants from herbivores: Locusts (*Schistocerca gregaria*), slugs (*Deroceras caruanae*), and caterpillars (*Pieris brassicae*); all showed significant preferences for *Thlaspi caerulescens* plants with lower foliar Zn^{2+} concentrations. Such differential feeding could result in selection pressures favouring the evolution of hyperaccumulation (Pollard & Baker 1997). In choice experiments, *Schistocerca gregaria* locusts preferred *Thlaspi caerulescens* plants containing low foliar Zn^{2+} concentrations (mean value 0.42 mg/g dry biomass) compared with intermediate (1.50 mg/g or high (5.77 mg/g) Zn^{2+} concentrations. In no-choice experiments, there was no difference in the duration of the first meal, suggesting that locusts could not taste Zn^{2+} directly, but total time spent feeding, amount of foliage eaten, and locust growth rate all decreased as foliar Zn^{2+} concentration increased (Behmer et al. 2005).

Not only plants, but also some other organisms, such as some yeast strains (Cha & Cho 2009) and some bacteria (Hussain et al. 2022; see Magnin et al. 2014 for further examples) may accumulate large amounts of zinc.

Plants that normally do not accumulate large amounts of zinc ions may nevertheless use zinc ions for their defense. When they are infected with a fungus they increase their uptake, and direct the zinc ions specifically to the infection sites. If they are prevented to do this, their resistance to the fungus is decreased (Escudero et al. 2022).

12.1.2 Zinc and mammal fertility

Zn^{2+} is very abundant throughout the male reproductive tract in humans. The prostate contains about 0.7 mg/g, and the testicles about a tenth of this (Zigo et al. 2022). Zinc ions are required for almost every step of sperm maturation. Zn^{2+} is involved in numerous sperm functions and is mainly taken up during sperm transport after movement through the epididymis. Once the sperm enters the female reproductive tract, the Zn^{2+} content of the seminal fluid is important for pacifying the responses against invasion and signaling that this invader should not be disturbed (Vickram et al. 2021).

Zinc is also an important component in many processes that regulate female germ cell growth, fertility, and pregnancy. An adequate supply of zinc is necessary for the oocyte to form a fertilization-competent egg (Garner et al. 2021). When the sperm enters the egg cell, the zinc spark (see below) induces egg activation in addition to facilitating zona pellucida hardening and reducing sperm motility to prevent polyspermy. The proper development of the fetus is compromised by zinc deficiency.

12.1.3 Fertilization and the zinc spark

During the last hours of maturation the mammal oocyte takes up over twenty billion zinc ions and arrests after the first meiotic division, until fertilization stimulates cell cycle progression toward a new embryo. Fertilization of the mature, zinc-enriched egg triggers the ejection of zinc into the extracellular milieu (zinc exocytosis) in a series of events termed “zinc sparks” (Kim et al. 2011). Zinc sparks can be monitored using a compound in the extracellular medium that becomes fluorescent upon association with zinc ions. The zinc sparks mediate a decrease in intracellular zinc content that is necessary for continued cell cycle progression. Each zinc spark is immediately preceded by an elevation in intracellular calcium.

The zinc spark arises from a system of thousands of zinc-loaded vesicles, each of which contains, on average, a million zinc ions, and it is estimated that each oocyte contains 8,000 vesicles (Que et al. 2015). Experimentally the inner concentration of free zinc ions can be achieved by chelation (by the zinc-specific chelator, N,N,N′ ,N′ -tetrakis(2-pyridylmethyl)ethane-1,2-diamine), and this is sufficient to induce cell cycle resumption and transition of a meiotic cell into a mitotic one (Duncan et al. 2016). Zhang et al. (2016) propose, based on experiments with mice, that the fertilization-induced zinc spark be used for checking embryo quality and early development.

It is essential that only one sperm cell fertilizes an egg cell, and Que et al. (2017, 2019) found that the zinc spark caused by the first-arriving sperm cell changes the properties of the zona pellucida, a layer surrounding the egg cell, in such a way that it becomes impossible for further sperms to enter. In the process, the number of zinc ions in the zona pellucida increases by ca. 300%. The zinc sparks have been observed not only in mammals, but also in other vertebrates. (Converse & Thomas 2020, Seeler et al. 2021).

12.1.4 COVID-19, SARS-CoV-2, and zinc

SARS-CoV-2 can compete with human proteins for Zn^{2+}. Fifteen Zn^{2+} sites have been identified in the proteome of coronavirus (Andreini et al. 2022). There was also an inverse correlation between zinc status and disease severity (from 814 ± 164 μg/L in the “mild” group to 574 ± 218 μg/L in the “ex. letalis” group). The normal Cu/Zn ratio is close to 1. Due to the well-known antagonistic effect between these metals, the Cu/Zn ratio is considered clinically more relevant than just the individual concentration of each element. The Cu/Zn ratio gradually increased in association with COVID-19 severity, from 1.55 ± 0.56 in the "mild" to 3.09 ± 2.85 in the "ex. letalis" group (Bego et al. 2022). The structural basis of ionic zinc binding to the SARS-CoV-2 main protease has been elucidated by X-ray crystallography (Grifagni et al, 2021).

Zinc deficiency might contribute to a poor prognosis of COVID-19 disease by altering the release of particular cytokines, and Zn supplementation reduces the risk (Boretti 2022, Dhawan et al. 2022, Engin et al. 2022). Low serum zinc and 25-hydroxyvitmain D levels appear to be risk factors for COVID-19 affliction (Ghanei et al. 2022).

Zinc supplementation is a viable therapeutic approach when used alone or in combination with other drugs to combat cytokine storm syndrome in COVID-19 patients. The supplementation may be especially effective for elderly patients or patients with co-morbidity already deficient in zinc (Hoang & Han 2021). However, according to Li et al. (2022) zinc might not decrease all-cause mortality in adults or the in-hospital mortality of COVID-19.

12.2 Cadmium

12.2.1 A dangerous relative of the essential metal zinc

In column 12 of the periodic system, as it is usually displayed, we find zinc (^{30}Zn), cadmium (^{48}Cd), mercury (^{80}Hg), and copernicium (^{112}Cn). The latter two have no biological function. As already described, zinc is a very important element for all organisms, as part of many enzymes and other proteins, such as insulin. Cadmium and mercury are both very toxic to us. We shall soon return to biological functions of cadmium. It is so toxic partly because it resembles zinc, and most organisms mistake it for zinc, the positive functions of which it cannot fulfill. It also forms complexes with various vital compounds (reviewed by Sigel et al., eds 2013).

Many plants, particularly in the family Brassicaceae (Lombi et al. 2000, Seregin et al. 2015, Dou et al. 2019), but also some others (Liu et al 2016, Tang et al. 2016, Tang et al. 2022, Dai et al. 2020, Yang et al. 2020, Wang et al. 2021, Liu et al 2022) store cadmium. At least in some cases this takes place using a protein specialized for importing cadmium (Lombi et al. 2001). Also some fungi (Sácký et al. 2014, 2021, Borovička et al. 2022), and even some animals store large amounts of cadmium. *Thlaspi caerulescens* is protected from the leaf-eating insect *Frankliniella occidentalis* when the plant has taken up cadmium (Jiang et al. 2005).

Different strains of *Thlaspi caerulescens* exhibit very different levels of cadmium tolerance and accumulation, and one strain requires 3 micromolar Cd^{2+} for optimum growth (Roosens et al. 2003). Sunflower (*Helianthus annuus*) takes up more cadmium after simulated herbivory (Grossman & Gruntman 2022).

Two species of *Crotalaria* react very differently to cadmium (Tang et al. 2022). *Crotalaria novae-hollandiae* accumulates remarkably high concentrations of zinc, cadmium, and copper in its shoots when growing on metalliferous (Zn-Cd 'calamine') soils in Australia. It accumulated up to 16.6 mg Zn g^{-1} and 1.25 mg Cd g^{-1}, with a 70% increase in biomass when exposed to Zn. *Crotalaria cunninghamii* accumulated up to 1.21 mg Zn g^{-1} and 0.0476 mg Cd g^{-1} in its leaves, with a 75% reduction in biomass in the Zn treatment.

The fact that there exist proteins specialized for taking up cadmium is a sign that an organism investing in such a protein gains some advantage from the cadmium. What can the advantage be? Dai et al. (2020) have shown that when *Ageratina adenophora* has acquired cadmium, it has also acquired a certain protection against the disease-causing fungus *Rhizoctonia solani* (*Thanatephorus cucumeris*). A similar experiment by Kazemi-Dinan et al. (2015) with *Arabidopsis halleri* is not easy to interpret, since both Zn and Cd were varied simultaneously.

But there are also organisms equipped with enzymes containing cadmium. An example is the diatom *Thalassiosira weissflogii* (Lane et al. 2005, Alterio et al. 2015) that has a carboanhydrase containing cadmium. Carboanhydrases are enzymes catalyzing the conversion $CO_2 + H_2O \leftrightarrow HCO_3^- + H^+$. (Other organisms have carboanhydrases containing zinc or manganese as active metals.) The composite plant *Picris divaricata* has a carboanhydrase that can use either zinc or cadmium, but the highest activity is obtained using zinc (Liu et al. 2016).

High levels of Cd in chocolate and other cacao-based products is a problem, and food safety regulations stipulating maximum permissible limits for Cd in chocolate and cacao-based products have been established or are being established. The injection of hydrated lime into the soil was found to reduce the cadmium content of the leaves of the coacoa plant, *Theobroma cacao* L. by 40% (Ramtahal et al. 2022).

12.2.2 Cadmium in fungi

Much of the older literature is reviewed by Kalač et al. (2003). They mention that some species, mainly from the genera *Agaricus*, *Macrolepiota*, *Lepista* and *Calocybe* accumulate a high content of cadmium and mercury even in unpolluted areas. The genus *Agaricus* has been rather well investigated with respect to cadmium. *Agaricus silvicola* can accumulate more than 50 mg Cd/(kg dry matter). Meisch et al. (1979) showed that the growth of *Agaricus abruptibulbus* is stimulated by cadmium acetate (Fig. 1). Also the growth of *Morchella spongiola* on agar is stimulated, although only a little, by cadmium ions (as $CdCl_2$), with an optimum between 0.15 and 0.90 mg/L (Xu et al. 2021).

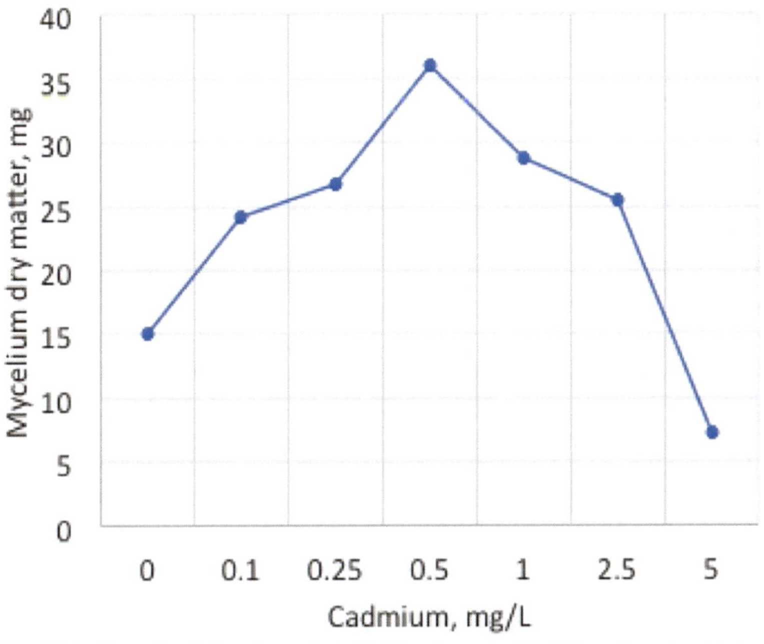

Fig. 12.1. Growth of *Agaricus abruptibulbus* in an artificial medium containing various concentrations of cadmium acetate. Data from Meisch et al. (1979).

In contrast to Fig. 12.1, Jackl et al. (1987) found that Cd up to 1 mg/L had no effect on the growth of *Agaricus arvensis* mycelium in liquid culture, and Münger et al. (1982) found only growth inhibition by cadmium of *Agaricus bisporus* grown on compost. The latter researchers also noted that cadmium stimulated the uptake of copper, and formation of tyrosinase (a copper-containing enzyme). According to Meisch & Schmitt (1986) in *Agaricus macrosporus* the cadmium becomes bound to a 12 kDa (kilodalton) protein. In contrast to this, Jackl et al. (1987) report that in *Agaricus arvensis* cadmium becomes bound to various low-molecular compounds.

Cystoderma carcharias is another fungus (closely related to *Agaricus*) is one of several fungi hyperaccumulating cadmium (Sácký et al. 2021), and also copper. The cadmium appears to be mostly bound to metallothionein (peptides) in vacuoles. *Thelephora penicillate* is one more fungus that hyperaccumulates cadmium (Borovička et al. 2022, 2023).

Huang et al. (2008) found that even the fruiting bodies of *Agaricus blazei* Murrill could accumulate cadmium to such an extent that caution has to be exercised in their use as food, and that cadmium content rises rapidly with phosphate content. Stoknes et al. (2019) propose the use of *Agaricus subrufescens* for protecting other crops from cadmium. This fungus extracted 80% of the Cd from the growing medium into the mushrooms, mostly at the start of fruiting. After disposal of the first mushroom harvest in a safe way, the soil was sufficiently depleted, so further mushroom crops could be safely used as food. The cadmium uptake by *Agaricus subrufescens* is so efficient that it has been considered for remediation of contaminated agricultural soils (Hu et al. 2022), while Khan et al. (2019) suggest a number of ascomycetes (strains of *Aspergillus niger*, *Aspergillus fumigatus*, and *Penicillium rubens*) for the same purpose (and also for chromium remediation). Din et al. (2022) isolated another Cd^{2+}-tolerant and Cd^{2+}-accumulating ascomycete strain, *Penicillium chrysogenum*. Fazli et al. (2015), after comparing various fungal strains (*Aspergillus versicolor, A. fumigatus, Paecilomyces sp., Terichoderma sp., Microsporum sp., Cladosporium sp.* and others), stated that there is little agreement between tolerance and accumulation of cadmium. Of the species investigated, a strain of *Aspergillus versicolor* was found to be the one most suitable for remediation.

12.2.3 Cadmium in bacteria

Also cadmium-tolerant bacteria have been considered for cadmium remediation. Dabir et al. (2019) found that *Microbacterium oxydans* CM3 and *Rhodococcus sp.* AM1 exhibited high ability to remove cadmium from coal and aluminium mines, and were also useful for lead. Wu et al. (2018) identified strains of *Bacillus cereus* that would be useful for removal of Cd^{2+}-ions from water.

12.3 Mercury

Gold extraction using mercury: A common gold extraction method uses liquid mercury, leading to a high exposure in workers. Gold-containing rocks are ground and afterwards pulverized into small pieces, and the material thus obtained is mixed with elemental mercury to form an amalgam out of gold and mercury. The mercury is then evaporated by heating. This procedure is associated with great intoxication risk for the workforce involved. For details, see Taux et al. (2022). Maps showing the distribution of mercury in the terrestrial environment are available in Budnik & Casteleyn (2019), and corresponding maps for the global ocean in Wu et al. (2021).

Of the commonly occurring forms of mercury, methyl mercury is the most poisonous. It can lead to neurological disorders and developmental troubles for unborn children (Bensefa-Colas et al. 2010). Soil contaminated by methyl mercury can, to some extent, be remedied by use of the fungus *Metarhizium robertsii*. It degrades methylmercury and reduces divalent mercury, decreasing mercury accumulation in plants and greatly increasing their growth in contaminated soils (Wu et al. 2022).

Liquid metallic mercury is relatively non-toxic, but inhalation of the vapour can be fatal, as described in several articles. Acute inhalation of metallic mercury vapours induces pulmonary disease, and chronic inhalation leads to neurological (encephalopathy) or renal (nephritis) disorders (Bensefa-Colas et al. 2010). Rowens et al. (1991) tell a story about how all four members of a family, and also their two dogs, died when they tried to reclaim silver from silver amalgam by heating it.

Caries-afflicted teeth have for a long time been treated with mercury-containing amalgam. There are indications that dental amalgam does not increase DNA damage but reduces the capacity to repair DNA (Al-Saleh et al. 2012).

Mercury may, under certain circumstances, possess carcinogenic properties. Low mercury exposure may induce a proliferative response in normal and cancerous cells by possessing estrogen-like activity and interfere with the estrogen receptor, and induce oxidative DNA modifications, thus increasing the risk of mutations. Mercury-induced inhibition of DNA repair mechanisms, increases the risk further (Skalny et al. 2022). For further information concerning mercury hazards to living organisms, I refer to the comprehensive treatise by Eisler (2006).

But mercury is not poisonous only to humans and other vertebrates. The fruitfly, *Drosophila melanogaster*, exposed to methyl mercury, showed a reduction in survival rate, associated with the inhibition of acetylcholinesterase and tyrosine hydroxylase activities in heads of flies and decreased dopamine and octopamine levels (Bianchini et al. 2022). While mercury in general is toxic also to molluscs, bivalves living at hydrothermal vents sometimes thrive in water that is enriched (sometimes to $2 \cdot 10^{-4}$ μM)

in mercury in comparison to normal seawater (<10^{-6} μM) (Kádár et al. 2005).

In a mushroom, *Neoboletus luridiformis*, mercury was found to accumulate specifically in the spore-forming part, the hymenium (Árvay et al. 2022).

References

Al-Saleh, I., Al-Sedairi, A. & Elkhatib, R, (2012) Effect of mercury (Hg) dental amalgam fillings on renal and oxidative stress biomarkers in children. Sci. Total Environm. 431, 188–196. https://doi.org/10.1016/j.scitotenv.2012.05.036

Alterio, V., Langella, E., De Simone, G. & Monti, S.M. (2015) Cadmium-containing carbonic anhydrase CDCA1 in marine diatom *Thalassiosira weissflogii*. Mar. Drugs 13, 1688–1697. https://doi.org/10.3390/md13041688

Andreini, C., Banci, L., Bertini, I. & Rosato, A. 2006. Counting the zinc-proteins encoded in the human genome. J. Proteome Res. 5, 196–201.

Andreini, C., Arnesano, F. & Rosato, A. (2022) The zinc proteome of SARS-CoV-2. Metallomics 14: mfac047 (10 pp.). https://doi.org/10.1093/mtomcs/mfac047

Árvay, J., Hauptvogl, M., Demková, L., Harangozo, Ľ, ˇSnirc, M., Bobuľská, L., ˇStef′aniková, J., Kováˇcik, A., Jakabová, S., Janˇco, I. & Kunca, V. & Relic′, D. (2022) Mercury in scarletina bolete mushroom (*Neoboletus luridiformis*): Intake, spatial distribution in the fruiting body, accumulation ability and health risk assessment. https://doi.org/10.1016/j.ecoenv.2022.113235

Balafrej, H., Bogusz, D., Triqui, Z.-E A., Guedira, A., Bendaou, N., Smouni, A. & Fahr, M. 2020. Zinc hyperaccumulation in plants: A review. Plants, 9, 562 (22 pp.). https://doi.org/10.3390/plants9050562

Bego, T., Meseldˇzic′, N., Prnjavorac, B., Prnjavorac, L., Marjanovic′, D., Azevedo, R., Pinto, E., Duro, M., Couto, C. & Almeida, A. 2022. Association of trace element status in COVID-19 patients with disease severity. J. Trace Elem. Medic. Biol. 74: 127055 (9 pp.). https://doi.org/10.1016/j.jtemb.2022.127055

Behmer, S.T., Lloyd, C.M., Raubenheimer, D. et al. 2005. Metal hyperaccumulation in plants: mechanisms of defence against insect herbivores. Functional Ecology 19, 55–66.

Bensefa-Colasa, L., Andujar, P. & Descathae, A. (2011) Intoxication par le mercure (Mercury poisoning). La Revue de médecine interne 32, 416–424. https://doi.org/10.1016/j.revmed.2009.08.024

Bianchini, M.C., Soares Jr., L.F.W., Sousa, J.M.F.M., Ramborger, B.P., Gayer, M.C., Bridi, J.C., Roehrs, R., Pinton, S., Aschner, M., , Ávila, D.S. & Puntel, R.L. (2022) Chemico-Biological Interact. 365: 110121 (10 pp.). https://doi.org/10.1016/j.cbi.2022.110121

Boretti, A. 2022. Zinc augments the antiviral potential of HCQ/CQ and ivermectin to reduce the risks of more serious outcomes from COVID-19 infection. J. Trace Elem. Medic. Biol. 71: 126954 (11 pp.). https://doi.org/10.1016/j.jtemb.2022.126954

Borovička, J., Braeuer, S., Walenta, M., Hršelová, H., Leonhardt, T., Sácký, J., Kaňa, A. & Goessler, W. (2022) A new mushroom hyperaccumulator: Cadmium and arsenic in the ectomycorrhizal basidiomycete *Thelephora penicillata*. Sci. Total Environm. 826: 154227 (10 pp.). http://dx.doi.org/10.1016/j.scitotenv.2022.154227

Borovička, J., Sácký, J., Kaňa, A., Walenta. M., Ackerman, L., Braeuer, S., Leonhardt, T, Hršelová, H., Goessler,W. & Kotrba, P. (2023) Cadmium in the hyperaccumulating mushroom *Thelephora penicillate*: Intracellular speciation and isotopic composition. Sci. Total Environm. 855: 159002 (9 pp.). http://dx.doi.org/10.1016/j.scitotenv.2022.159002

Budnik, L.T. & Casteleyn, L. (2019) Mercury pollution in modern times and its socio-medical consequences. Sci. Total Environm. 654, 720–734. https://doi.org/10.1016/j.scitotenv.2018.10.408

Butterworth, B.E.; Korant, B.D. Characterization of the large picornaviral polypeptides produced in the presence of zinc ion. 1974. J. Virol. 14, 282–291.

Castillo-González, J., Ojeda-Barrios, D., Hernández-Rodríguez, A., González Franco, A.C., Robles-Hernández, L. & López-Ochoa, G.R. Zinc 2018. Metalloenzymes in Plants. Interciencia 43, 242–248.

Cha, J.-Y. & Cho, Y.-S. (2009) Determination of Optimal Conditions for Zinc-Hyperaccumulation by *Saccharomyces cerevisiae* FF-10. J. Korean Soc. Appl. Biol. Chem. 52, 227–233. https://doi.org/10.3839/jksabc.2009.041

Coleman, J.E. 1992. Zinc proteins: enzymes, storage proteins, transcription factors, and replication proteins. Annu. Rev. Biochem. 61, 897–946.

Converse, A. & Thomas, P. 2020. The zinc transporter ZIP9 (Slc39a9) regulates zinc dynamics essential to egg activation in zebrafish. Sci. Rep. 10:15673 (14 pp.). https://doi.org/10.1038/s41598-020-72515-4

Dabir, A., Heidari, P., Ghorbani, H. & Ebrahimi, A. (2019) Cadmium and lead removal by new bacterial isolates from coal and aluminum mines. Intern. J. Environm. Sci. Technology 16, 8297–8304. https://doi.org/10.1007/s13762-019-02303-9

Dai, Z.-C., Cai, H.-H., Qi, S.-S., Li, J., Zhai, D.-L., Wan, J.S.H. & Du, D.-L. (2020) Cadmium hyperaccumulation as an inexpensive metal armor against disease in Crofton weed. Environm. Pollut. 267: 115649 (9 pp.). https://doi.org/10.1016/j.envpol.2020.115649

Din, G., Hassan, A., Dunlap, J., Ripp, S. & Shah, A.A. (2022) Cadmium tolerance and biore-mediation potential of filamentous fungus *Penicillium chrysogenum* FMS2 isolated from soil. Internat. J. Environ. Sci.Technol. 19, 2761–2770. https://doi.org/10.1007/s13762-021-03211-7

Dou, X., Dai, H., Twardowska, I. & Wei, S. (2019) Hyperaccumulation of Cd by *Rorippa globosa* (Turcz.) Thell. from soil enriched with different Cd compounds, and impact of soil amendment with glutathione (GSH) on the hyperaccumulation efficiency. Environm. Pollut. 255: 113270 (10 pp.). https://doi.org/10.1016/j.envpol.2019.113270

Dhawan, M., Emran, T.B., Priyanaka & Choudhary, O.P. 2022. Immunomodulatory effects of zinc and its impact on COVID-19 severity. Ann. Medic. Surg. 77: 103638 (4 pp.). https://doi.org/10.1016/j.amsu.2022.103638

Duan, M., Li, T., Liu, B., Yin, S., Zang, J., Lv, C., Zhao, G. & Zhang, T. 2021. Zinc nutrition and dietary zinc supplements. Crit. Revs Food Sci. Nutrition (16 pp.). https://doi.org/10.1080/10408398.2021.1963664

Duncan, F.E., Que, E.L., Zhang, N., Feinberg, E.C., O'Halloran, T.V. & Woodruff, T.K. 2016. The zinc spark is an inorganic signature of human egg activation. Sci. Rep. 6:24737 (8 pp.). https://doi.org/10.1038/srep24737

Fazli, M.M., Soleimani, N., Mehrasbi, M., Darabian, S., Mohammadi, J. & Ramazani, A. (2015) J. Environm. Health Sci. Engineer. 13:19 (9 pp.). https://doi.org/10.1186/s40201-015-0176-0

Eisler, R. (2006) Mercury hazards to living organisms. 175 pp., CRC, Taylor & Francis Group, Boca Raton, London, New York.

Engin, A.B., Engin, E.D. & Engin, A. 2022. Can iron, zinc, copper and selenium status be a prognostic determinant in COVID-19 patients? Environm. Toxicol. Pharmacol. 95: 103937 (14 pp.). https://doi.org/10.1016/j.etap.2022.103937

Escudero, V., Ferreira Sánchez, D., Abreu, I., Sopeña-Torres, S., Makarovsky-Saavedra, N., Bernal, M., Krämer, U., Grolimund, D., González-Guerrero, M. & Jordá, L. 2022. *Arabidopsis thaliana* Zn^{2+}-efflux ATPases HMA2 and HMA4 are required for resistance to the necrotrophic fungus *Plectosphaerella cucumerina* BMM. J. Exp. Bot. 73, 339–350. https://doi.org/10.1093/jxb/erab400

Ghanei, E., Baghani, M., Moravvej, H., Talebi, A., bahmanjahromi, A. & Abdollahimajd, F. 2022. Low serum levels of zinc and 25-hydroxyvitmain D as potential risk factors for COVID-19 susceptibility: a pilot case-control study. Eur. J. Clin. Nutrit. 76, 1297– 1302. https://doi.org/10.1038/s41430-022-01095-5

Grifagni, D., Calderone, V., Giuntini, S., Cantini, F., Fragai, M. & Banci, L. 2021. SARS-CoV-2 Mpro inhibition by a zinc ion: structural features and hints for drug design. Chem. Commun. 57, 7910–7913. https://doi.org/10.1039/d1cc02956h

Grossman, E. & Gruntman, J. (2022) Simulated herbivory enhances Cd phytoextraction efficiency of sunflowers. Plant Soil (online Aug. 2022) https://doi.org/10.1007/s11104-022-05643-7

Hoang, B.X. & Han, B. 2021. Micronutrient zinc roles in adjunctive therapy for COVID-19 by enhancing patients immunoregulation and tolerance to the pathogen. Revs Medical Microbiol. 32, 149–157. https://doi.org/10.1097/MRM.0000000000000263

Hu, Y., Bandara, A.R., Xu, J., Kakumyan, P., Hyde, K.D. & Mortimer, P.E. The use of *Agaricus subrufescens* for rehabilitation of agricultural soils. Agronomy 2022, 12, 2034 (17 pp.). https://doi.org/10.3390/agronomy12092034

Huang, J.C., Li, K.B., Yu,Y.R., Wu, H. & Liu, D.L. (2008) Cadmium accumulation in *Agaricus blazei* Murrill. J. Sci. Food Agric. 88, 1369–1375. https://doi.org/10.1002/jsfa.3225

Hussain, S., Khan, M., Sheikh, T.M.M., Mumtaz, M.Z., Chohan, T.A., Shamim, S. & Liu, Y. 2022. Zinc essentiality, toxicity, and Its bacterial bioremediation: A comprehensive Insight. Front. Microbiol. 13:900740 (15 pp.). https://doi.org/10.3389/fmicb.2022.900740

Li, J., Cao, D., Huang, Y., Chen, B., Chen, Z., Wang, R., Dong, Q., Wei, Q. & Liu, L. 2022. Zinc intakes and health outcomes: An umbrella review. Front. Nutr. 9:798078 (14 pp.). https://doi.org/10.3389/fnut.2022.798078

Jackl, G.A., Reidel, G. & Kollmer, W.E. (1987) Compounds in *Agaricus arvensis* hyphae using ^{109}Cd. Appi. Radiat. Isot. 38, 431–435.

Jiang, R.F., Ma, D.Y., Zhao, F.J. & McGrath, S.P. (2005) Cadmium hyperaccumulation protects *Thlaspi caerulescens* from leaf feeding damage by thrips (*Frankliniella occidentalis*). New Phytol. 167, 805–814. https://doi.org/10.1111/j.1469-8137.2005.01452.x

Joulaei, H., Keshani, P., Foroozanfar, Z., Zamanian, D., Hassani, A., Parvizi, F., Khadem, Y., Omidifar, N. & Davarpanah, M.A. 2022. Serum zinc associated with immunity and inflammatory markers in Covid-19. Open Medicine 17, 702–711. https://doi.org/10.1515/med-2022-0469

Kádár, E., Costa, V., Martins, I., Serrao Santos, R. & Powell, J.J. (2005) Enrichment in trace metals (Al, Mn, Co, Cu, Mo, Cd, Fe, Zn, Pb and Hg) of macro-invertebrate habitats at hydrothermal vents along the Mid-Atlantic Ridge. Hydrobiologia 548, 191–205. https://doi.org/10.1007/s10750-005-4758-1

Kalač, P., Svoboda, L. & Havlíčková, B. (2003) Contents of cadmium and mercury in edible mushrooms. J. Appl. Biomed. 2, 15–20.

Kazemi-Dinan, A. Barwinski, A., Stein, R.J., Krämer, U. & Müller, C. (2015) Metal hyperaccumulation in Brassicaceae mediates defense against herbivores in the field and improves growth. Entomol. Exp. Applic. 157, 3–10.

Khan, I., Aftab, M., Shakir, S.U., Ali, M., Qayyum, S., Rehman, M.U., Haleem, K.S. & Touseef, I. (2019) Mycoremediation of heavy metal (Cd and Cr)–polluted soil through indigenous metallotolerant fungal isolates. Environ. Monit. Assess. 191: 585 (11 pp.). https://doi.org/10.1007/s10661-019-7769-5

Khan, S.T., Malik, A., Alwarthan, A. & Shaik, M.R. 2022. The enormity of the zinc deficiency problem and available solutions; an overview. Arabian J. Chem. 15, 103668 (31 pp.). https://doi.org/10.1016/j.arabjc.2021.103668

Kim, A.M, Bernhardt, M.L., Kong, B.Y., Ahn, R.W., Vogt, S., Woodruff, T.K. & O'Halloran, T.V. 2011. Zinc sparks are triggered by fertilization and facilitate cell cycle resumption in mammalian eggs. ACS Chem. Biol. 6, 716–723.

Kluska, K., Veronesi, G., Deniaud, A., Hajdu, B., Gyurcsik, B., Bal, W. & Krężel, A. 2022. Structures of silver fingers and a pathway to their genotoxicity. Angew. Chem. Int. Ed. 61, e202116621 (8 pp). https://doi.org/10.1002/anie.202116621

Lane, T.W., Saito, M.A., George, G.N., Pickering, I.J., Prince, R.C. & Morel, F.M.M. (2005) A cadmium enzyme from a marine diatom. Nature 435, 42–42.

Liu, S., Tang, Y., Qiu, R., Ying, R., Ge, R. & Ji, X. (2016) Responses of carbonic anhydrase to cadmium in the zinc/cadmium hyperaccumulator *Picris divaricata* Vant. Pedosphere 26, 709–716. A cadmium enzyme from a marine diatom. https://doi.org/10.1016/S1002-0160(15)60081-9

Lombi, E., Zhao, F.J., Dunham, S.J. & McGrath, S.P. (2000) Cadmium accumulation in populations of *Thlaspi caerulescens* and *Thlaspi goesingense*. New Phytol. (2000), 145, 11–20.

Lombi, E. Zhao, F.J., McGrath, S.P., Young, S.D. & Sacchi, G.A. (2001) Physiological evidence for a high-affinity cadmium transporter highly expressed in a *Thlaspi caerulescens* ecotype. New Phytol. 149, 53–60.

Maares, M., Hackler, J., Haupt, A., Arved Heller, R.A., Bachmann, M., Diegmann, J., Moghaddam, A., Schomburg, L. & Haase, H. 2022. Free zinc as a predictive marker for COVID-19 mortality risk. Nutrients 14: 1407 (13 pp.). https://doi.org/10.3390/nu14071407

Magnin, J.-P., Gondrexon, N. & Willison, J.C. 2014. Zinc biosorption by the purple non-sulfur bacterium *Rhodobacter capsulatus*. Can. J. Microbiol. 60, 829–837. https://doi.org//10.1139/cjm-2014-0231

Meisch, H.-U., Schmitt, J.A. & Scholl, A.-R. (1979) Growth stimulation by cadmium in the mushroom *Agaricus abruptibulbus*. Naturwiss. 66, 209–209.

Meisch, H.-U- & Schmitt, J.A. (1986) Characterization studies on cadmium-mycophosphatin from the mushroom *Agaricus macrosporus*. Environm. Health Perspect. 65, 29–32. https://www.jstor.org/stable/3430158

Münger K., Lerch K. and Tschierpe H. J. (1982) Metal accumulation in Agaricus bisporus: influence of Cd and Cu on growth and tyrosinase activity. Experientia 38, 1039–1041.

Murdoch, C.C. & Skaar, E.P. 2022. Nutritional immunity: the battle for nutrient metals at the host–pathogen interface. Nature Revs in press (13 pp.). https://doi.org/10.1038/s41579-022-00745-6

Norouzi, S., Adulcikas, J., Sohal, S.S. & Myers, S. 2018. Zinc stimulates glucose oxidation and glycemic control by modulating the insulin signaling pathway in human and mouse skeletal muscle cell lines. PLoS ONE 13(1): e0191727 (15 pp.). https://doi.org/10.1371/journal.pone.0191727

Pollard, A.J. & Baker, A.J.M. 1997. Deterrence of herbivory by zinc hyperaccumulation in *Thlaspi caerulescens* (Brassicaceae). New Phytol. 135, 655–658.

Poudel, R.R., Bhusal, Y., Tharu, B. & Kafle, N.K. (2017) Role of zinc in insulin regulation and diabetes. J. Social Health Diabetes 5, 83–87.

Pritts, J.D. & Michel, S.L.J. 2022. Fe-S clusters masquerading as zinc finger proteins. J. Inorg. Biochem, 230: 111756 (20 pp.). https://doi.org/10.1016/j.jinorgbio.2022.111756

Ramtahal, G., Umaharan, P., Davis, C., Roberts, C., Hanuman, A. & Ali, L. (2022) Mitigation of cadmium uptake in Theobroma cacao L: efficacy of soil application methods of hydrated lime and biochar. Plant Soil 477, 281–296. https://doi.org/10.1007/s11104-022-05422-4

Roosens, N., Verbruggen, N., Ximénez-Embún, P. & Smith, J.A.C. (2003) Natural variation in cadmium tolerance and its relationship to metal hyperaccumulation for seven populations of *Thlaspi caerulescens* from western Europe. Plant, Cell Environm. 26, 1657–1672.

Sácký, J., Leonhardt, T., Borovicˇka, J., Gryndler, M., Briks, A. & Kotrba, P. (2014) Intracellular sequestration of zinc, cadmium and silver in *Hebeloma mesophaeum* and characterization of its metallothionein genes. Fungal Genetics and Biology 67 (2014) 3–14. http://dx.doi.org/10.1016/j.fgb.2014.03.003

Sácký, J., ˇCerný, J., ˇSantrůˇcek, J., Boroviˇcka, J., Leonhardt, T. & Kotrba, P. (2021) Cadmium hyperaccumulating mushroom *Cystoderma carcharias* has two metallothionein isoforms usable for cadmium and copper storage. Fungal Genetics and Biology 153: 103574 (9 pp.). https://doi.org/10.1016/j.fgb.2021.103574

Seregin, I.V., Kozhevnikova, A.D.Zhukovskaya, N.V. & H. Schat, H. (2015) Cadmium tolerance and accumulation in excluder Thlaspi arvense and various accessions of hyperaccumulator *Noccaea caerulescens*. Russian Journal of Plant Physiology, 2015, Vol. 62, No. 6, pp. 837–846. Russian original in Fiziologiya Rastenii, 62, 854–864.

Sigel, A., Sigel, H. & Sigel, R.K.O., eds (2013) Cadmium: From toxicity to essentiality. Metal ions in life sciences, vol. 11 (Series eds Sigel, A., Sigel, H. & Sigel, R.K.O.), pp. xxxvi+560. Springer, Dordrecht, Heidelberg, New York, London. ISSN 1559-0836, ISSN 1868-0402 (electronic) ISBN 978-94-007-5178-1, ISBN 978-94-007-5179-8 (eBook), https://doi.org/10.1007/978-94-007-5179-8

Que, E.L, Bleher, R., Duncan, F.E., Kong, B.Y., Gleber, S.C., Vogt, S., Chen, S., Garwin, S.A., Bayer, A.R., Dravid, V.P., Woodruff, T.K. & O'Halloran, T.V. 2015. Quantitative mapping of zinc fluxes in the mammalian egg reveals the origin of fertilization-induced zinc sparks. Nature Chemistry 7, 130–139. https://doi.org/10.1038/NCHEM.2133

Que, E.L., Duncan, F.E., Bayer, A.R., Philips, S.J., Roth, E.W., Bleher, R., Gleber, S.C., Vogt, S., Woodruff, T.K. & O'Halloran, T.V. 2017. Zinc sparks induce physiochemical changes in the egg zona pellucida that prevent polyspermy. Integr. Biol. 9, 135–144. https://doi.org/10.1039/c6ib00212a

Que, E.L., Duncan, F.E., Chang Lee, H.C., Hornick, J.E., Vogt, S., Fissore, R.A., O'Halloran, T.V. & Woodruff, T.S. 2019. Bovine eggs release zinc in response to parthenogenetic and sperm-induced egg activation. Theriogenology 127, 41–48. https://doi.org/10.1016/j.theriogenology.2018.12.031

Restrepo Caro, C.R., Coronell, M.D.C., Arrollo, J., Martinez, G., Sanchez Majana, L., Sarmiento-Rubiano, L.A. 2016. Zinc deficiency: A global problem that affect the health and cognitive development. Arch. Lationoam. de nutricion. 66, 165–175.

Rowens, B., Guerrero-Betancourt, D., Gottlieb, C.A., Boyes, R.J. & Eichenhorn, M.S.(1991) Acute Inhalation of mercury vapor. A clinical and histologic perspective. Chest 99, 185–190.

Seeler, J.F., Sharma. A., Zaluzec, N.J., Bleher, R., Lai, B., Schultz, E.G., Hoffman, B.M., LaBonne, C., Woodruff, T.K. & O'Halloran, T.V. 2021. Metal ion fluxes controlling amphibian fertilization. 13, 683–691. https://doi.org/10.1038/s41557-021-00705-2

Skalny, A.V., Aschner, M., Sekacheva, M.I., Santamaria, A., Barbosa, F., Ferrer, B., Aaseth, J., Paoliello, M.M.B., Rocha, J.B.T. & Tinkov, A.A. (2022) Mercury and cancer: Where are we now after two decades of research? FoodChem. Toxicol. 164: 113001 (11 pp.). https://doi.org/10.1016/j.fct.2022.113001

Smith, G.D., Swenson, D.C., Dodson, E.J. & Reynolds, C.D. 1984. Structural stability in the 4-zinc human insulin hexamer. Proc. Natl Acad. Sci. USA 81, 7093–7097.

Stanton C., Sanders D., Krämer U., and Podar D. 2022. Zinc in plants: Integrating homeostasis and biofortification. Mol. Plant. 15, 65–85. https://doi.org/10.1016/j.molp.2021.12.008

Stoknes, K., Scholwin, F., Jasinska, A., Wojciechowska, E., Mleczek, M., Hanc, A. & Niedzielski, P. (2019) Cadmium mobility in a circular food-to-waste-to-food system and the use of a cultivated mushroom (Agaricus subrufescens) as a remediation agent. J. Environ. Manag. 245, 48–54.

Tang, L., Yao, A., Yuan, M., Tang, Y., Liu, J., Liu, X. & Qiu, R. (2016) Transcriptional up-regulation of genes involved in photosynthesis of the Zn/Cd hyperaccumulator *Sedum alfredii* in response to zinc and cadmium. Chemosphere 164, 190–200. http://dx.doi.org/10.1016/j.chemosphere.2016.08.026

Tang, R.H., Nkrumah, P.N., Erskine, P.D. & van der Ent, A. (2022) Polymetallic (zinc and cadmium) hyperaccumulation in the Australian legume *Crotalaria novae-hollandiae* compared to Crotalaria cunninghamii. Plant Soil, in press. https://doi.org/10.1007/s11104-022-05547-6

Taux, K., Kraus, T. & Kaifie, A. (2022) Mercury exposure and Its health effects in workers in the artisanal and small-scale gold Mining (ASGM) sector—A systematic review. Int. J. Environ. Res. Public Health 19, 2081 (34 pp.). https://doi.org/10.3390/ijerph19042081

te Velthuis, A. J., van den Worm, S. H., Sims, A. C., Baric, R. S., Snijder, E. J., van Hemert, M. J. 2010. Zn^{2+} inhibits coronavirus and arterivirus RNA polymerase activity in vitro and zinc ionophores block the replication of these viruses in cell culture. PLoS Pathog. 6: e1001176 (10 pp.). https://doi.org/10.1371/journal.ppat.1001176

UNEP, 2015. Global Mercury Modelling: Update of Modelling Results in the Global Mercury Assessment 2013. Arctic Monitoring and Assessment Programme, Oslo, Norway/UNEP Chemicals Branch, Geneva, Switzerland.

Verbruggen, N., Hermans, C. & Schat, H. (2009) Molecular mechanisms of metal hyperaccumulation in plants. New Phytol. 181, 759–776. https://doi.org/10.1111/j.1469-8137.2008.02748.x

Wang, X., Li, H., Liu, Y., Cui, T., Liao, M., Lin, L., Wang, Z., Deng, H., Liang, D., Xia, H. & Tang, Y. (2021) Mowing enhances the phytoremediation ability of cadmium-contaminated soil in the post-grafting generations of potential cadmium-hyperaccumulator *Solanum photeinocarpum.* Internat. J. Phytoremediation 23, 652–657. https://doi.org/10.1080/15226514.2020.1847036

Wozniak, K.L., Bainbridge, R.E., Summerville, D.W., Tembo, M., Phelps, W.A., Sauer, M.L., Wisner, B.W., Czekalski, M.E., Pasumarthy, S., Hanson, M.L., Linderman, M.B., Luu, C.H., Boehm, M.E., Sanders, S.M., Buckley, K.M., Bain, D.J., Nicotra, M.L., Lee, M.T. & Carlson, A.E. (2020) Zinc protection of fertilized eggs is an ancient feature of sexual reproduction in animals. PLoS Biol 18: e3000811 (17 pp.). https://doi.org/10.1371/journal.pbio.3000811

Wu, C., Tang, D., Dai, J., Tang, X., Bao, Y., Ning, J., Zhen, Q., Song, H., St. Leger, R.J., Fang, W. (2022) Bioremediation of mercury-polluted soil and water by the plant symbiotic fungus *Metarhizium robertsii.* Proc. Acad. Sci. USA 119: e2214513119 (9 pp.). https://doi.org/10.1073/pnas.2214513119

Wu, H., Wu, Q., Wu, G., Gu, Q. & Wei, L. (2016) Cd-resistant strains of B. cereus S5 with endurance capacity and their capacities for cadmium removal from cadmium-polluted water. PLoS ONE 11:e0151479 (25 pp.). https://doi.org/10.1371/journal.pone.0151479

Wu, P., Dutkiewicz, S., Monier, E. & Zhang, Y. (2021) Bottom-heavy trophic pyramids impair methylmercury biomagnification in the marine Plankton ecosystems. Environ. Sci. Technol. 55, 15476−15483. https://doi.org/10.1021/acs.est.1c04083

Xu, H., Xie, Z., Guo, J. & Men, Q. (2021) Morphological changes and bioaccumulation in response to cadmium exposure in *Morchella spongiola*, a fungus with potential for detoxification. Can. J. Microbiol. 67, 789–798. https://doi.org//10.1139/cjm-2020-0571

Yang, L., Liu, Q., Zhou, A., Liu, Y., Xia, X., Lin, L., Huan, Y., Sun. J., Wu, C., Liao, M. & Xia, H. 2020. Effects of intercropping with three ploidies of hyperaccumulator *Solanum* species on the growth and cadmium uptake of *Cyphomandra betacea* seedlings. Internat. J. Environ. Anal. Chem. 2020, 1–11. https://doi.org/10.1080/03067319.2020.1774569

Zhang, N., Duncan, F.E., Que, E.L., Thomas V. O'Halloran, T.V. & Woodruff, T.K. (2016) The fertilization-induced zinc spark is a novel biomarker of mouse embryo quality and early development. Sci. Rep. 6:22772 (9 pp.). https://doi.org/10.1038/srep22772

CHAPTER THIRTEEN

BORON, ALUMINIUM, GALLIUM, AND THALLIUM

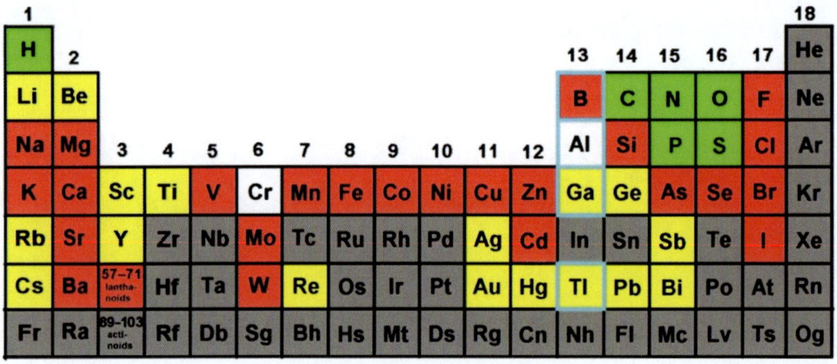

13.1 Boron

13.1.1 Plants

It has been known for a long time (a century and a half) that boron is an essential nutrient element for plants. One of the crop problems with boron deficiency that was discovered long ago is heart rot in sugar beets (Rowe 1936, Brenchley &Watson 1937, Cox 1940). Shorrocks (1997) has a long list of plant species and countries in which boron deficiency has been observed. The map below (Fig. 13.1.1.1) is redrawn from his article.

Fig. 13.1.1.1. Boron deficient areas (red colour). Areas identified on the basis of crop responses to boron and by soil B analysis. Modified from Shorrocks (1997).

Until recently it has been a mystery exactly what the function of boron is in plant cells. It has been known that it is necessary for pollen germination and pollen tube growth, and needed for growth of root meristems, and more generally for cell extension, but only recently have the molecular details started to become a little more clear. As boron starts to become deficient in a plant, a greater proportion of it is allocated to cell walls, while the cytoplasmic concentration drops. Both plastic and elastic extensibility drop (Hu & Brown 1994). There is a very narrow range between boron deficiency and toxicity as more than 5.00 mg L^{-1} available boron can be toxic to many agronomic crops (Ozturk et al. 2010). One function of borate is to link two molecules of rhamnogalacturonan II in the cell wall together (Ishii et al. 2001). This is one of many examples of how boric acid exerts its action by combining with two neighbouring hydroxy groups (Bolaños et al. 2002).

The concentration of boron in plants is regulated. When enough has been taken up, import is shut down, at at excess export is activated (Yoshinari & Takano 2017). In the phloem of celery (*Apium graveolens*) boron is transported as a mannitol complex (Hu et al. 1997, Fig. 13.1.1.2).

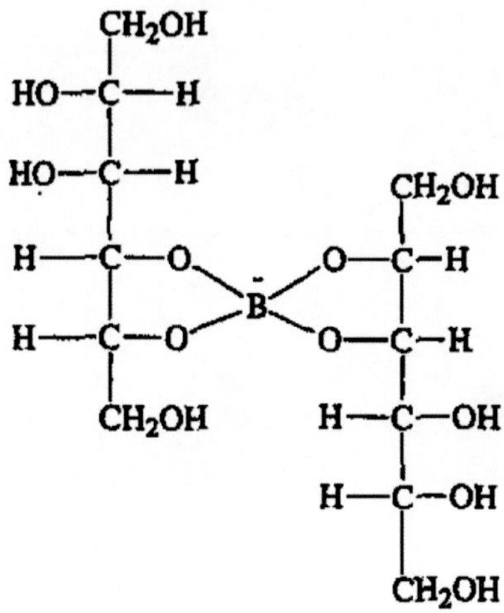

Fig. 13.1.1.2. Dimannitol-borate complex from phloem, also present in extrafloral nectar of *Prunus*. From Hu et al. (1995).

13.1.2 Algae

Several studies on red algae were conducted in the 1940s, especially by Harald Kylin and Anders Kylin (see Henkel 1952). More detailed measurements of the boron requirement were carried out by Henkel (1952); see Fig. 13.2.1.

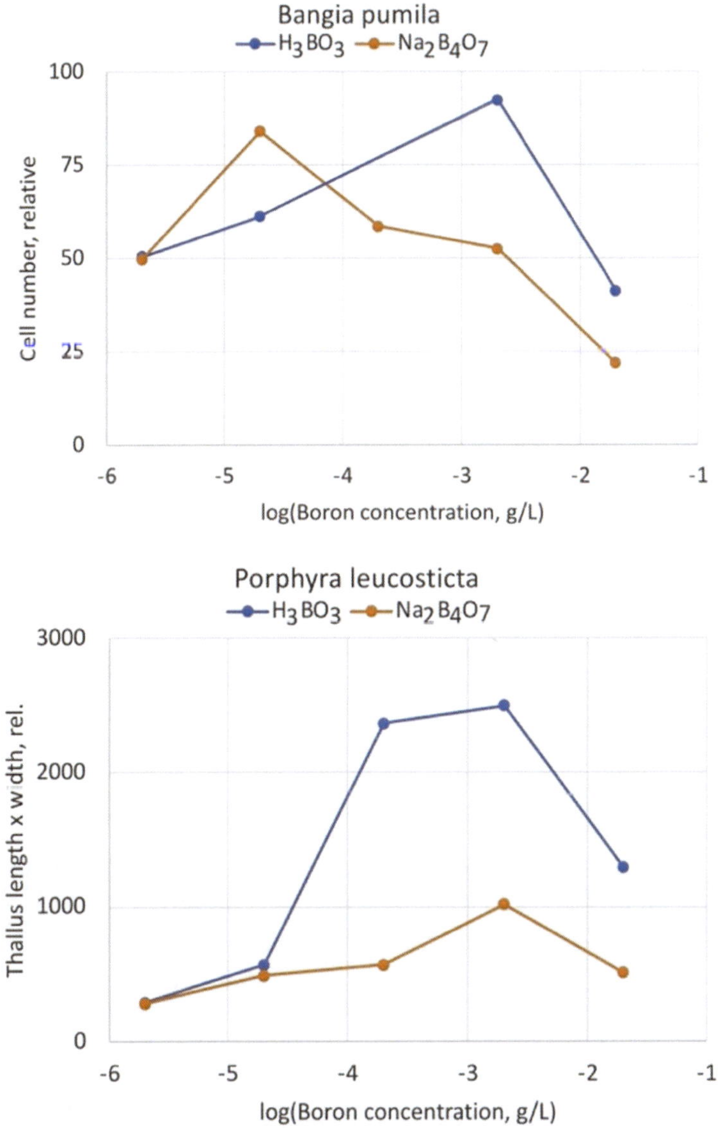

Fig. 13.1.2.1. The boron (as boric acid, H_3BO_3 or borax, $Na_2B_4O_7$) requirement of the red algae *Bangia pumila* and *Porphyra leucosticta*. Except for borax and *Bangia pumila* the optimum boron concentration is around 2 mg L^{-1}. The lowest concentration in the diagram is uncertain. Data replotted from Henkel (1952).

The necessity of boron for the growth of a diatom was first demonstrated by Lewin (1965) (Fig. 13.1.2.2). Lewin (1966) demonstrated the boron requirement for 12 diatom species from different groups, so there seems to be a general boron requirement for diatoms.

Boron is not required for the green algae *Chlorella pyrenoidosa, Draparnaldia plumosa,* or *Stigeoelonium tenue* (Bowen et al. 1965, Gerloff 1968).

The essentiality of boron for the brown alga *Fucus edentatus* was demonstrated by McLachlan (1977).

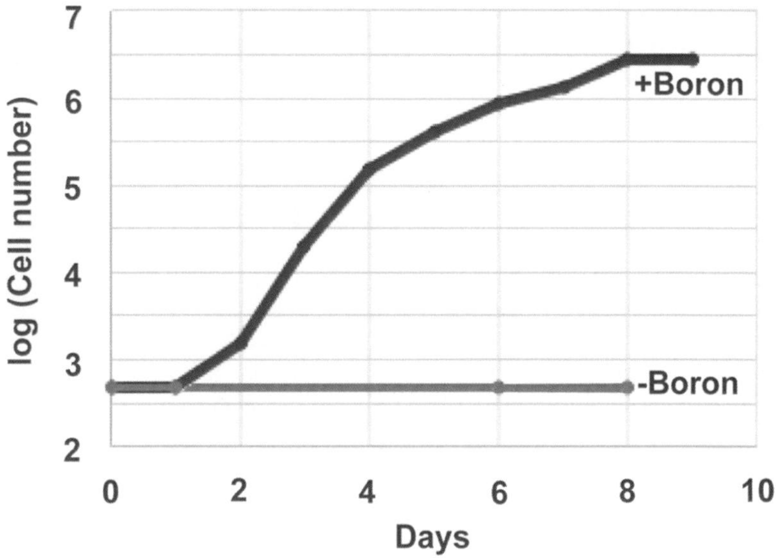

Fig. 13.1.2.2. Growth of of the diatom *Cylindrotheka fusiformis* in an artificial medium with and without boron, 0.5 mg/1 (added as H_3BO_2). Data from Lewin (1965).

13.1.3 Fungi

For fungi the reactions differ between groups, but there seems to be no absolute requirement. Boron very much stimulated the growth rate of the yeast *Saccharomyces cerevisiae* (Bennett et al. 1999), but not of that of *Penicillium chrysogenum* (Gerloff 1968). Estevez-Fregoso et al. (2021) review the literature on boron compounds in fungi.

13.1.4 Bacteria

Gerretsen & de Hoop (1954) found boron to very much stimulate growth and nitrogen fixation by the bacterium *Azotobacter chroococcum*. Growth of free-living Frankia BCU110501 was delayed in B-deficient, and completely inhibited in N-free media (Bolaños et al. 2002). Since boron is necessary also for nitrogen fixation by the cyanobacterium *Anabaena* (Mateo et al. 1986), boron may be an almost general requirement for nitrogen fixation, although at least some non-heterocystous cyanobacteria do not need boron (Bonilla et al. 1991). One possibility is that boron helps to form a barrier that keeps the nitrogenase-inhibiting oxygen out.

Borate anion seems to be involved in ferric ion binding by a protein in the marine bacterium *Marinobacter algicola* (Weerasinghe et al. 2013).

The bacterium *Vibrio harveyi* uses a boron-containing chemical signal (Fig. 5) for feeling the population density of their cells, the so-called quorum sensing. For this bacterium high population density results in bioluminescence (Colthurst et al. 2002).

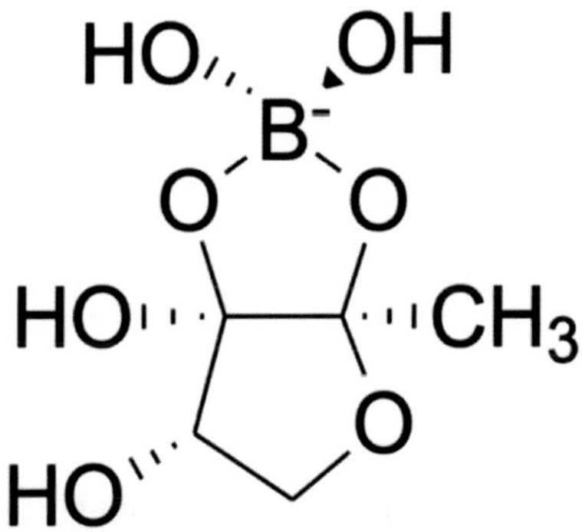

Fig. 13.1.4.1. The quorum-sensing boron compound (termed Al-2) from *Vibrio harveyi*. Redrawn from Colthurst et al. (2002).

Boromycin (Fig. 13.1.4.2) is an antibiotic active against, e.g., human immunodeficiency virus, that is excreted by *Streptomyces antibioticus* (Kohno et al. 1996), and a related boron antibiotic compound, aplasmomycin, is excreted by another bacterium (Blue et al. 2022).

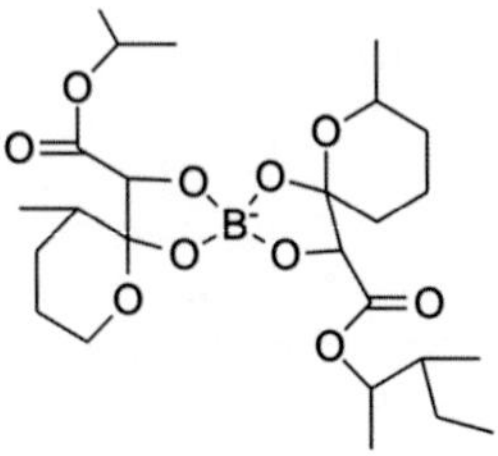

Fig. 13.1.4.2. The central part of a boromycin molecule. Adapted from Kohno et al. (1996).

13.1.5 Humans and other vertebrates

Pizzorno (2015) has provided an excellent review of the importance of boron in human nutrition. She lists eleven beneficial effects of boron on human health, the first two being

(1) essentiality for the growth and maintenance of bone;
(2) greatly improving wound healing.

Boron counteracts osteoporosis (literature review by Vielma et al. 2017), and has a positive effect on alveolar bone (the bone holding the teeth) by enhancing osteoblast formation (Shalehin et al. 2020). It inhibits the formation of osteoclasts, bone-disassembling cells (Xu et al. 2022). An overview of the literature regarding human bone health was published by Rondanelli et al. (2020). Supplemental boron in drinking water promotes bone development in ostrich chicks (Zhu et al. 2021). In experiments with zebrafish (*Danio rerio*) Sojan et al. (2022) found synergism between boron and vitamin D on bone development.

Boron also seems to have importance for immunity (Hunt 2003).

13.2 Aluminium

13.2.1 Aluminium in plants

Aluminium is the most common metallic element in the Earth's crust, estimated to contribute 8.23% of its mass. It occurs in the environment in its trivalent form, at low pH as Al^{3+}, at high pH as aluminate, $Al(OH)_3$ or $Al(OH)_4^-$. There are many other ionic forms: $[Al(H_2O)_6]^{3+}$, $Al(OH)_2^+$, $AlOH^{2+}$. Aluminium can also assume other oxidation levels, but they seem to be without biological or geological importance. It conjugates with other elements to generate complexes like aluminosilicate.

In general Al^{3+} is toxic to plants, animals, and humans, but there are important exceptions. One important plant that seems to benefit from aluminium ions is the tea plant (Fung et al. 2008). For this plant it may even be essential (Sun et al. 2020). Aluminium ion promoted new root growth in five representative tea varieties of *Camellia sinensis* with dose-dependent responses to Al^{3+} availability. In the absence of Al^{3+}, the tea plants failed to generate new roots (Fig. 3.2.1.1), and the root tips were damaged within 1 d of Al deprivation. Similar results were obtained by Liu et al. (2020) for *Camellia japonica*.

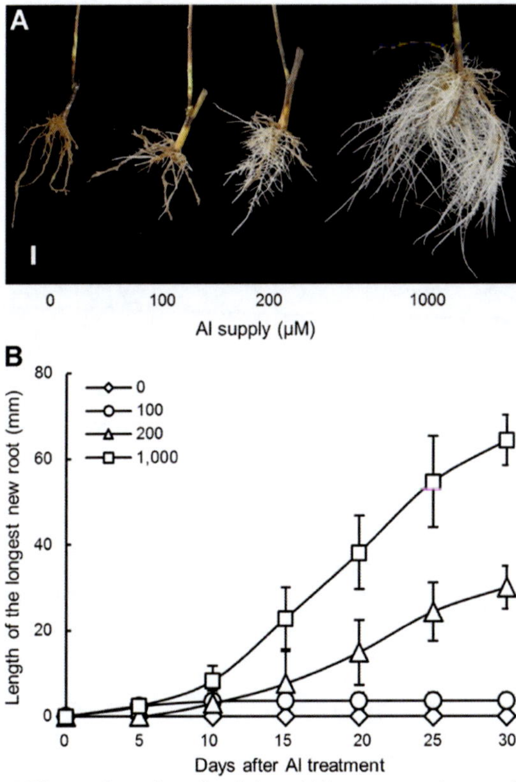

Fig. 13.2.1.1. Effects of varying aluminium (Al) concentrations on Tieguanyin tea (*Camellia sinensis*) root growth. Tieguanyin tea plants were harvested from field acid soil and then transferred to nutrient solution containing 0, 100, 200, or 1,000 μmol/L $AlCl_3$, at pH 4.5. (A) Photographs of root growth phenotypes. Bar = 10 mm. (B) Length of the longest new root. Data are means with standard error (n = 5). From Sun et al. (2020). Reproduced under Creative Commons Attribution License (http://creativecommons.org/licenses/by/4.0/).

Also growth of cell cultures of the tea plant *Camellia sinensis* in nutrient solution is stimulated by Al^{3+} ions.

Similar effects on root development are reported also for some plants not closely related to the tea plant, e.g., *Symplocos paniculata* (Fig. 13.2.1.2).

Fig. 13.2.1.2. Another plant, *Symplocos paniculata* (Symplocaceae, Ericales), for which root development is stimulated by Al^{3+}. From Schmitt et al. (2016). Reproduced under Creative Commons Attribution License (http://creativecommons.org/licenses/by/4.0/).

Still another, and from the view of human nutrition very important plant, rice, reacts in a similar way to aluminium ions (Fig. 3.2.1.3).

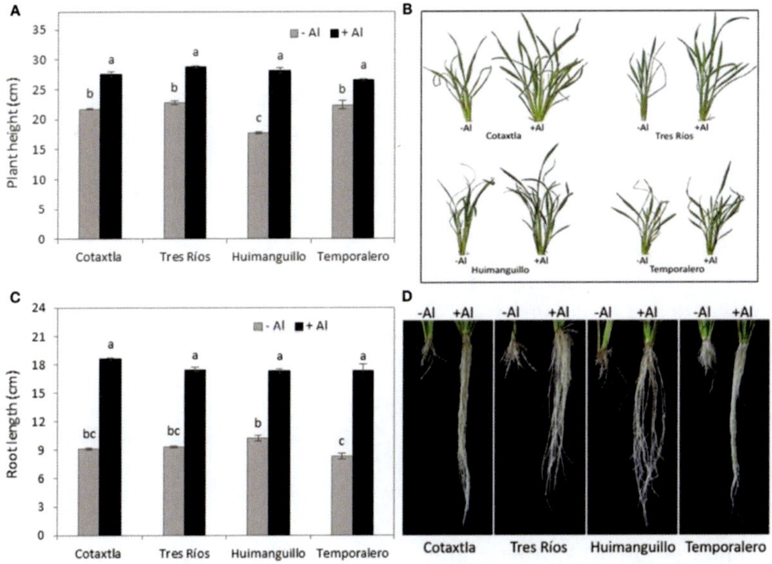

Fig. 3.2.1.3. Response of four varieties of rice to 200 μM Al^{3+} for 20 days. From Moreno-Alvorado et al. (2017). Reproduced under Creative Commons Attribution License (CC BY) (http://creativecommons.org/licenses/by/4.0/).

A plant is classified as a hyperaccumulator of aluminium if the amount of Al in the shoot is at least 0.1% by weight (Branquinho et al. 2007). For the tea plant, *Camellia sinensis*, it can reach over 0.04% in twigs and over 0.8% in leaves on a dry matter basis (van der Ent 2020); it is an aluminium hyperaccumulator. In the relative *Camellia oleifera* Abel. Only 0.011% accumulated in the leaves, but it was still called hyperaccumulator (Zeng et al. 2011).

As already mentioned, aluminium ions (Al^{3+}) are toxic to most plants. In barley they reduced the mitotic activity of the root tip cells and also induced micronuclei and damaged nuclei. They decreased the frequency of S phase cells and increased the frequency of cells in G2/M phase (Jaskowiak et al. 2018). In soybean Al^{3+} induced programmed cell death in the root tips (Huang et al. 2022), and further examples can be found in these publications.

13.2.2 Aluminium hyperaccumulators

Aluminium is generally taken up as Al^{3+} from a substrate with low pH. A plant is classified as a hyperaccumulator of aluminium if the amount of Al in the shoot is at least 0.1% by weight (Branquinho et al. 2007). The older literature about aluminium hyperaccumulators is summarized by Jansen et al. (2002). Very few aluminium hyperaccumulators are monocotyledons; most are eudicots, but apart from that, they are not closely related. They can be asterids, as *Rudgea viburnoides* (Gonçalves Malta et al. 2016), *Camellia* (the tea genus) or *Plantago* (Serrano et al. 2017), or superasterids without being asterids (belonging to Caryophyllales), as *Rumex obtusifolius* (Vondráčková et al. 2015) and *Phytolacca americana* (Liu et al. 2022), or rosids, as *Qualea grandiflora, Callisthene major, and Vochysia pyramidalis* (Miranda de Andrade et al. 2011) and various members of Melastomataceae (Milanez et al. 2019, Timpone & Habermann 2022, Zaia et al. 2022), or basal eudicots belonging to Proteales: *Gevuina avellana*, *Orites myrtoidea*, *Lomatia dentata* (Delgado et al. 2019). Aluminium accumulation was also found in nine out of ten plants in Lycopodiales and many, but not all ferns (Schmitt et al. 2017). Liu et al. (2021) studied the Al-accumulating fern *Dicranopteris linearis* in more detail.

As for causes for hyperaccumulation of the generally toxic Al^{3+} ion, it can be supposed that it protects against herbivores and/or parasites. Rebula et al. (2008) showed that leaves with high aluminium content cause the death of a fungus that leaf-cutter ants (*Atta sexdens*) cultivate on a substrate made from the leaves, and also death of the ants. Ribeiro et al. (2017) found that aluminum (Al) reduced herbivory in *Lychnophora ericoides* (Fig. 13.2.2.1). However, this reduction happened only in a population that had a scent that varied with aluminum content.

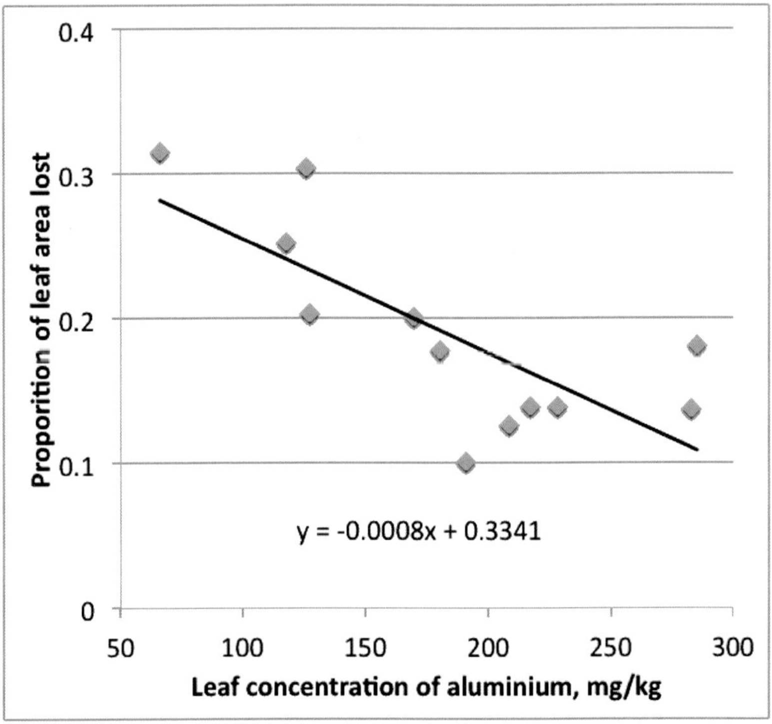

Fig. 13.2.2.1. The proportion of leaf area lost by herbivory was reduced with the increase in aluminum (Al) concentration in populations of *Lychnophora ericoides* at the Ouro Branco Mountain in the Ferriferous Quadrangle, Minas Gerais, Brazil. Modified from Ribeiro et al. (2017); plotted without arcsin transformation, as this makes very little difference.

Dueli et al. (2021) studied herbivory on the tree *Eremanthus erythropappus* (Asteraceae) in Brazil. Since here not only content of aluminum but also of copper and zinc was varied, the results are a less straightforward, but nevertheless the conclusion is that herbivory shows negative effects on plant growth if leaves have a high aluminium content.

13.2.2 Aluminium and flower colours

Trivalent aluminium cations can have a large effect on anthocyanin colouration of flowers. The best studied example is *Hydrangea macrophylla* flowers. Many researchers have written about this; most recently Yoshida et al. (2021). But already Chenery (1937) found that that in acidic soil *Hydrangea* sepals tend to be blue, while in alkaline soil the sepals become red in color. Asen et al. (1957) found that the blue colour under acid conditions is due to aluminium ions. Okada & Okawa (1974) confirmed that in acid soil the color shifts from scarlet to dark blue or mauve under the influence of aluminium ions. Houghton et al. (2021) shows the formula, with one molecule of delphinidin (an anthocyanin) and two molecules of neochlorogenic acid complexed to the aluminium ion (Fig. 13.2.3.1, Fig. 13.2.3.2). Kaliya (2020) has a similar formula, but with only one molecule of neochlorogenic acid.

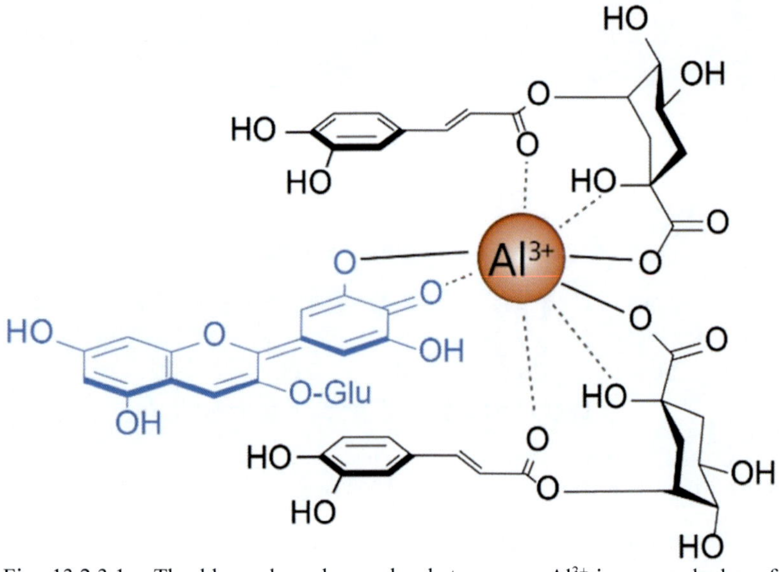

Fig. 13.2.3.1. The blue-coloured complex between an Al^{3+}-ion, a moleulce of delphinidin (an anthocyanin (Glu = glucose) and two molecules of neochlorogenic acid. Reproduced from Houghton et al. (2021) under the Creative Commons Attribution (CC BY) license (https://creativecommons.org/licenses/by/4.0/).

Fig. 13.2.3.2. Flowers of *Hydrangea macrophylla*, in both cases with low pH (3.1–3.6) in the vacuoles. The Al3+ concentration was 0.3 m*M* for the left picture, 2.4 m*M* for the right picture. Reproduced from Houghton et al. (2021) under the Creative Commons Attribution (CC BY) license (https://creativecommons.org/licenses/by/4.0/).

The article by Houghton et al. (2021) is recommended for more related information.

Another flower for which the aluminium content plays an important role for the colour of its flowers *Camellia japonica*, a relative of the tea plant. The aluminium content of the purplish red and the purple petals was found to be 4–10 times higher and 14–21 times higher than that of red petals, respectively (Tanikawa et al. 2016). Other ions that can affect the colour of flowers with anthocyanins are Fe^{3+} and molybdate (MoO_4^{2-}).

13.2.3 Aluminium and mammal reproduction

A comprehensive review of the earlier literature is provided by Yokel (2020). The main conclusions are as follows: For rodent females the exposure giving noticeable effects were more than 100-fold above the typical human daily Al consumption equivalent. Male rodents seem to be more susceptible to aluminium-induced reproductive toxicity than females. Increased Al intake increases aluminium to a greater extent in the placenta than fetus, and perhaps to a greater extent in the testes than placenta. One human study showed reduction in normal and mobile sperm by aluminium ions and positive correlations between high spermatozoa aluminium ion concentration and decreased sperm motility. According to Khalaf et al.

(2022) endoplasmic reticulum stress and mitochondrial injury are critical molecular drivers of $AlCl_3$-induced testicular and epididymal distortion and dysfunction. Even during the peripuberal period aluminium ions induce prostate damage in male rats (Leal et al. 2022). Some of the toxic effects of aluminium ions can be ameliorated by zinc oxide nanoparticles (Lokman et al. 2022).

13.2.4 Aluminium and Alzheimer's disease

Alzheimer's disease is a neurodegenerative disease that affects the brain. Usually it starts slowly, progresses over a long time, and causes about two thirds of the cases of dementia. An introduction is provided by Knopman et al. (2021). A typical sign is difficulty in remembering recent events. There is a hereditary and a sporadic version of the disease. Familial Alzheimer's disease is an early-onset and aggressive form, driven through autosomal dominant mutations in genes encoding the amyloid precursor protein and presenilins 1 and 2 (Mold et al. 2020a). Rondeau et al. (2000), after an 8-year study, concluded that a concentration of aluminum in drinking water above 0.1 mg/liter may be a risk factor of dementia and, especially, Alzheimer's disease. Although there are indications of increased brain aluminium content also in a number of other neurodegenerative diseases: congophilic amyloid angiopathy, epilepsy and autism (Exley & Mold 2020), we shall here focus on Alzheimer's disease, which is the most common cause of dementia in the elderly. Several others have later confirmed the connection between aluminium and this disease. Overviews of the early literature are provided by Flaten (2001). Many later references are provided by Colomina & Peris-Sampedro (2017) and Hara (2018).

A typical histological trait typical of Alzheimer's disease is the presence of tau-containing intracellular neurofibrillary tangles in the brain, and they have been shown to be co-localized with aluminium (Mold et al. 2020b). Tau is a protein with multiple functions in healthy brain cells; a very important one is stabilization of the microtubules. Another brain trait typical of Alzheimer patients is the presence of β-amyloid-containing plagues (Murphy & LeVine III 2010, Knopman et al. 2021), and several investigations have shown that aluminium ions associates also with it. The interaction results in the production of reactive oxygen species (Bolognin et al. 2013). Molecular details of the interaction were explored by Petersingham et al. (2022).

Dey et al. (2022) summarize the effects of exposure to aluminium ions thus: "It accumulates in various brain regions including the prefrontal cortex, hippocampus, corpus striatum, cerebellum, and brain stem.

Aluminium dysregulates the level of essential metals and forms insoluble precipitates, which gets accumulated in the exposed tissues resulting into molecular structural damage and aggregation of cellular proteins, lipids, and DNA."

Aluminium chloride is often used in water treatment as a coagulant, to reduce the number of small particles and to remove colouration. The main mechanism is that Al^{+3} ions, having a high positive electrical charge, bind to the negatively charged particles and coloured humic compounds and form connections between them. In this way larger particles are formed that can to be filtered out, and most of the aluminium is removed by filtration and sedimentation together with the particles and humic compounds. This sometimes results in increased water concentrations of aluminium ions, but if the treatment process is functioning optimally, the addition may decrease the aluminium content of the water.

Another worry is the use of aluminium for beverage containers and for cooking pots. Aluminium may be released if berries or other sour foods or drinks are heated in aluminium pots. Aluminium ions are also used in antiperspirants and cosmetics. Aluminium intoxication during haemodialysis was first noticed by Alfrey et al. (1976) and Parkinson et al. (1981).

Tietz et al. (2019) estimate that when taking into account the overall aluminium exposure from foods, cosmetic products, pharmaceuticals and FCM from uncoated aluminium, a significant exceedance of the EFSA-derived tolerable weekly intake of 1 mg/(kg body mass)/week, and even the provisional tolerable weekly intake of 2 mg/(kg body mass)/week, derived by the Joint FAO/WHO Expert Committee on Food Additives, may occur.

13.3 Gallium

Gallium occurs in solution as Ga^{3+} at pH up to about 4, and from pH=4 upwards as $Ga(OH)_4^-$. Around pH=4 it is mostly in the form $Ga(OH)^{2+}$, but to some extent also $Ga(OH)_2^+$ and $Ga(OH)_3$ (Verron et al. 2012a). It inhibits several ATPases. The inhibition of magnesium-dependent ATPase was suspected to be due to competition between Ga^{3+} and Mg^{2+}.

Gallium has a considerable effect on the skeleton. Bockman et al. (1986) report that it a potent inhibitor of bone resorption that is clinically effective in controlling cancer-related hypercalcemia as well as the accelerated bone resorption associated with bone metastases. They found that bones of gallium-treated animals had higher calcium content than control animals. Verron et al. (2010, 2012a) found, in *in vitro* experiments, that gallium reduces osteoclastic resorption, differentiation and formation without negatively affecting osteoblasts. (Osteoblasts build bone, while osteoclasts are cells that break down bone tissue). Gallium accumulates preferentially in metabolically active regions of bone (metaphysis, growth cartilage and fracture zones (literature cited by Verron et al. (2012a). Gallium inhibits many enzymes involved in bone tissue metabolism (Verron et al. 2012b). Hydroxyapatite doped with gallium is not toxic (Melnikov et al 2019), and gallium has a positive effect on reconstruction of bone (Geljic et al. 2018, Gómez-Cerezo et al. 2018).

Gallium can be used in several ways for medical treatments. There are several radioactive isotopes of the element, and since it accumulates in some malignant tumors some isotopes are suitable for combatting them. The positron emitter ^{68}Ga, decaying to ^{68}Zn with a half-life of 67.71 min (and to 11% by electron capture, also resulting in ^{68}Zn), is suitable for positron emission tomography (PET) of tumors; it is used as ^{68}Ga-labelled peptides (Maecke et al. 2005). But also stable gallium has medical use. Given as gallium maltolate it disrupts tumor iron metabolism and retards the growth of glioblastoma by inhibiting mitochondrial function and palmitoleic, stearic, linoleic, arachidic, and eicosenoic acids and threo-18-fluoro-9,10-dihydroxystearic acid (Chitambar et al. 2018). Glioblastoma is a very serious disease with a median survival time of a little over a year from diagnosis.

Gallium can also be used to combat bacteria. As with tumors, much of the effect is thought to be due to the similarity (as well as the difference) between the Ga^{3+} and Fe^{3+} ions (e.g., Olakanmi et al. 2000, Kaneko et al. 2007). The uptake system of cells cannot effectively distinguish the two ions. The ion radii are 0.620 Å for Ga^{3+} and 0.645 Å for Fe^{3+} in octahedral high spin complexes, and 0.47 Å for Ga^{3+} and 0.49 Å for Fe^{3+} in tetrahedral complexes (Shannon 1976 as cited by Kircheva & Dudev 2021). Bacterial resistance to antibiotics is a growing problem in many contexts. Valappil et al. (2012) suggest that the use of gallium–silver-doped phosphate-based glasses may offer an effective alternative to antibiotics treatments or could be used to complement current therapies. Lessa et al. (2020) discuss the use of various organic gallium compounds for fighting bacteria and cancer.

13.4 Thallium

Thallium, a metal with atomic number 81, has its place in the periodic system of elements between mercury (atomic number 80) and lead (atomic number 82). In common with its neighbours it lacks known biological functions, but is more poisonous than both, as monovalent Tl^{+} as well as in the form of Tl^{3+}. These ions lack colour, taste and odour, and are for this reason used as rat poisons as well as for homicide. As trivalent cation, Tl^{3+}, the lethal dose is 8 mg kg^{-1} body weight (Moeschlin 1980). Also as monovalent cation, Tl^+, it is poisonous, since the organism mistakes it for potassium ions, which have the same charge and ion size, and use the potassium uptake system. Tl^+ interferes with K^+-metabolism and, in many cases, with the formation of ATP in respiration (Sager 1994). Monovalent thallium is the dominant form in water solution in equilibrium with atmospheric oxygen and in the absence of complexing agents. The symptoms of thallium intoxication are described by Galván-Arzate & Santamaria (1998), but are rather diffuse, except for loss of hair. The most reliable early clue for thallium toxicity is an ascending, rapidly progressive painful sensory peripheral neuropathy, often with pain in the feet (literature in Lawrence & Kirk 2007). There is no good antidote for thallium poisoning, but Prussian blue ($Fe^{III}_4[Fe^{II}(CN)_6]_3$) may help. Thallium was found to inhibit protein synthesis and growth in human embryonic kidney cells (Chou & Lo 2019) already at a concentration of 5 ppm.

Even a single 48 h exposure to Tl^+ doses as low as 1, 10, and 100 g/L have been found to have significant effects on neuronal growth rate and morphology (Colombaioni et al. 2017, Bramanti et al. 2019). Campanella et al. (2019) have reviewed the toxicity of thallium at low doses.

For a general introduction to this element I recommend Rodríguez-Mercado (2013), Lennartson (2015) and Corzo Remigio et al. (2022a,b).

Only rarely can thallium fulfill the functions of the alkali elements that it mimics. One case where it does, is for the mammalian glutamate transporter EEAC1 (Tao et al. 2008).

Of all elements that are not needed by plants, thallium has one of the highest leaf to soil ratio. In particular many plants in the family Brassicaceae have high such ratios, some being real hyperaccumulators, i.e. reaching 0.1 mg thallium per gram shoot dry matter (Reeves et al. 2018). Three thallium hyperaccumulators in Brassicaceae are *Biscutella laevigata* (Pošćić et al. 2013, Corzo-Remigio et al. 2022b) (recorded with up to 32.7 mg Tl g^{-1} in the leaves from an Italian ecotype), Wierzbicka et al. 2016), *Iberis intermedia* (synonym of *Iberis linifolia*) (with up to 4 mg Tl g^{-1} in the leaves from a French ecotype), and *Sinapis alba* (Holubík et al. 2021, Krasnodębska-Ostręga et al. 2022). Scheckel et al. (2007) showed that in *Iberis intermedia* the thallium in the leaves is deposited in the veins. There are also thallium hyperaccumulators in other families, e.g., Caryophyllaceae, *Silene latifolia* (with up to 1.5 mg Tl g^{-1} in the leaves from a French ecotype) (Escarré et al. 2011; LaCoste et al. 1999, Fellet et al. 2012). A long list of thallium hyperaccumulating plant species has been compiled by Krasnodębska-Ostręga et al. (2022). At least in *Brassica oleracea* the thallium is accumulated primarily in the monovalent form (Jia et al. 2018). Thallium in rice fields is a considerable problem in China. According to Jiang et al. (2022) most of the thallium coming in contact with rice plants is retained before it reaches the kernels, but despite this the kernels acquire amounts that should not be neglected.

Hyperaccumulation of thallium occurs mostly in seed plants, but there are also examples of other thallium hyperaccumulators, such as a fern (Wei et al. 2020). In Brassicaceae hyperaccumulation of thallium and of other element seems to have evolved repeatedly (Manara et al. 2020).

Why have some plants evolved thallium hyperaccumulation? It is not because it is mistaken for potassium, as clearly shown by Renkema et al. (2015). Canola (*Brassica napus* L.) takes up Tl^+ much more rapidly than K^+ from a solution of similar concentration (while wheat takes up the two ions at the same rate). Although there are not solid data for this for thallium, one might speculate that the thallium protects the plant from infection, insect attack, or competitors as is the case with some other elements, such as cerium (Adisa et al. 2018), gadolinium (Batista-Oliveira et al. 2021), cadmium (Dai et al. 2020), cobalt, nickel, and selenium (Statwick et al. 2016).

Thallium is mined on a small scale only, with a production of about 15 tons per year (Rodríguez-Mercado et al. 2013) for use in electrical devices, such as superconductors (Sheng et al. 1988), as rat poison, and earlier also in medicine (the radioactive nuclide ^{210}Tl is still used for myocardial perfusion scintigraphy, one of the most used techniques for performing non-invasive cardiac imaging tests in the diagnosis of acute onset chest pain as well as in the evaluation of subjects with coronary artery disease in conditions of exercise stress or pharmacological stress to increase coronary blood flow (Genchi et al. 2021). More than a hundred times this amount is released into the environment as a byproduct of mining of other metals, and combustion of coal and oil.

However, alarmingly high amounts of thallium are released at an increasing rate into the environment and biosphere as a by-product of other mining activities. This is clearly illustrated by a comparison of the pyrite mining yield and the thallium content of tree-rings (Wang et al. 2022).

References

Adisa, I.O., Reddy, V.L., Pullagurala, R., Rawat, S., Hernandez-Viezcas, J.A., Dimkpa, C.O., Elmer, W.H., White, J.C., Peralta-Videa, J.R. & Gardea-Torresdey, J.L. 2018. Role of cerium compounds in *Fusarium* wilt suppression and growth enhancement in tomato (*Solanum lycopersicum*). J. Agric. Food Chem. 66, 5959−5970.

Alfrey, A.C., LeGendre, G.R. & Kaehny, W.D. 1976. The dialysis encephalopathy syndrome ± possible aluminium intoxication. New England J. Medicine, 294, 184–188.

Allen, R.C. 1943. Influence of aluminum on the flower color of *Hydrangea macrophylla* DC. Contr. Boyce Thompson Inst. Pl. Res. 13, 221–242.

Asen, S. and Siegelman, H.W. 1957. Effect of aluminum on absorption spectra of the anthocyanin and flavonols from sepals of *Hydrangea macrophylla* var. Merveille. Proc. Am. Soc. Hort. Sci. 70, 478–481.

Bennett, A., Rowe, R.I., Soch, N. & Eckhert, C.D 1999. Boron stimulates yeast (*Saccharomyces cerevisiae*) growth. J. Nutrition, 129, 2236–2238. https://doi.org/10.1093/jn/129.12.2236

Blue, R.M., Macho, J.M., Lee, H.-W. & MacMillan, J.B. 2022. ^{11}B and ^{1}H–^{11}B HMBC NMR as a tool for identification of a boron-containing nucleoside dimer. J. Nat. Prod. 85, 2682−2686. https://doi.org/10.1021/acs.jnatprod.2c00745

Bockman, R.S., Boskey, A.L., Blumenthal, N.C., Alcock, N.W. & Warrell Jr, R.P. 1986. Gallium increases bone calcium and crystallite perfection of hydroxyapatite. Calcified tissue Internat, 39, 376–381.

Bolaños, L., Lukaszewski, K., Bonilla, I. & Blevins, D. (2002) Plant Physiol. Biochemi. 42, 907–912.

Bolognin, S., Zatta, P., Lorenzetto, E., Valenti, M.Y. & Buffelli, M. (2013) b-Amyloid-aluminum complex alters cytoskeletal stability and increases ROS production in cortical neurons. Neurochem. Internat. 62, 566–574. Ltd. http://dx.doi.org/10.1016/j.neuint.2013.02.008

Bonilla, I., Garcia-Gonzalez, M. & Mateo, P. (1990) Boron requirement in cyanobacteria – Its possible role in the early evolution of photosynthetic organisms. Plant Physiol. 94, 1554–1560.

Bowen J.E. & Gauch, H.G. (1966) Nonessentiality of boron in fungi and the nature of its toxicity. Plant Physiol. 41, 319–324.

Bowen J.E, Kraussr, W. & Gallowary, A. (1965) The nonessentiality of boron for *Chlorella*. J. Phycol. 1, 151–154.

Branquinho, C., Serrano, H.C., Pinto, M.J. & Martins-Loucão, M.A. (2007) Revisiting the plant hyperaccumulation criteria to rare plants and earth abundant elements. Environm. Pollut. 146, 437–443. https://doi.org/10.1016/j.envpol.2006.06.034

Brenchley, W.E. & Watson, D.J. (1937) The influence of boron on the second year's growth of sugar beet affected with heart-rot.

Campanella. B., Colombaioni, L., Benedetti, E., Di Ciaula, A., Ghezzi, L., Onor, M., D'Orazio, M., Giannecchini, R., Petrini, R. & Bramanti, E. (2019) Toxicity of thallium at low doses: A review. Int. J. Environ. Res. Public Health 16, 4732 (6 pp.). https://doi.org/10.3390/ijerph16234732

Chenery, E.M. (1937) The problem of the blue hydrangea. J. R. Hort. Soc. 62, 320–604.Chitambar, C.R., Al-Gizawiy, M.M., Alhajala, H.S., Pechman, K.R., Wereley, J.P., Wujek, R., Clark, P.A., Kuo, J.S., Antholine, W.E. & Schmainda, K.M. (2018) Gallium maltolate disrupts tumor iron metabolism and retards the growth of glioblastoma by inhibiting mitochondrial function and ribonucleotide reductase. Mol. Cancer Ther. 17, 1240–1250. https://doi.org/10.1158/1535-7163.MCT-17-1009

Chou, Y.-T. & Lo, K.-Y. (2019) Thallium(I) treatment induces nucleolar stress to stop protein synthesis and cell growth. Sci. Rep. 9:6905 (14 pp.). https://doi.org/10.1038/s41598-019-43413-1

Colomina, M.T. & Peris-Sampedro, F. (2017) Aluminum and Alzheimer's Disease, pp. 183–197 *in* Neurotoxicity of metals (M. Aschner & L.G. Costa, eds.), Adv. Neurobiol., Vol. 18 (Ser. Ed. Schousboe, E.). Springer International Publ. AG. ISSN 2190-5215 ISSN 2190-5223 (electronic), ISBN 978-3-319-60188-5 ISBN 978-3-319-60189-2 (eBook). https://doi.org/10.1007/978-3-319-60189-2

Corzo Remigio, A., Nkrumah, P.N., Pošćić, F., Edraki, M., Baker, A,J.M., van der Ent, A. (2022a) Thallium accumulation and distribution in *Silene latifolia* (Caryophyllaceae) grown in hydroponics. Plant Soil in press. https://doi.org/10.1007/s11104-022-05575-2

Corzo Remigio, A., Pošćić, F., Nkrumah, P.N., Edraki, M., Spiers, K.M., Brueckner, D. & van der Ent, A. (2022b) Comprehensive insights in thallium ecophysiology in the hyperaccumulator *Biscutella laevigata*. Sci. Total Environm. 838: 155899 (18 pp.). http://dx.doi.org/10.1016/j.scitotenv.2022.155899

Coulthurst, S.J., Whitehead, N.A., Welch, M. & Salmond, G.P.C. (2002) Can boron get bacteria talking? Trends Biochem. Sci. 27, 217–219.

Cox, T.R. (1940) Relation of boron to heart rot in the sugar beet. J. Am. Soc. Agronomy 354–370.

Dai, Z.-C., Cai, H.-H., Qi, S.-S., Li, J., Zhai, D.-L., Wan, J.S.H. & Du, D.-L. (2020) Cadmium hyperaccumulation as an inexpensive metal armor against disease in Crofton weed. Environm. Pollut. 267: 115649 (9 pp.). https://doi.org/10.1016/j.envpol.2020.115649

Dey, M. & Singh, E.K. (2022) Neurotoxic effects of aluminium exposure as a potential risk factor for Alzheimer's disease. Pharmacol. Rep. 74, 439–450. https://doi.org/10.1007/s43440-022-00353-4

Delgado, M., Valle, S., Barra, P.J., Reyes-Díaz,M. & Zúñiga-Feest, A. (2019) New aluminum hyperaccumulator species of the Proteaceae family from southern South America. Plant Soil 444, 475–487. https://doi.org/10.1007/s11104-019-04289-2

Dueli, G.F., DeSouza, O. & Ribeiro, S.P (2021) Metal bioaccumulation alleviates the negative effects of herbivory on plant growth. Sci. Reports 11:19062 (10 pp.). https://doi.org/10.1038/s41598-021-98483-x

Escarré, J., Lefèbvre, C., Raboyeau, S., Dossantos, A., Gruber, W., Cleyet Marel, J.C., Frérot, H., Noret, N., Mahieu, S., Collin, C. & van Oort, F. (2011) Heavy metal concentration survey in soils and plants of the Les Malines Mining District (southern France): implications for soil restoration. Water Air Soil Pollut. 216, 485–504. https://doi.org/10.1007/s11270-010-0547-1

Estevez-Fregoso, E., Farfán-García, E.D., García-Coronel, I.H., Martínez-Herrera, E., Alatorre, A., Scorei, R.I. & Soriano-Ursúa, M.A. (2021) Effects of boron-containing compounds in the fungal kingdom. J. Trace Elem. Med. Biol. 65: 126714 (10 pp.). https://doi.org/10.1016/j.jtemb.2021.126714

Exley, C. & Mold, M.J. (2020) Imaging of aluminium and amyloid β in neurodegenerative disease. Heliyon 6: e03839 (5 pp.). https://doi.org/10.1016/j.heliyon.2020.e03839

Flaten, T. (2001) Aluminium as a risk factor in Alzheimer's disease, with emphasis on drinking water. Brain Res. Bull. 55, 187–196.

Fellet, G., Pošćić, F., Casolo, V. & Marchiol, L. (2012) Metallophytes and thallium hyperaccumulation at the former Raibl lead/zinc mining site (Julian Alps, Italy). Plant Biosyst 146, 1023–1036. https://doi.org/10.1080/11263 504.2012.703250

Fung, K.F., Carr, H.P., Zhang, J. & Wong, M.H. (2008) Growth and nutrient uptake of tea under different aluminium concentrations. J. Sci. Food Agric. 88, 1582–1591. https://doi.org/10.1002/jsfa.3254

Gerloff, G.C. (1968) The comparative boron nutrition of several green and blue-green algae. Physiol. Plantarum 21, 369–377.

Galván-Arzate, S. & Santamaría, A. (1998) Thallium toxicity. Toxicology Letters 99, 1–13.

Genchi, G., Carocci, A., Lauria, G., Sinicropi, M.S. & Catalano, A. (2021) Thallium use, toxicity, and detoxification therapy: An overview. Appl. Sci. 11, 8322 (15 pp). https://doi.org/10.3390/app11188322

Gerretsen, F.C. & de Hoop, H. (1954) Boron, an essential micronutrient for *Azotobacter chroococcum*. Plant Soil. 5, 349–367.

Gómez-Cerezo, N., Verron, E., Montouillout, V., Fayon, F., Lagadec, P., Bouler. J.M., Bujoli, B., Arcos, D. & Vallet-Regí, M. (2018) The response of pre-osteoblasts and osteoclasts to gallium containing mesoporous bioactive glasses. Acta Biomaterialia 76, 333–343. https://doi.org/10.1016/j.actbio.2018.06.036

Gonçalves Malta, P., Arcanjo-Silva, S., Cleberson Ribeiro, C., Campos, N.V. & Azevedo, A.A. (2016) *Rudgea viburnoides* (Rubiaceae) overcomes the low soil fertility of the Brazilian Cerrado and hyperaccumulates aluminum in cell walls and chloroplasts. Plant Soil 408, 369–384. https://doi.org/10.1007/s11104-016-2926-x

Hara, Y. (2018) Is there a link between aluminum and Alzheimer's? https://www.alzdiscovery.org/cognitive-vitality/blog/is-there-a-link-between-aluminum-and-alzheimers

Holubík, O., Vaněk, A., Mihaljevič, M. & Vejvodov, K. (2021) Thallium uptake/tolerance in a model (hyper)accumulating plant: Effect of extreme contaminant loads. Soil Water Res. 16, 129–135. https://doi.org/10.17221/167/2020-SWR

Houghton, A., Appelhagen, I. & Martin, C. (2021) Natural blues: Structure meets function in anthocyanins. Plants 10: 726 (22 pp.). https://doi.org/10.3390/plants10040726

Hu, H., Penn, S.G., Lebrilla, C.B. & Brown, P.H. (1997). Plant Physiol. 113, 649–655.

Huang, J., Han, R., Ji, F., Yu, Y., Wang, R., Hai, Z., Liang, W. & Wang, H. (2022) Glucose-6-phosphate dehydrogenase and abscisic acid mediate programmed cell death induced by aluminum toxicity in soybean root tips. J. Hazard. Mat 425: 127964 (11 pp.). https://doi.org/10.1016/j.jhazmat.2021.127964

Hunt, C.D. (2003) Dietary boron: An overview of the evidence for Its role in immune function. J. Trace Elements Exp. Med. 16, 291–306. https://doi.org/10.1002/jtra.10041

Ishii, T., Matsunaga, T. & Hayash, N. (2001) Formation of rhamnogalacturonan II-borate dimer in pectin determines cell wall thickness of pumpkin tissue. Plant Physiol. 126, 1698–1705.

Jansen, S., Broadley, M.R., Robbrecht, E. & Smets, E. (2002) Aluminum hyperaccumulation in angiosperms: A review of its phylogenetic significance. Botanical Rev. 68, 235–269.

Jaskowiak, J., Tkaczyk, O., Slota, M., Kwasniewska, J. & Szarejko, I. (2018) Analysis of aluminum toxicity in *Hordeum vulgare* roots with an emphasis on DNA integrity and cell cycle. PLoS ONE 13: e0193156 (18 pp.). https://doi.org/10.1371/journal.pone.0193156

Jia, Y., Xiao, T., Sun, J., Yang, F. & Baveye, P.C. (2018) Microcolumn-based speciation analysis of thallium in soil and green cabbage. Science of the Total Environment 630, 146–153. https://doi.org/10.1016/j.scitotenv.2018.02.147

Jiang, Y., Wei, X., He, H., She, J., Liu, J., Fang, F., Zhang, W., Li, Y., Wang, J., Xiao, T. & Tsang, D.C.W. (2022) Transformation and fate of thallium and accompanying metal(loid)s in paddy soils and rice: A case study from a large-scale industrial area in China. Journal of Hazardous Materials 423 (2022) 126997 (9 pp.). https://doi.org/10.1016/j.jhazmat.2021.126997

Josane A. Lessa, J.A., Parrilha, G.L. & Beraldo, H. (2012) Gallium complexes as new promising metallodrug candidates. Inorg. Chim. Acta 393, 53–63. http://dx.doi.org/10.1016/j.ica.2012.06.003

Kajiya, D. (2020) Demonstrating purple color development to students by showing the highly visual effects of aluminum ions and pH on aqueous anthocyanin solutions. J. Chem. Educ. 97, 4084–4090. https://dx.doi.org/10.1021/acs.jchemed.0c00476

Kandimalla, R., Vallamkondu, J., Corgiat, E.B & Gill, K.D. (2016) Understanding aspects of aluminum exposure in Alzheimer's disease development. Brain Pathol. 26, 139–154. https://doi.org/10.1111/bpa.12333

Kaneko, Y., Thoendel, M., Olakanmi, O., Britigan, B.E. & Singh, P.K. (2007) The transition metal gallium disrupts *Pseudomonas aeruginosa* iron metabolism and has antimicrobial and antibiofilm activity. J. Clin. Investig. 117, 877–888. https://doi.org/10.1172/JCI30783

Khalaf, H.A., Elsamanoudy, A.Z., Abo-Elkhair, S.M., Hassan, F.E., Mohie, P.M. & Ghoneim, F.M. (2022) Endoplasmic reticulum stress and mitochondrial injury are critical molecular drivers of $AlCl_3$-induced testicular and epididymal distortion and dysfunction: protective role of taurine. Histochem, Cell Biol. 158, 97–121. https://doi.org/10.1007/s00418-022-02111-2

Kircheva, N. & Dudev, T. (2021) Competition between abiogenic and biogenic metal cations in biological systems: Mechanisms of gallium's anticancer and antibacterial effect. J. Inorg. Biochem. 214: 111309 (10 pp.). https://doi.org/10.1016/j.jinorgbio.2020.111309

Knopman, D.S., Amieva, H., Petersen, R.C., Chételat, G., Holtzman, D.M., Hyman, B.T., Nixon, R.A. & Jones, D.T. (2021) Alzheimer disease. Nat. Rev. Dis. Primers. 7: 33. https://doi.org/10.1038/s41572-021-00269-y. Author manuscript available at https://www.ncbi.nlm.nih.gov/pmc/articles/PMC8574196/pdf/nihms-1749938.pdf

Kohno, J., Kawahata, T., Otake, T., Morimoto, M., Mori, H., Ueba, N., Nishio, M., Kinumaki, A., Komatsubara, S. & Kawashima, K. (1996) Boromycin, an anti-HIV antibiotic. Biosci. Biotechnol. Biochem. 60, 1036–1037.

Krasnodębska-Ostręga, B., Sadowska, M., Biaduń, E., Mazur, R. & Kowalska, J. (2022) *Sinapis alba* as a useful plant in bioremediation – studies of defense mechanisms and accumulation of As, Tl and PGEs. Internat. J. Phytoremediation 24, 1475–1490. https://doi.org/10.1080/15226514.2022.2036098

LaCoste, C., Robinson, B., Brooks R, Anderson, C., Chiarucci, A. & Leblanc, M. (1999) The phytoremediation potential of elements and associated health risk assessment in the industrial area of Volos, Greece. Environ. Int. 124, 79–88. https://doi.org/10.1016/j.envint.2018.12.053

Lawrence, D.T. & Kirk, M.A. (2007) Chemical terrorism attacks: Update on antidotes. Emerg Med. Clin. N. Am. 25, 567–595. https://doi.org/10.1016/j.emc.2007.02.002

Leal, F.A.V.D., Gonçalves, G.D., Soncini, J.G.M., Staurengo-Ferrari, L., Fattori, V., Verri Jr, W.A., Scarano, W.R. & Fernandes, G.S.A. (2022) Exposure to aluminium chloride during the peripuberal period induces prostate damage in male rats. Acta Histochem. 124: 151843 (7 pp.). https://doi.org/10.1016/j.acthis.2022.151843

Lennartson, A. (2015) Toxic thallium. Nature Chem. 7, 610. https://doi.org/10.1038/nchem.2286

Lewin, J.C. (1965) The boron requirement of a marine diatom. Naturwiss. 52, 70–70.

Lewin, J. (1966) Boron as a growth requirement for diatoms. J. Phycol. 2, 160–163).

Liu, C., Sun, D., Zheng, H.-X., Wang, G.-B., Liu, W.-S., Cao, Y., Tang, Y.-T. & Qiu, R.L. (2022) The limited exclusion and efficient translocation mediated by organic acids contribute to rare earth element hyperaccumulation in *Phytolacca americana*. Sci. Total Environm. 805: 150335 (10 pp). https://doi.org/10.1016/j.scitotenv.2021.150335

Liu, W.-S., Zheng, H.-X., Liu, C., Guo, M.N., Zhu, S.C., Cao, Y., Qiu, R.-L., Morel, J.L., van der Ent, A., & Tang, Y.-T. (2021) Variation in rare earth element (REE), aluminium (Al) and silicon (Si) accumulation among populations of the hyperaccumulator *Dicranopteris linearis* in southern China. Plant Soil 461, 565–578. https://doi.org/10.1007/s11104-021-04835-x

Liu, Y., Tao, J., Cao, J., Zeng, Y., Li, X., Ma, J., Huang, Z., Jiang, M. & Sun, L. (2020) The beneficial effects of aluminum on the plant growth in *Camellia japonica*. J. Soil Sci. Plant Nutrit. 20. 1799–1809. https://doi.org/10.1007/s42729-020-00251-9

Lokman, M., Ashraf, E., Kassab, R.B., Moneim, A.E.A. & El-Yamany, N.A. (2022) Aluminum chloride–induced reproductive toxicity in rats:

the protective role of zinc oxide nanoparticles. Biolog. Trace Element Res. 200, 4035–4044. https://doi.org/10.1007/s12011-021-03010-8

Maecke, H.R., Hofmann, M. & Haberkorn, U. (2005) ^{68}Ga-labeled peptides in tumor imaging. J. Nuclear Med. 46, 172S–178S.

McLachlan, J. (1976) Effects of nutrients on growth anddevelopment of embryos of *Fucus edentatus* Pyl. (Phaeophyceae, Fucales). Phycologia 16, 329–338. https://doi-org.ludwig.lub.lu.se/10.2216/i0031-8884-16-3-329.1

Manara, A., Fasani, E., Furini, A. & DalCorso, G. (2020) Evolution of the metal hyperaccumu-lation and hypertolerance traits. Plant Cell Environ. 43, 2969–2986. https://doi.org/10.1111/pce.13821

Mateo, P., Bonilla, I., Fernández-Valiente, E. & Sanchez-Maeso, E. (1986) Essentiality of boron for dinitrogen fixation in *Anabaena sp*. PCC 7119. Plant Physiol. 81, 430–433.'

Milanez, C.R.D., Marcati, C.R. & Rodrigues Machado, S. (2017) Trabeculae and Al-accumulation in the wood cells of Melastomataceae species from Brazilian savanna. Botany 95, 521–530. https://doi.org//10.1139/cjb-2016-0135

Melnikov, P., de Fatima Cepa Matos, M., Malzac, A., Teixeira, A.R. & Muniz de Albuquerque, D. (2019) Evaluation of in vitro toxicity of hydroxyapatite doped with gallium. Materials Lett. 253, 343–345. https://doi.org/10.1016/j.matlet.2019.06.095

Miranda de Andrade, L.R., Barros, L.M.G., Echevarria, G.F., Velho do Amaral, L.I., Cotta, M.G., Rossatto, D.R., Haridas, M. & Franco, A.C. (2011) Al-hyperaccumulator Vochysiaceae from the Brazilian Cerrado store aluminum in their chloroplasts without apparent damage. Environm. Exp. Bot. 70, 37–42. https://doi.org/10.1016/j.envexpbot.2010.05.013

Moeschlin, S. (1980) Thallium poisoning. Clin. Toxicol. 17, 133–146. https://doi.org/10.3109/15563658008985073

Mold, M., Linhart, C., Gomez-Ramirez, J., Villegas-Lanau, A. & Exley, C. (2020a) Aluminum and amyloid-beta in familial Alzheimer's disease. J. Alzheimers Dis. 73, 1627–1635. https://doi.org/10.3233/JAD-191140

Mold, M.J., O'Farrell, A., Morris, B. & Exley, C. (2020b) Aluminum and neurofibrillary tangle co-localization in familial Alzheimer's disease and related neurological disorders. J. Alzheimer's Dis. 78, 139–149. https://doi.org/10.3233/JAD-200838

Moreno-Alvarado, M., García-Morales, S., Trejo-Téllez, L.I., Hidalgo-Contreras, J.V. & Gómez-Merino, F.C. (2017) Aluminum enhances growth and sugar concentration, alters macronutrient status and regulates the expression of NAC transcription factors in rice. Front. Plant Sci. 8:73 (16 pp.). https://doi.org/10.3389/fpls.2017.00073

Murphy, M.P & LeVine III, H. (2010) Alzheimer's disease and the β-amyloid peptide. J. Alzheimers Dis. 19: 311 (17 pp.). doi:10.3233/JAD-2010-1221.

Olakanmi, O., Britigan, B.E. & Schlesinger, L.S. (2000) Gallium disrupts iron metabolism of mycobacteria residing within human macrophages. Infect. Immun. 68, 5619–5627.

Ozturk, M., Sakcali, S, Gucel, S. & Tombuloglu, H. (2010) Boron and plants. Ch.16, pp. 275–311 in M. Ashraf et al. (eds.), Plant Adaptation and Phytoremediation, Springer. https://doi.org/10.1007/978-90-481-9370-713,

Parkinson, I.S., Ward, M.K. & Kerr, D.N.S. (1981) Dialysis encephalopathy, bone disease and anaemia: the aluminium intoxication syndrome during regular haemodialysis. J. Clin. Pathol. 34, 1285–1294.

Petersingham, G., Zaman, M.S., Johnson, A.J., Reddy, N., Torres, A.M. & Wu, M.J. (2022) Molecular details of aluminium-amyloid β peptide interaction by nuclear magnetic resonance. Biometals 35, 759–769. https://doi.org/10.1007/s10534-022-00399-0

Pizzorno, L. (2015) Nothing boring about boron. Integrative Medicine 14, 35–48.

Poščić, F., Marchiol, L. & Schat, H. (2013) Hyperaccumulation of thallium is population-specific and uncorrelated with caesium accumulation in the thallium hyperaccumulator, *Biscutella laevigata*. Plant Soil 365, 81–91. https://doi.org/10.1007/s11104-012-1384-3

Reeves, R.D., Baker, A.J.M., Jaffré, T., Erskine, P.D., Echevarria, G. & van der Ent, A. (2018) A global database for plants that hyperaccumulate metal and metalloid trace elements. New Phytol. 218, 407–411. https://www.jstor.org/stable/10.2307/90019919

Rebula, C.A., Haridasan, M., Kitayama, K. & Morais, HC. (2008) Toxic effects of leaves of an aluminum-accumulating plant on laboratory colonies of *Atta sexdens* (Hymenoptera: Formicidae). Sociobiology 52, 13–24.

Renkema, H., Koopmans, A., Hale, B. & Berkelaar, E. (2015) Thallium and potassium uptake kinetics and competition differ between durum wheat and canola. Environ. Sci. Pollut. Res. 22, 2166–2174. https://doi.org/10.1007/s11356-014-3490-7

Rodríguez-Mercado, J.J. & Altamirano-Lozano, M.A. (2013) Genetic toxicology of thallium: A review. Drug Chem. Toxicol. 36, 369–383.

Rondanelli, M., Faliva, M.A., Peroni, G., Infantino, V., Gasparri, C., Iannello, G., Perna, S., Riva, A., Petrangolini, G. & Tartara, A. (2020)

Pivotal role of boron supplementation on bone health: A narrative review. J. Trace Elem. Med. Biol. 62: 126577 (11 pp.). https://doi.org/10.1016/j.jtemb.2020.126577

Rondeau, V., Commenges, D., Jacqmin-Gadda, H. & Dartigues, J.F. (2000b) Relation between aluminum concentrations in drinking water and Alzheimer's disease: An 8-year follow-up study. Am. J. Epidemiol. 152, 59–66. https://doi.org/10.1093/aje/152.1.59

Rowe, E.A. (1936) A study of heart-rot of young sugar-beet plants grown in culture solutions. Ann. Bot. 50, 735–746.

Sager, M. (1994) Thallium. Toxicol. Environ. Chem. 45, 11–32. https://doi.org/10.1080/02772249409358067)

Santos Batista-Oliveira, J., Formey, D., Torres, M., Aragón, W., Romero-Contreras, Y.J., Maruri-López, I., Tromas, A., Schwan-Estrada, K.R.F. & Serrano, M. (2021) Gadolinium protects *Arabidopsis thaliana* against *Botrytis cinerea* through the activation of JA/ET-induced defense responses. Int. J. Mol. Sci. 22, 4938 (20 pp.). https://doi.org/10.3390/ijms22094938

Scheckel, K.G., Hamon, R., Jassogne, L., Rivers, M. & Lombi, E. (2007) Synchrotron X-ray absorption-edge computed microtomography imaging of thallium compartmentalization in *Iberis intermedia*. Plant Soil 290, 51–60. https://doi.org/10.1007/s11104-006-9102-7

Schmitt, M., Watanabe, T. & Jansen, S. (2016) The effects of aluminium on plant growth in a temperate and deciduous aluminium accumulating species. AoB PLANTS 8: plw065. https://doi.org/10.1093/aobpla/plw065

Schmitt, M., Mehltreter, K., Sundue, M., Testo, W., Watanabe, T. & Jansen, S. (2017) The evolution of aluminum accumulation in ferns and lycophytes. Am. J. Bot. 104, 573 – 583. https://doi.org/10.3732/ajb.1600381

Serrano, H.C., Cotrim, H., Pinto, M.J., Martins-Loução, M.A. & & Branquinho, C. (2017) Metal hyperaccumulation patterns within *Plantago* phylogeny (Plantaginaceae). Plant Soil 411, 227–241. https://doi.org/10.1007/s11104-016-3024-9

Shalehin, N., Hosoya, A., Takebe, H., Hasan, M.R. & Irie, K. (2020) Boric acid inhibits alveolar bone loss in rat experimental periodontitis through diminished bone resorption and enhanced osteoblast formation. J. Dental Scie. 15, 437–444. https://doi.org/10.1016/j.jds.2019.09.009

Shannon, R.D. (1976) Revised effective ionic radii and systematic studies of interatomic distances in halides and chalcogenides, Acta Crystallogr. Sect. A. 32, 751–767.

Sheng, Z.Z., Hermann, A.M., El Ali, A., Almasan, C., Estrada, J. & Datta, T. (1988) Superconductivity at 90 K in the Tl-Ba-Cu-0 System. Phys, Rev. Lett. 60, 937–940.

Shorrocks, V.M. (1997) The occurrence and correction of boron deficiency. Plant Soil 193, 121–148.

Sojan, J.M., Gundappa, M.K., Carletti, A., Gaspar, V., Gavaia, P., Maradonna, F. & Carnevali, O. (2022) Zebrafish as a model to unveil the pro-osteogenic effects of boron-vitamin D_3 synergism. Frontiers Nutrition 9: 868805 (15 pp.). https://doi.org/10.3389/fnut.2022.868805

Statwick, J., Majestic, B.J. & Sher, A.A. (2016) Characterization and benefits of selenium uptake by an *Astragalus* hyperaccumulator and a non-accumulator. Plant Soil 404, 345–359. https://doi.org/10.1007/s11104-016-2842-0

Strazic Geljic, I., Melis, N., Boukhechba, F., Schaub, S., Mellier, C., Janvier, P., Laugier, J.-P., Bouler, J.-M., Verron, E. & Scimeca, J.-C. (2018) Gallium enhances reconstructive properties of a calcium phosphate bone biomaterial. J. Tissue Eng. Regen. Med. 12, e854-e866. https://doi.org/10.1002/term.2396

Sun, L., Zhang, M., Liu, X., Mao. Q., Shi. C., Kochian, L.V. & Liao, H. (2020) Aluminium is essential for root growth and development of tea plants (*Camellia sinensis*). J. Integrative Plant Biol. 62, 984–997. https://doi.org/10.1111/jipb.12942

Tanikawa, N., Inoue, H. & Nakayama, M. (2016) luminum Ions Are Involved in Purple Flower Coloration in *Camellia japonica* 'Sennen-fujimurasaki'. Horticult. J. 85, 331–339. https://doi.org/10.2503/hortj.MI-114

Tao, Z., Gameiro, A. & Grewer, C. (2008) Thallium ions can replace both sodium and potassium ions in the glutamate transporter EAAC1. Biochemistry 47, 12923–12930. https://doi.org/10.1021/bi8017174.

Tietz, T., Lenzner, A., Kolbaum, A.E. and 18 more authors (2019) Aggregated aluminium exposure: risk assessment for the general population. Arch. Toxicol. 93, 3503–3521. https://doi.org/10.1007/s00204-019-02599-z

Timpone, L.T. & Habermann, G. (2022) Is aluminum (Al) eliminated by senescent structures of *Miconia albicans*, an Al-accumulating species from Brazilian savanna? Flora 289: 152036 (8 pp.). https://doi.org/10.1016/j.flora.2022.152036

Valappil, S.P., Coombes, M., Wright, L., Owens, G.J., Lynch, R.J.M., Hope, C.H. & Higham, S.M. (2012) Role of gallium and silver from phosphate-based glasses on in vitro dual species oral biofilm models of

Porphyromonas gingivalis and *Streptococcus gordonii*. Acta Biomaterialia 8, 1957–1965. https://doi.org/10.1016/j.actbio.2012.01.017

van der Ent, A., Kopittke, P.M., Paterson, D.J., Casey, L.W. & Nkrumah, P.N. (2020) Distribution of aluminium in hydrated leaves of tea (*Camellia sinensis*) using synchrotron- and laboratory-based X-ray fluorescence microscopy. Metallomics 12, 1062–1069. https://doi.org/10.1039/c9mt00300b

Verron, E., Masson, M., Khoshniat, S., Duplomb, L., Wittrant, Y., Baud'huin, M., Badran, Z., Bujoli, B., Janvier, P., Scimeca, J.-C., Bouler, J.-M. & Guicheux, J. (2010) Gallium modulates osteoclastic bone resorption in vitro without affecting osteoblasts. British J. Pharmacol. 159, 1681–1692.

Verron, E., Bouler, J.M. & Scimeca, J.C. (2012a) Gallium as a potential candidate for treatment of osteoporosis. Drug Discovery Today 17, 1127–1132. http://dx.doi.org/10.1016/j.drudis.2012.06.007

Verron, E., Loubat, A., Carle, G.F., Vignes-Colombeix, C., Strazic, I., Guicheux, J., Rochet, N., Bouler, J.M. & Scimeca, J.C (2012b) Molecular effects of gallium on osteoclastic differentiation of mouse and human monocytes. Biochem. Pharmacol. 83, 671–679. https://doi.org/10.1016/j.bcp.2011.12.015

Vielma, J.R., Picón-Borregales, D., Vergara, M.A., Carrero, P.E. & Gutiérrez- Peña. L.V. (2017) El boro, un elemento benéfico que ayuda a prevenir la osteoporosis en el humano: una revisión de literatura. Avances en Biomedicina 6, 216–233.

Vogt (1986) Water quality and health: Study of a possible relation between Al in drinking water and dementia. Sosiale og Økonomiske Studier. Oslo: Statistics Norway.61.

Vondráčková, S., Száková, J., Drábek, O., Tejnecký, V., Hejcman, M., Müllerovà, V. & Tlustoš, P. (2015) Aluminium uptake and translocation in Al hyperaccumulator *Rumex obtusifolius* is affected by low-molecular-weight organic acids content and soil pH. PLoS ONE 10: e0123351 (18 pp.). https://doi.org/10.1371/journal.pone.0123351

Wang, J., Huang, Y., Beiyuan, J., Wei, X., Qi, J., Wang, L., Fang, F., Liu, J., Cao, J. & Xiao, T. (2022) Thallium and potentially toxic elements distribution in pine needles, tree rings and soils around a pyrite mine and indication for environmental pollution. Sci. Total Environm. 828: 154346 (9 pp.). http://dx.doi.org/10.1016/j.scitotenv.2022.154346

Weerasinghe, A.J., Amin, S.A., Barker, R.A., Othman, T., Romano, A.N., Parker Siburt, C.J., Tisnado, J., Lambert, L.A., Huxford, T., Carrano, C.J. & Crumbliss, A.L. (2013) Borate as a synergistic anion for

Marinobacter algicola ferric binding protein, FbpA: A role for boron in iron transport in marine life. J. Am. Chem. Soc. 135, 14504−14507. https://doi.org//10.1021/ja406609s

Wei, X., Zhou, Y., Tsang, D.C.W., Song, L., Zhang, C., Yin, M., Liu, J., Xiao, T., Zhang, G. & Wang, J. (2020) Hyperaccumulation and transport mechanism of thallium and arsenic in brake ferns (*Pteris vittata* L.): A case study from mining area. J. Hazardous Materials 388: 121756 (10 pp.). https://doi.org/10.1016/j.jhazmat.2019.121756

Wierzbicka, M., Pidlicowska, M., Abratowska, A., WiŁkomirska, B., Wyscocka, I., Panufnik-MĘdrzycka. D. & Bulska, E. (2016) Thallium hyperaccumulation in Polish populations of *Biscutella laevigata* (Brassicaceae). Acta Biol. Cracoviensia Ser. Botanica 58/2: 7–19. https://doi.org/10.1515/abcsb-2016-0012

Yokel, R.A, (2020) Aluminum reproductive toxicity: a summary and interpretation of scientific reports. Crit. Revs Toxicol. 50, 551–593. https://doi.org/10.1080/10408444.2020.1801575

Yoshida, K., Oyama, K. & Kondo, T. (2021) Insight into chemical mechanisms of sepal color development and variation in hydrangea. Proc. Jpn. Acad., Ser. B 97, 51–68. https://doi.org/10.2183/pjab.97.003

Yoshinari, A. & Takano, J. (2017) Insights into the mechanisms underlying boron homeostasis in plants. Front. Plant Sci. 8: 1951 (8 pp.). https://doi.org/10.3389/fpls.2017.01951

Xu, B., Dong, F., Yang, P., Wang, Z., Yan, M., Fang, J. & Zhang, Y. (2023) Boric acid inhibits RANKL-stimulated osteoclastogenesis in vitro and attenuates LPS-induced bone loss in vivo. Biol. Trace Elem. Res. 201, 1388–1397 https://doi.org/10.1007/s12011-022-03231-5

Zaia, M., Timpone, L.T. & Habermann, G. (2022) Do aluminum (Al)-accumulating species from the Brazilian savanna accumulate Al in the roots? Trees 36, 1677–1685.https://doi.org/10.1007/s00468-022-02301-4

Zeng, Q. L., Chen, R. F., Zhao, X. Q., Wang, H. Y. and Shen, R. F. (2011) Aluminium uptake and accumulation in the hyperaccumulator *Camellia oleifera* Abel. Pedosphere 21, 358–364.

Zhu, D., Abdur R., Xiao, K., Wang, W., Wang, L., Qiu, W. Zheng, X., Song, H., (Liu, H., Zhong, J. & Kemei, P. (2021) Boron supplementation promotes osteogenesis of tibia by regulating the bone morphogenetic protein-2 expression in African ostrich chicks. Biol. Trace Element Res. 199, 1544–1555.
https://doi.org/10.1007/s12011-020-02258-w

CHAPTER FOURTEEN

SILICON, GERMANIUM, AND LEAD

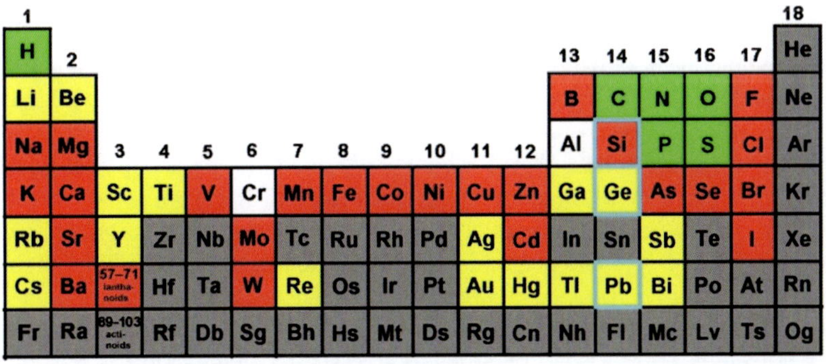

14.1 Silicon

Silicon is, second only to oxygen, the most abundant element in the Earth's crust. Compared to its neighbour carbon in column 14 of the periodic table it is of less, but far from negligible biologic importance. Because it mostly remains in its inorganic form, not forming characteristic compounds, its role at the molecular level has been hard to pinpoint. It is not known to be an essential element for many organisms. Diatoms and siliceous sponges (Uriz 2003, Müller et al. 2007) are important exceptions; nevertheless silicon is of great biological importance. Wang et al. (2012) describe how the siliceous spicules of sponges are formed, as well as the proteins involved.

It is thought that early ocean contained more dissolved silicon as silicic acid, $Si(OH)_4$, than what is the case today. From the Cambrian to the Triassic radiolarians and siliceous sponges slowly depleted the ocean of dissolved silicon (Conley et al. 2017, Ye et al. 2021), and then, accelerating throughout the Cretaceous, depletion by diatoms took over. The silicon is taken up in the form of silicic acid (Annenkov & Gordon 2020), and deposited essentially as SiO_2.

Diatoms are currently the dominating silicon-storing organisms. Radiolarians and siliceous sponges are still important. Other aquatic organisms with siliceous skeletons are the Chrysophyta (e.g., Gusev & Martynenko 2022, Fig. 14.1.1). Silicon is also accumulated by some macroalgae, especially the red algae *Cystoclonium* and *Polysiphonia* (Yacano et al. 2022).

But also some bacteria precipitate silicon, either intracellularly or extracellularly (Ikeda 2021). A silicon-containing layer of nanometer-sized particles in the spore-coat of several *Bacillus* strains, including *Bacillus cereus*, have been shown to increase the resistance of the spores to low pH (Hirota et al. 2010). Intracellular silicon-accumulation has been shown for magnetotactic bacteria (Li et al. 2022) and the cyanobacterium *Synechococcus* (Baines et al. 2012, Ohnemus et al. 2018, Wei et al. 2022).

On land the dominating silicon accumulators are horsetails (*Equisetum*), grasses, and plants belonging to Cyperaceae and Commelinaceae have relatively large silicon content of about 4% (Hodson et al. 2005, Currie and Perry 2007). Also other plants, such as *Selaginella* (Corrêa Lopes & Feio 2020), *Isoëtes echinospora* (Polevova & Moiseenko 2022) or the date palm, *Phoenix dactylifera* (Bokor et al. 2019), may deposit hydrated silica, an intermediate between silicic acid and silicon dioxide, as phytoliths in their cells. The fern *Dicranopteris linearis* accumulates silicon and aluminium, and precipitates them together (Liu et al. 2019). With the possible exception for horsetails, silicon is not regarded as an essential element, but as Epstein (2009) has forcefully argued, this does not mean that it is of no importance to plants.

One gets a first indication of its importance from the observation that plants may completely deplete a nutrient solution of silicon (Epstein 2009).

The older literature on the roles of silicon in plants is reviewed by Epstein (2009). No silicon-containing enzymes or other complex silicon compounds are known in plants, so why do plants take up silicon? Is there any advantage in containing phytoliths and other forms of silicon? As Islam et al. (2020) state: In many cases "the underlying Si-mediated mode of action to enhance plant resistance to disease and insect pests is unclear". But several different advantages have been demonstrated for various plants. In some cases the mechanical effect of phytoliths, making plants more difficult to chew, and wearing teeth or mandibles down may be important (Jeer et al. 2017). For instance, silicon offers some protection against moth (*Cnaphalocrocis medinalis*) larvae feeding on rice (Rizwan et al. 2022). But some insects are specialized to deal with this defense, e.g., diptera of the genus *Pegomya* on horsetail (Michelsen & Palmer 2020). In rice silicon supplementation alters the composition of herbivore induced plant volatiles and enhances attraction of parasitoids (the enemies' enemies) to infested plants (Liu et al. 2017). Palatability to insects is probably reduced by silicon-containing trichomes; they may hurt thin-skinned caterpillars (Hall et al. 2020). Silicon may also alleviate drought, salt stress, and UV stress in plants (Rizwan et al. 2015, Čermelj et al. 2021). The concentration of silicon (as K_2SiO_3) in the nutrient solution greatly influenced the uptake of some other ions. With increasing silicon in the medium, the contents of Si and K increased, while that of Ca decreased in both roots and shoots of barely plants, and the contents of Fe, Mn, and Zn decreased in the roots (Čermelj et al. 2022). Although it appears clear that silicon protects against a number of abiotic stresses, infections and herbivores, the mechanisms for these effects are not clear (Coskun et al. 2019).

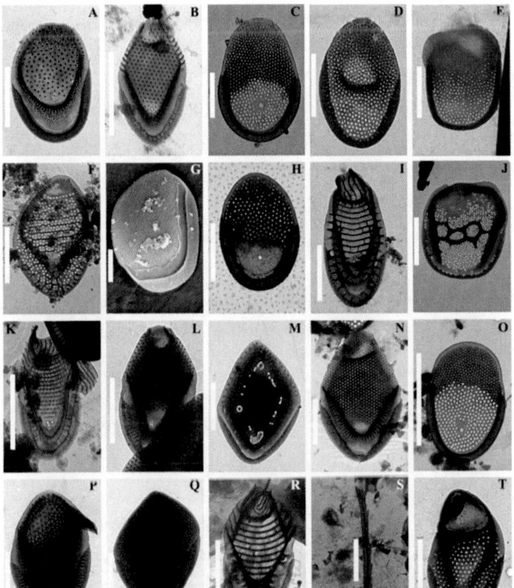

Fig. 14.1.1. Scales and bristles of members of the genus *Mallomonas*, Chrysophyta. A *M. cf. ouradion*, B *M. papillosa*, C *M. paragrandis*, D *M. peronoides*, E *M. plumosa*, F *M. portae-ferreae*, G *M. pseudobronchartiana*, H *M. pseudomatvienkoae*, I *M. pseudocratis*, J *M. punctifera*, K *M. punctostriata*, L *M. rasilis*, M *M. sidereal*, N *M. cf. skvortsovii*, O *M. sorohexareticulata*, P *M. spinosa*, Q *M. splendens*, R,S *M. striata var. serrata*, T *M. tonsurata*. Scale bars: 0: 5 μm; (A–L,N,P–T): 2 μm; (M): 1 μm. TEM: (A–F,H–T); SEM: (G). From Gusev & Martynenko (2022). Reproduced under the Creative Commons Attribution (CC BY) license (https:// creativecommons.org/licenses/by/4.0/)

Silicon is required for skeletal and connective tissue health in humans and other mammals (for older literature se Refitt et al. 2003). Orthosilicic acid stimulates collagen type 1 synthesis and osteoblastic differentiation in human osteoblast-like cells in vitro (Refitt et al. 2003). Jugdaohsingh (2007) states that in the Western world "we ingest between 20–50 mg/day, greater than two-fold our intake of iron and zinc". After reviewing eight studies, Rondanelli et al. (2021) find that "an adequate intake in order to promote beneficial effects for bone could be considered to be around 25 mg silicon/day". They further write, however, that "there are still no precise indications regarding a possible role of silicon on bone health in humans.""

A warning against silicon consumption is given by Titov et al. (2019) who write that silicon dioxide of plant origin can penetrate into the human body unchanged and be transferred through the blood to the heart, where it can cause mineralized heart valves.

Aluminium is a very common, and biologically mostly useless, element in the environment, and especially at low pH it may intoxicate both animals (e.g., fish) and plants. Silicon may alleviate this toxicity (Jiang et al. 2022).

14.2 Germanium

Germanium dioxide (GeO_2) has been used for many years in the cultivation of red and green, and brown algae as a means of controlling the growth of diatoms (Tarakhovskaya et al. 2012, Kerrison et al. 2016). Diatoms are sensitive to germanium dioxide, but red and green algae seem to be more resistant. Brown algae are inhibited or even damaged (Fig. 14.2.1) by high concentration, but their growth may be stimulated by low concentrations of germanium (Shea & Chopin 2007).

Use of germanium dioxide has been proposed for preventing "biofouling" of underwater constructions not only by diatoms, but also solitary and colonial ascidians, tubeworms and barnacles (Pérez et al. 2015).

The effects of germanium appear somewhat enigmatic. Germanium follows after carbon and silicon in the periodic table of elements as it is usually displayed. Therefore one might suspect that germanium would interfere with the use of silicon. At least some brown algae take up silicon (Mizuta & Yasui 2012), but at least some brown algae are stimulated by germanium, while others are inhibited if exposed to it for a longer time (Praeger et al. 2022) Also on certain red algae silicon has a positive effect (Le et al. 2019), but red algae are only affected by germanium at very high concentration (Tatewaki and Mizuno 1979 as cited by Tarakhovskaya et al. 2012).

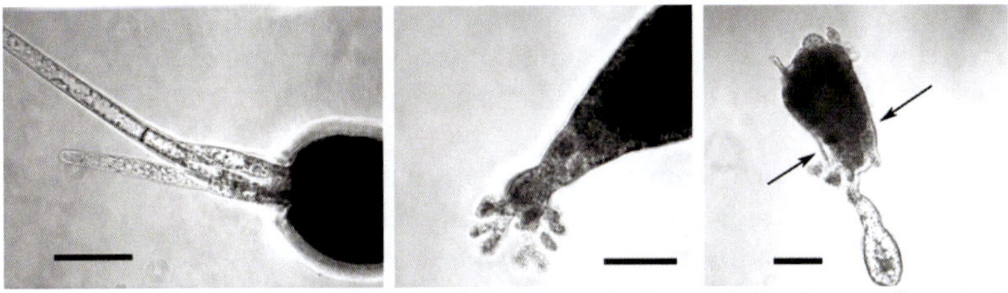

Fig. 14.2.1. Developing embryos of *Fucus vesiculosus*. Left: Apex of control embryo at 8 days with three normal apical hairs. Centre: Base of 8 days control embryo with typical rhizoid development. Right: Whole 8 days GeO_2 (5 mg/L) treated embryo with irregular shape, inflated rhizoid tip, poor development of apical hairs and secondary rhizoid development from thallus base; arrows indicate common point of breakage of GeO_2 treated embryos. Scale bar 50 μm. Reproduced from Tarakhovskaya et al. (2012) under the Creative Commons Attribution Non-Commercial License (http://creativecommons.org/licenses/by-nc/3.0/).

Marron et al. (2016) describe how germanium, supplied as GeO_2, interferes with silicate in formation of the silicate shells (loricae). The effects can be ameliorated by increasing the concentration of silicate in the medium.

A number of germanium compounds have been proposed for various medical problems, but we shall not deal with that here.

14.3 Lead

14.3.1 Not required, only toxic

Lead is not known to be a required element for any organism. Instead it is very poisonous to most organisms. In humans lead as a cumulative toxicant affects cardiovascular, renal, neurological, skeletal, hematological, immune, respiratory, gastrointestinal, reproductive, and endocrine systems (Mitra et al. 2017). A significant source of lead intake is drinking water. For toxicity to humans, see also https://www.who.int/news-room/fact-sheets/detail/lead-poisoning-and-health.

The prime target to lead toxicity in both plants and animals is the heme synthesis enzyme δ-aminolevulinic acid dehydratase (porphobilinogen synthase). It forms a precursor of heme, chlorophylls, cytochromes, and vitamin B_{12}. Other affected molecules are thiol-containing antioxidants and enzymes: superoxide dismutase, catalase, glutathione peroxidase, glucose 6-phosphate dehydrogenase and antioxidant molecules like glutathione (Nemsadze et al. 2009). Part of the toxicity of lead results from its ability to substitute for calcium and zinc, and to a lesser extent magnesium in various proteins, without fulfilling the functions of these metal ions. Lead affects nerve action by decreasing the amplitude of calcium currents (Garza et al. 2006). Other mechanisms involve damage of nucleic acids and induction of oxidative stress.

Some plants and microorganisms accumulate lead; possibly it gives some protection from being eaten. Very high amounts are accumulated in both root and shoot of the shrub *Sesbania drummondii*, Fabaceae (Sahi et al. 2002, Fig. 14.3.1).

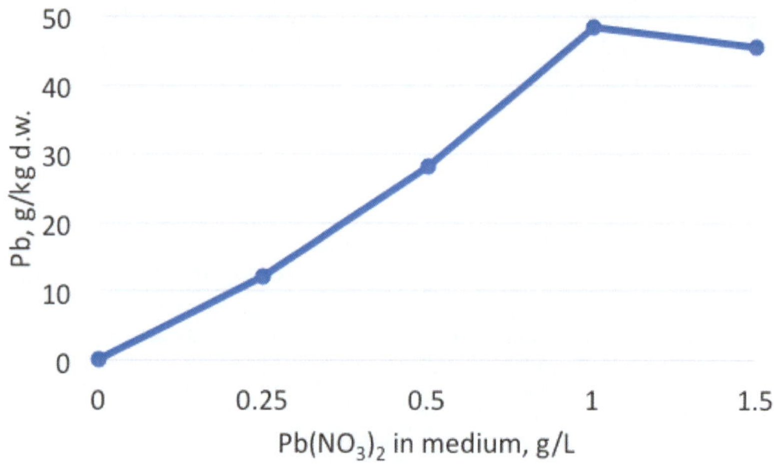

Fig. 14.3.1. The concentration of lead in the shoot of *Sesbania drummondii* grown in media with various concentrations of lead nitrate. To convert medium concentration to g/L Pb, multiply by 0.626. Data from Sahi et al. (2002).

Also some other plants accumulate considerable amounts, e.g., the common *Plantago major*, which can have over 9 g/kg in the roots (Romeh et al. 2016) and the "centipede grass", *Eremochloa ophiuroides*, which can reach over 5 g/kg (Pu et al. 2022). The photosynthetic flagellate *Euglena gracilis* can be considered a hyperaccumulator of lead. After five days exposure to Pb^{2+} it contained 4.6 g Pb^{2+} per kg dry matter (Hernández-Garnica et al. 2021).

Lead poisoning of the grass *Lolium perenne* is counteracted by the lead-resistant fungus *Trichoderma asperellum* (Sun et al. 2021), and the fungus *Funneliformis mosseae*, forming arbuscular mycorrhiza with *Sophora viciifolia* (Fabaceae) helps its host in the same way (Xu et al. 2016). The bacterium *Pseudomonas fluorescence* decreased the uptake of lead by pea (*Pisum sativum*) plants (Shabaan et al. 2021). For instance, when the soil had been amended with 750 mg lead per kg, the lead content of roots was decreased from 20 to 5 mg/kg by the bacteria, that of the shoots from 17 to 4 mg/kg, and that of the grains from 9.5 to 2 mg/kg. Bacterial tolerance of lead has been reviewed by Mitra et al. (2021).

The rest of this chapter will focus on the relation of lead to humans, and its toxicity to humans.

14.3.2 Ancient history

The first reference to lead toxicity can be traced back to Egyptian medical papyri. Apparently, it was often used with homicidal intentions at that time (Montes-Santiago 2013). This author also gives examples of how lead products were in the past used in dangerous ways without intention to harm. For instance, "lead sugar", i.e., lead acetate, was used as a sweetener and preservative for wine. The physician Eberhard Gockel noticed that monks who drank no wine seemed healthy, while those who drank were ill with classic abdominal colic pain. Gockel discovered that the addition of a sweetener, litharge or white lead oxide to the wine, was responsible. Lead acetate was used also much more recently for treatment of various medical conditons: cholera, dysentery and other diarrheal diseases, typhoid pneumonia, nosebleeds, abdominal swelling, trachoma, bronchitis, pneumonia, yellow fever, uterine hemorrhage, aortic aneurism, epilepsy, chorea, hydrophobia, heavy menstrual flow, tuberculosis (literature cited by Jonasson & Afshari (2018).

14.3.3 Lead and the Romans

The old Romans were aware of the toxicity of lead, at least the toxicity of "cerussa", i.e., lead carbonate, $PbCO_3$ (Hodge 1981). Perhaps the people in general did not associate it with metallic lead used in water pipes (Fig. 14.3.3.1), beakers and other metal objects in contact with food and drink; according to Delile et al. (2014) they were unaware of the risk. It is estimated that the water distribution system increased lead contents in drinking water of the capital city by up to a hundred times. The lead used by the Romans came from many parts of their empire (Italy, Spain, and Greece, to France or possibly Britain.), as revealed by isotope analysis (Carroll et al. 2021). Some people have suggested that the decline and fall of the Roman empire was caused by lead poisoning, but according to Delile et al. (2014) this is unlikely. An ice-core from the Alps show a very large and sharp peak centered at 250 B.C (Preunkert et al.2019).

Fig. 14.3.3.1. Roman water pipes (fistulae) made of lead. Rome's water pipe history can be inferred from lead-contaminated waters trapped in its ancient harbor basins. From Mail online 1 September 2017 (https://www.dailymail.co.uk/sciencetech/article-4843640/Romans-using-poisonous-lead-pipes-200-years-earlier.html); cf. Delile et al. (2017).

The danger of water pipes made of lead only slowly became generally recognized. Troesken (2008) describes how still in 1897 about half of all American municipalities used lead pipes to distribute water. Using data from Massachusetts, he found that in the average town in 1900, the use of lead pipes increased infant mortality by 25 to 50 percent. However, in cities using new pipes and distributing acidic water, lead pipes increased infant mortality three- to four-fold. Already Ernst Milligan (1931), cited by Troesken (2008) found a connection between infant mortality and the lead content of the water supplied. There is an inverse relationship between water hardness and lead dissolved from water tubes. The more wealthy Romans had another way of becoming poisoned by lead: They used lead compounds in their cosmetics (Robles-Osorio & Sabath 2014).

An estimation of the worldwide production of lead, partly as a byproduct of silver production, is shown in Fig. 14.3.3.2. It is mirrored in the lead content of petrous bones (bones at the temples) of human bodies (Erel et al. 2021) and lead in Greenland ice (Fig. 14.3.3.3).

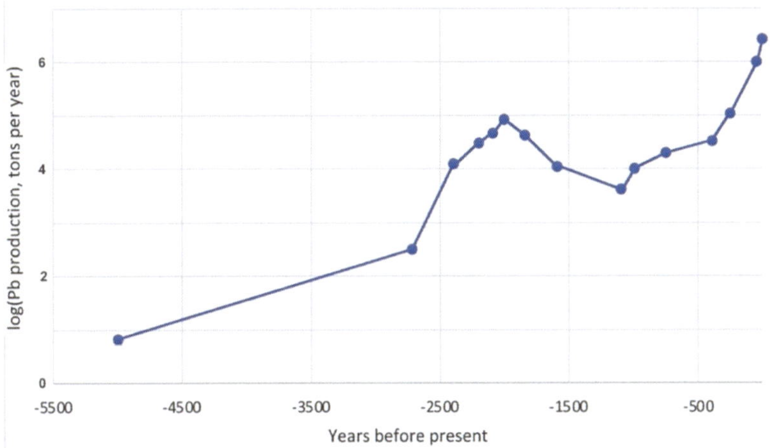

Fig. 14.3.3.2. Worldwide Pb production.The decline from about 2000 years before present to 1100 before present is due to exhaustion of the Roman mines. The more recent rise is due mainly to Spanish silver production in the America and the Industrial revolution. Based on Settle & Patterson (1980) as cited by Hong et al. (1994).

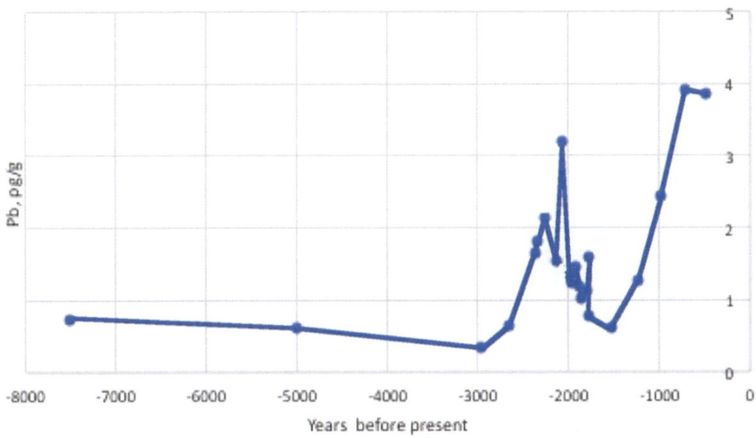

Fig. 14.3.3.3. Lead content in Greenland ice (pg/g = µg/ton) of different age. Based on Hong et al. (1994).

14.3.4 Lead and physical performance

In modern times not only people who come in contact with lead during their work, or ingest it with food or drink, are at risk of lead poisoning. A lead compound, tetraethyl lead ($Pb(C_2H_5)_4$) has in the past been added as antiknock agent to gasoline. Because of its highly lipophilic nature, tetraethyl lead passes the blood-brain barrier readily and accumulates in the limbic forebrain, frontal cortex, and hippocampus. It has now been prohibited in all countries. Algeria used up its last supply in 2021; from most other countries it disappeared much earlier. As seen in Fig. 14.3.4.1, there has been (in the US) a close parallel between the amount of lead in gasoline used and blood concentration of lead of the general population.

Nowak-Szczepanska et al. (2022) found that the handgrip strength was lower for children with higher blood level of lead than in children with a lower level of lead in the blood.

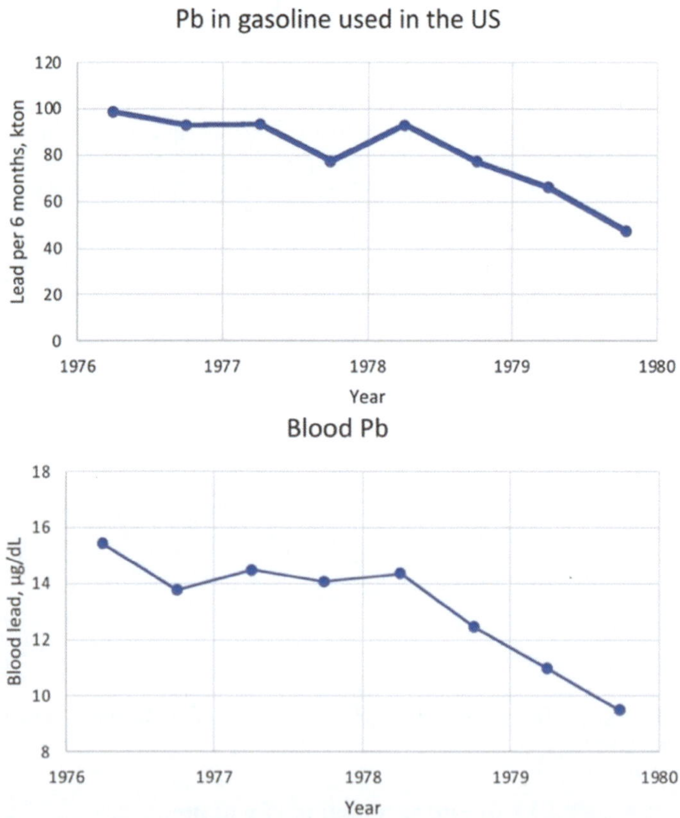

Fig. 14.3.4.1. Comparison between the lead content in gasoline and in the blood of the general population in the United states during the late 1970s. Redrawn from data published by Dissanayake & Erickson (2012).

14.3.5 Lead and neurological problems

High blood level of lead is associated with neurological problems, such as seizures, as described by Greig et al. (2014). These authors also give references to earlier literature on the subject.

14.3.6 Lead and mental ability

High blood level of lead is associated with disturbed cognition and other mental problems. Reuben et al. (2017) compared blood concentrations of lead at age 11 years with intelligence test results at age 38 years. After adjustment for confounding factors (maternal IQ, childhood IQ, and childhood socioeconomic status) each 5-μg/dL higher level of blood lead in childhood was associated with a 1.61-point lower adult IQ (the Wechsler Adult Intelligence Scale–IV) score, a 2.07-point lower score in perceptual reasoning, and a 1.26-point lower score in working memory. Thus childhood lead exposure may have long-term consequences. Vlasak et al. (2019), after doing a systematic literature search covering all articles published before September 2018, concluded that subjects with higher blood lead levels show lower visuospatial abilities as well as lower manual dexterity, and thus that lead significantly impairs cognitive performance and sensorimotor functioning in adults.

Heidari et al. (2022) performed a systematic review and meta-analysis of case-control comparison of IQ test score between a group of children exposed to a concentration of lead above 10 μg/dL, and the control group was exposed to a concentration of less than 10 μg/dL. In the subgroup with duration of exposure less than the average of 4.5 years, the difference of IQ test mean score (-3.53) between two groups was significant (P-value <0.05). In the subgroup with more than 4.5 years of duration, the difference of IQ test mean score (-22.63) had even higher significance (P-value < 0.001).

Even more alarming results were published by McFarland et al. (2022). In 2015, over 170 million persons in the United States (more than 53% of the population) had blood lead levels above 5 μg/dL in early life, over 54 million above 15 μg/dL, and over 4.5 million above 30 μg/dL. Blood lead levels greater than 5 μg/dL were nearly universal (>90%) among those born 1951 to 1980, while the levels were considerably lower than 5 μg/dL among those born since 2001. The average lead-linked loss in cognitive ability was 2.6 IQ points per person as of 2015. This amounts to a total loss of 824,097,690 IQ points.

House and industrial painters have been at risk of lead poisoning, but perhaps more interesting is that it has been shown that some of the great artists became lead-intoxicated by the paint they used. Since the

Renaissance period, lead intoxication has been called painter's colic or painter's madness (Montes-Santiago 2013). Michelangelo, Goya and Portinari almost certainly became intoxicated by lead, perhaps also van Gogh. Other artists, as well as Händel and Beethoven, are also discussed by Montes-Santiago as possibly having been poisoned by lead.

Analyses of long-lived corals indicate that lead pollution of the Mediterranean started soon after the end of World War II, and peaked in the late 1970s (Ricolleau et al. 2019).

A meta-analysis by Meng et al. (2020) showed that lead was a risk factor for meningioma and brain cancer, and when parents were exposed to lead, their children had a higher incidence of brain tumors.

He et al. (2022) fond a negative correlation between relationships between blood lead level and serum folate (vitamin B_9) concentration.

It should be observed that the blood level of lead reflects only primarily lead that has been taken up recently. Over long time more than 90% of body lead is stored in bone (Rosin 2009). But bone lead is mobilized into the blood, e.g., during osteoporosis.

Because of the cessation of adding tetraethyl lead and other measures, the environmental lead situation has improved in recent years. There is some evidence that a recent fall in homicide rate in Britain is related to this (Higney et al. 2022). MacDonald Gibson et al. (2022) found that higher blood Pb was significantly associated with reported juvenile delinquency.

It seems that some persons are more susceptible to lead than others. One reason could be polymorphism of aminolevulinic acid dehydratase, one of the possible targets of lead toxicity. Children carrying the less frequent allele had five times higher risk for high blood lead levels than those with the more frequent allele (Irawati et al. 2022). However, this is certainly not the only, probably not even the major, mechanism of lead toxicity. Experiments with mice (Javorac et al. 2022) indicate that its neurotoxicity is due to decrease of acetylcholinesterase activity (an effect similar to that of nerve gas).

A WHO committee has concluded that there is no safe exposure limit for lead, and the chronic dietary exposure corresponding to a decrease of 1 IQ point was estimated to be 0.6 μg/(kg body weight)/day (https://apps.who.int/food-additives-contaminants-jecfa database/Home/Chemical/3511).

Lead was detected in 37% of samples of baby food sold in the US; maximum=183.6 μg/kg (Gardener et al. 2019). According to Shvanhiy et al. (2022) physiological impacts induced by prenatal, preweaning, and post-weaning Pb exposure persists until adulthood, but neurobehavioral deficits are reversible, suggesting that only a continuous exposure to Pb during early life impacts cognitive performance in adulthood. Wang et al. (2022a) found that children's cognitive abilities negatively correlate with blood lead concentration (P=0.03737). Higher blood lead level was found to be related to higher risk of subclinical myocardial injury (Wang et al. 2022b).

References

Annenkov, V.V. & Gordon, R. 2020. The probable mechanism for silicon capture by diatom algae: Assimilation of polycarbonic acids with diatoms – is endocytosis a key stage in building of siliceous frustrules? J. Phycol. 56, 1729–1737. https://doi.org/10.1111/jpy.13062-20-094

Baines, S.B., Twining, B.S., Brzezinski, M.A., Krause, J.W., Vogt, S., Assael, D. & McDaniel, H. 2012. Significant silicon accumulation by marine picocyanobacteria. Nature Geosci. 5, 886–891. https://doi.org/10.1038/NGEO1641

Bokor, B., Soukup, M., Vaculík, M., Vd'acˇný, P., Weidinger, M., Lichtscheidl, I., Vávrová, S., Šoltys, K., Sonah, H., Deshmukh, R., Bélanger, R.R., White, P.J., El-Serehy. H.A. & Lux, A. 2019. Silicon uptake and localisation in date palm (*Phoenix dactylifera*) – A unique association with sclerenchyma. Front. Plant Sci. 10:988 (17 pp.).: https://doi.org/10.3389/fpls.2019.00988

Carroll, M., Evans, J., Pashley, V. & Prowse, R. 2021. Tracking Roman lead sources using lead isotope analysis. A case study from the imperial rural estate at Vagnari (Puglia, Italy). J. Archaeol. Sci.: Reports 36: 102821 (9 pp.). https://doi.org/10.1016/j.jasrep.2021.102821

Conley, D.J., Frings, P.J., Fontorbe, G., Clymans, W., Stadmark, J., Hendry, K.R., Marron, A.O. & De La Rocha, C.L. 2017. Biosilicification drives a decline of dissolved Si in the oceans through geologic time. Front. Mar. Sci. 4:397 (19 pp.). https://doi.org/10.3389/fmars.2017.00397

Coskun, D., Deshmukh, R., Sonah, H., Menzies, J.G., Reynolds, O., Ma, J.F., Kronzucker, H.J. & Bélanger, R.R. 2019. The controversies of silicon's role in plant biology. New Phytol. 221, 67–85. https://doi.org/10.1111/nph.15343

Delile, H., Blichert-Toft, J., Goiran, J.-P., Keay, S. & Albarède, F. 2014. Lead in ancient Rome's city waters. Proc. Natl Acad. Sci. USA 111, 6594–6599.

Delile, H., Keenan-Jones, D., Blichert-Toft, J., Goiran, J.-P., Arnaud-Godet, F. & Albarède, F. 2017. Rome's urban history inferred from Pb-contaminated waters trapped in its ancient harbor basins. Proc. Natl Acad. Sci. USA 114, 10059–10064. www.pnas.org/cgi/doi/10.1073/pnas.1706334114

Dissanayake, V. & Erickson, T.B. 2012. Ball and chain: The global burden of lead poisoning. Clinical Toxicol. 50, 528–531. https://doi.org/10.3109/15563650.2012.698018

Epstein, E. 2009. Silicon: its manifold roles in plants. Ann. Appl. Biol. 155, 155–160. https://doi.org/10.1111/j.1744-7348.2009.00343.x

Erel, Y., Pinhasi, R., Coppa, A., Ticher, A., Tirosh, O. & Carmel, L. 2021. Lead in archeological human bones reflecting historical changes in lead production. Environ. Sci. Technol. 55, 14407−14413. https://doi.org/10.1021/acs.est.1c00614

Gardener, H., Bowen, J. & Callan, S.P. 2019. Lead and cadmium contamination in a large sample of United States infant formulas and baby foods. Sci. Total Environm. 651, 822–827. https://doi.org/10.1016/j.scitotenv.2018.09.026

Garza, A., Vega, R. & Soto, E. 2006. Cellular mechanisms of lead neurotoxicity. Med. Sci. Monit. 12, RA57–RA65. http://www.medscimonit.com/abstract/index/idArt/447121

Greig, J., Thurtle, N., Cooney, L., Ariti, C., Ahmed, A.O., Ashagre, T., Ayela, A., Chukwumalu, K., Criado-Perez, A., Gómez-Restrepo, C., Meredith, C., Neri, A., Stellmach, D., Sani-Gwarzo, N., Nasidi, A., Leslie Shanks, L. & Dargan, P.I. 2014. Association of blood lead level with neurological features in 972 children affected by an acute severe lead poisoning outbreak in Zamfara State, Northern Nigeria. PLoS ONE 9: e93716 (9 pp.). https://doi.org/10.1371/journal.pone.0093716

Gusev, E. & Martynenko, N. 2022. Diversity of silica-scaled Chrysophytes in central Vietnam. Water 14: 65 (20 pp.). https://doi.org/10.3390/w14010065

Hall, C.R., Dagg, V., Waterman, J.M. & Johnson, S.N. 2020. Silicon alters leaf surface morphology and suppresses insect herbivory in a model grass species. Plants 9: 643 (12 pp.). https://doi.org/10.3390/plants9050643

He, H., Zhang, Z, & Li, M. 2022. Association between serum folate concentrations and blood lead levels in adolescents: A cross-sectional study. Front. Pediatr. 10:941651 (9 pp.). https://doi.org/10.3389/fped.2022.941651

Heidari, S., Mostafaei, S., Razazian, N., Rajati, M., Saeedi, A. & Rajati, F. 2022. The effect of lead exposure on IQ test scores in children under 12 years: a systematic review and meta-analysis of case-control studies. Systematic Revs 11:106 (8 pp.). https://doi.org/10.1186/s13643-022-01963-y

Hernández-Garnica, M., García-García, J.D., Moreno-Sánchez, R. & Sánchez-Thomas, R. 2021. Lead accumulation in photosynthetic

Euglena gracilis depends on polyphosphates and calcium. Environm,l Pollut, 272: 116007 (8 pp.). https://doi.org/10.1016/j.envpol.2020.116007

Higney, A., Hanley, N. & Moro, M. 2022. The lead-crime hypothesis: A meta-analysis. Regional Sci. Urban Econ. 97: 103826 (18 pp.). https://doi.org/10.1016/j.regsciurbeco.2022.103826

Hirota, R., Hata, Y., Ikeda, T., Ishida, T. & Kuroda, A. 2010. The Silicon layer supports acid resistance of *Bacillus cereus* spores. J. Bacteriol. 192, 111–116. https://doi.org/10.1128/JB.00954-09

Hodson, M. & Evans, D.E. 2020. Aluminium–silicon interactions in higher plants: an update. J. Exp. Bot. 71, 6719–6729. https://doi.org/10.1093/jxb/eraa024

Ikeda, T. 2021. Bacterial biosilicification: a new insight into the global silicon cycle. Biosci. Biotechnol. Biochem. 85, 1324–1331. https://doi.org/10.1093/bbb/zbab069

Irawati, Y., Kusnoputranto, H., Achmadi, U.F. et al. 2022. Aminolevulinic acid dehydratase allelic frequency and lead toxicity in children under five in a former used lead-acid battery area. Kesmas: J. Kesehatan Masyarakat Nasional 17, 66–73. https://doi.org/10.21109/kesmas.v17i1.5525

Islam, W., Tayyab, M., Khalil, F., Zhang Hua, Z., Huang, Z. & Chen, H.Y.H. 2020. Silicon-mediated plant defense against pathogens and insect pests. Pesticide Biochem. Physiol. 168: 104641 (21 pp.). https://doi.org/10.1016/j.pestbp.2020.104641

Javorac, D., Anđelkovic´, M., Repic´, A., Tatovic´, S., Djordjevic, A.B., Antonijevic´ Miljakovic´, E., Batinic´, B., Boriˇcic´, N., Đukic¨-C´osic´, D., Antonijevic´, B. & Bulat, Z. 2022. Comprehensive insight into the neurotoxic mechanisms of low dose Pb exposure in Wistar rats: Benchmark dose analysis. Chemico-Biological Interactions 360: 109932 (10 pp.). https://doi.org/10.1016/j.cbi.2022.109932

Jeer, M., Telugu, U.M., Voleti, S.R. & Padmakumari, A.P. 2017. Soil application of silicon reduces yellow stem borer, *Scirpophaga incertulas* (Walker) damage in rice. J. Appl. Entomol. 141, 189–201. https://doi.org/10.1111/jen.12324

Jiang, D., Wu, H., Cai, H. & Chen, G. 2022. Silicon confers aluminium tolerance in rice via cell wall modification in the root transition zone. Plant Cell Environ. 45, 1765–1778. https://doi.org/10.1111/pce.14307

Jonasson, M.E. & Afshari, R. 2018. Historical documentation of lead toxicity prior to the 20th century in English literature. Human and Exp. Toxicol. 37, 775–788. https://doi.org/10.1177/0960327117737146

Jugdaohsingh, R. 2007. Silicon and bone health. J. Nutr. Health Aging 11, 99–110 (22 pp.)

Kerrison, P.D., Stanley, M.S., Kelly, M., MacLeod, A., Black, K.D. & Hughes, A.D. 2016. Optimising the settlement and hatchery culture of *Saccharina latissima* (Phaeophyta) by manipulation of growth medium and substrate surface condition. J. Appl. Phycol. 28, 1181–1191. https://doi.org/10.1007/s10811-015-0621-6

Le, B., Nadeem, M., Yang, S.H., Shin, J.A., Kang, M.-G., Chung, G. & Sun, S. 2019. Effect of silicon in *Pyropia yezoensis* under temperature and irradiance stresses through antioxidant gene expression. J. Appl. Phycol. 31, 1297–1302. https://doi.org/10.1007/s10811-018-1605-0

Liu, J., Zhu, J., Zhang, P., Han, L., Reynolds, O.L., Zeng, R., Wu, J., Shao, Y., You, M. & Gurr, G.M. 2017. Silicon supplementation alters the composition of herbivore induced plant volatiles and enhances attraction of parasitoids to infested rice plants. Front. Plant Sci. 8:1265 (8 pp.). https://doi.org/10.3389/fpls.2017.01265

Liu, W.-S., Zheng, H.-X., Guo, M.N., Liu, C., Huot, H., Morel, J.L., van der Ent, A., Tang, Y.-T. & Qiu, R.-L. 2019. Co-deposition of silicon with rare earth elements (REEs) and aluminium in the fern *Dicranopteris linearis* from China. Plant Soil 437, 427–437. https://doi.org/10.1007/s11104-019-04005-0

MacDonald Gibson, J., MacDonald, J.M., Fisher, M., Chen, X., Pawlick, A. & Cook, P.J. 2022. Early life lead exposure from private well water increases juvenile delinquency risk among US teens. Proc. Natl Acad. Sci. USA 119: e2110694119 (11 pp.). https://doi.org/10.1073/pnas.2110694119

Mavricˇ Cˇermelj, A., Golob, A., Vogel-Mikuš, K. & Germ, M. 2021. Silicon mitigates negative impacts of drought and UV-B radiation in plants. Plants 11, 91 (16 pp.). https://doi.org/10.3390/plants11010091

Mavricˇ Cˇermelj, A., Fideršek, E., Golob, A., Maršic´, N.K., Vogel Mikuš, K. & Germ, M. 2022. Different concentrations of potassium silicate in nutrient solution affects selected growth characteristics and mineral composition of barley (*Hordeum vulgare* L.). Plants 11, 1405 (12 pp.). https://doi.org/10.3390/plants11111405

McFarland, M.J., Hauer, M.E. & Reuben, A. 2022. Half of US population exposed to adverse lead levels in early childhood. Proc. Natl Acad. Sci. USA 119: e2118631119 (7 pp.). https://doi.org/10.1073/pnas.2118631119

Meng, Y., Tang, C., Yu, J., Meng, S. & Zhang, W. 2020. Exposure to lead increases the risk of meningioma and brain cancer: A metaanalysis. J. Trace Elem. Med. Biol.60: 126474 (4 pp.).

https://doi.org/10.1016/j.jtemb.2020.126474
Marron, A.O., Chappell, H., Ratcliffe, S. & Goldstein, R.E. 2016. A model for the effects of germanium on silica biomineralization in choanoflagellates. J. R. Soc. Interface 13: 20160485 (10 pp.). http://dx.doi.org/10.1098/rsif.2016.0485
Michelsen, V. & Palmer, M.W. 2020. *Pegomya disticha* Griffiths and *P. cedrica* Huckett (Diptera: Anthomyiidae)—first documented case of insects trespassing the silica barrier of Common scouring-rush, *Equisetum hyemale* L. Zootaxa 4718: 355–370 (16 pp.). https://doi.org/10.11646/zootaxa.4718.3.4
Mitra, A., Chatterjee, S., Kataki, S., Rastogi, R.P. & Gupta, D.K. 2021. Bacterial tolerance strategies against lead toxicity and their relevance in bioremediation application. Environm. Sci. Pollut. Res. 28, 14271–14284. https://doi.org/10.1007/s11356-021-12583-9
Mitra, P., Sharma, S., Purohit, P. & Sharma, P. 2017. Clinical and molecular aspects of lead toxicity: An update. Crit. Revs Clin. Lab. Sci. 54, 506–528. https://doi.org/10.1080/10408363.2017.1408562
Mizuta, H. & Yasui, H. 2012. Protective function of silicon deposition in *Saccharina japonica* sporophytes (Phaeophyceae). J. Appl. Phycol. 24, 1177–1182. https://doi.org/10.1007/s10811-011-9750-8
Montes-Santiago, J. 2013. The lead-poisoned genius: Saturnism in famous artists across five centuries. Progr. Brain Res. 203, 223–240. http://dx.doi.org/10.1016/B978-0-444-62730-8.00009-8
Müller, W.E.G., Li, J., Schröder, H.C., Qiao, L. & Wang, X. 2007. The unique skeleton of siliceous sponges (Porifera; Hexactinellida and Demospongiae) that evolved first from the Urmetazoa during the Proterozoic: a review. Biogeoscience 4, 219–232. www.biogeosciences.net/4/219/2007/
Nemsadze, K., Sanikidze, T., Ratiani, L., Gabunia, L. & Sharashenidze, T. 2009. Mechanisms of lead-induced poisoning. Georgian Med. News 172–173, 92–6. https://pubmed.ncbi.nlm.nih.gov/19644200/
Nowak-Szczepanska, N., Gomula, A., Sebastjan, A., Ignasiak, Z., Malina, R.M. & Kozieł, S. 2022. Blood lead level and handgrip strength in preadolescent Polish schoolchildren. Toxics 10: 646 (8 pp.). https://doi.org/10.3390/toxics10110646
Ohnemus, D.C., Krause, J.W., Brzezinski, M.A., Collier, J.L., Baines, S.B. & Twining, B.S. 2018. The chemical form of silicon in marine *Synechococcus*. Marine Chemistry 206, 44–51. https://doi.org/10.1016/j.marchem.2018.08.004

Pérez, M., García, M., Roldán, J.P. & Blustein, G. 2018. Efecto inhibidor del dióxido de germanio sobre el biofouling del puerto de Mar del Plata. Revista de Matéri Artículo e-11955. http://dx.doi.org/10.1590/S1517-707620170001.0291

Polevova, S. & Moiseenko, A. 2023. Silicon in sporoderms of micro- and megaspores of *Isoëtes echinospora* Durieu registered by EDS and EELS. Protoplasma 260, 663–667. https://doi.org/10.1007/s00709-022-01791-w

Praeger, C., Magnusson, M. & Lawton, R. 2022. Optimising the zoospore release, germination, development of gametophytes and formation of sporophytes of *Ecklonia radiata*. J. Appl. Phycol. 34, 2535–2549. https://doi.org/10.1007/s10811-022-02806-y

Preunkert, S., McConnell, J,R., Hoffmann, H., Legrand, M., Wilson, A.I., Eckhardt, S., Stohl, A., Chellman, N.J., Arienzo, M.M & Friedrich, R. 2019. Lead and Antimony in Basal Ice From Col du Dome (French Alps) Dated With Radiocarbon: A Record of Pollution During Antiquity. Geophysical Research Letters, 46, 4953–4961. https://doi.org/10.1029/2019GL082641

Pu, S., Cai, X., Wang, W., Liu, X., Li, S., Fu, J., Sun, L., Ma, J., Jiang, M. & Li, X. 2022. NTA-assisted mineral element and lead transportation in *Eremochloa ophiuroides* (Munro) Hack. Environm. Sci. Pollut. Res. 29, 20650–20664. https://doi.org/10.1007/s11356-021-17306-8

Ricolleau, A., Floquet, N., Devidal, J.-L., Bodnar, R.J., Perrin, J., Garrabou, J., Harmelin, J.-G., Costantini, F., Boavida, J.R. & Vielzeuf, D. 2019. Lead (Pb) profiles in red coral skeletons as high resolution records of pollution in the Mediterranean Sea. https://doi.org/10.1016/j.chemgeo.2019.07.005

Rizwan, M., Ali, S., Ibrahim, M., Farid, M., Adrees, M., Bharwana, S.A., Zia-ur-Rehman, M., Qayyum, M.F. & Abbas, F. 2015. Mechanisms of silicon-mediated alleviation of drought and salt stress in plants: a review. Environ. Sci. Pollut. Res. 22, 15416–15431. https://doi.org/10.1007/s11356-015-5305-x

Rizwan, M., Atta, B., Rizwan, M., Sabir, A.M., Tahir, M., Sabar, M., Ali, M. & Ali, M.Y. 2022. Integrated pest management against rice leaffolder *Cnaphalocrocis medinalis* Guenée (Lepidoptera: Pyralidae). Pakistan J. Zool. 54, 569–575. https://dx.doi.org/10.17582/journal.pjz/20200330090344

Robles-Osorio, M.L. & Sabath, E. 2014. Breve historia de la intoxicación por plomo: de la cultura egipcia al Renacimiento. Rev. Investig. Clínica 66, 88 –91.

Romeh, A.A., Khamis, M.A. & Metwally, S.M. 2016. Potential of *Plantago major* L. for phytoremediation of lead-contaminated soil and water. Water Air Soil Pollut. 227: 9 (9 pp.). https://doi.org/10.1007/s11270-015-2687-9

Rondanell, M., Faliva, M.A., Peroni, G., Gasparri, C., Perna, S., Riva, A., Petrangolini, G. & Tartara, A. 2021. Silicon: A neglected micronutrient essential for bone health. Exp. Biol. Medicine 246, 1500–1511. https://doi.org/10.1177/1535370221997072

Rosin, A. (2009) The long-term consequences of exposure to lead. Israel Med. Assoc. J. (IMAJ) 11, 689–694.

Sahi, S., Bryant, N., Sharma, N. & Singh, S. 2002. Characterization of a Lead hyperaccumulator ahrub, *Sesbania drummondii*. Environ. Sci. Technol. 36, 4676-4680.

Shabaan, M., Asghar, H.N., Akhtar, M.J., Ali, Q. & Ejaz, M. 2021. Role of plant growth promoting rhizobacteria in the alleviation of lead toxicity to *Pisum sativum* L. Intern. J. Phytoremediation. 23, NO. 8, 837–845. https://doi.org/10.1080/15226514.2020.1859988

Shea, R. & Chopin, T. 2007. Effects of germanium dioxide, an inhibitor of diatom growth, on the microscopic laboratory cultivation stage of the kelp, *Laminaria saccharina*. J Appl. Phycol. 19, 27–32. https://doi.org/10.1007/s10811-006-9107-x

Shvachiy, L., Amaro-Leal, Â., Outeiro, T.F., Rocha, I. & Geraldes, V. 2022. From molecular to functional effects of different environmental lead exposure paradigms. Biology 11: 1164 (18 pp.). https://doi.org/10.3390/biology11081164

Sun, X., Sun, M., Chao, Y., Wang, H., Pan, H., Yang, Q., Cui, X., Lou, Y. & Zhuge, Y. 2021. Alleviation of lead toxicity and phytostimulation in perennial ryegrass by the Pb-resistant fungus *Trichoderma asperellum* SD-5. Func. Plant Biol. 48, 333–341. https://doi.org/10.1071/FP20237

Tarakhovskaya, E.R., Kang, E.J., Kim, K.Y. & Garbary, D.J. 2012. Effect of GeO_2 on embryo development and photosynthesis in *Fucus vesiculosus* (Phaeophyceae). Algae 27, 125–134. http://dx.doi.org/10.4490/algae.2012.27.2.125

Tatewaki, M. & Mizuno, M. 1979. Growth inhibition by germanium dioxide in various algae, especially in brown algae. Jpn. J. Phycol. 27, 205-212.

Titov, A., Zaikovskii, V. & Larionov, P.M. 2019. Silicon dioxide in mineralized heart valves. ICAM, SPEES, pp. 432–435. https://doi.org/10.1007/978-3-030-22974-0_105

Troesken, W. 2008. Lead water pipes and infant mortality at the turn of the twentieth century. J. Human Resources 43, 553–575.

Uriz, M.-J., Turon, X., Becerro, M.A. & Agell, G. 2003. Siliceous spicules and skeleton frameworks in sponges: Origin, diversity, ultrastructural patterns, and biological functions. Microscopy Res. Technique 62, 279–299.

Vlasak, T., Jordakieva, G., Gnambs, T., Augner, C., Crevenna R., Winker, R. & Barth, A. 2019. Blood lead levels and cognitive functioning: A meta-analysis.Sci. Total Environm. 668, 678–684. https://doi.org/10.1016/j.scitotenv.2019.03.052

Wang, X., Schloßmacher, U., Wiens, M., Batel, R., Schröder, H.C. & Müller, W.E.G. 2012. Silicateins, silicatein interactors and cellular interplay in sponge skeletogenesis: formation of glass fiber-like spicules. FEBS J. 279, 1721–1736. https://doi.org/10.1111/j.1742-4658.2012.08533.x

Wang, Y., Wang, Y. & Yan, C.-H. 2022a. Gender differences in trace element exposures with cognitive abilities of school-aged children: a cohort study in Wujiang city, China. Environm. Sci. Pollut. Res- 29, 64807–64821. https://doi.org/10.1007/s11356-022-20353-4

Wang, Z., Huang, X., Li, J., Liu, N. & Wei, Q. 2022b. Lead exposure is non-linearly associated with subclinical myocardial injury in the general population without cardiovascular disease. Front. Public Health 10:975413 (9 pp.). https://doi.org/10.3389/fpubh.2022.975413

Wei, Y., Sun, J., Li, L. & Cui, Z. 2022. *Synechococcus* silicon accumulation in oligotrophic oceans. Limnol. Oceanogr. 67, 552–566. https://doi.org/10.1002/lno.12015

Xu, Z., Ban, Y., Yang, R., Zhang, X., Chen, H. & Tang, M. 2016. Impact of *Funneliformis mosseae* on the growth, lead uptake, and localization of *Sophora viciifolia*. Can. J. Microbiol. 62, 361–373. https://doi.org//10.1139/cjm-2015-0732

Yacano, M.R., Foster, S.Q., Ray, N.E., Oczkowski, A., Raven, J.A. & Fulweiler, R.W. 2022. Marine macroalgae are an overlooked sink of silicon in coastal systems. New Phytol. 233, 2330–2336.

Ye, Y., Frings, P.J., vonBlanckenburg, F. & Feng, Q. 2021. Silicon isotopes reveal a decline in oceanic dissolved silicon driven by biosilicification: A prerequisite for the Cambrian Explosion? Earth Planet. Sci. Lett. 566: 116959 (12 pp.). https://doi.org/10.1016/j.epsl.2021.116959

CHAPTER FIFTEEN

PNICTOGENS: ARSENIC, ANTIMONY, AND BISMUTH

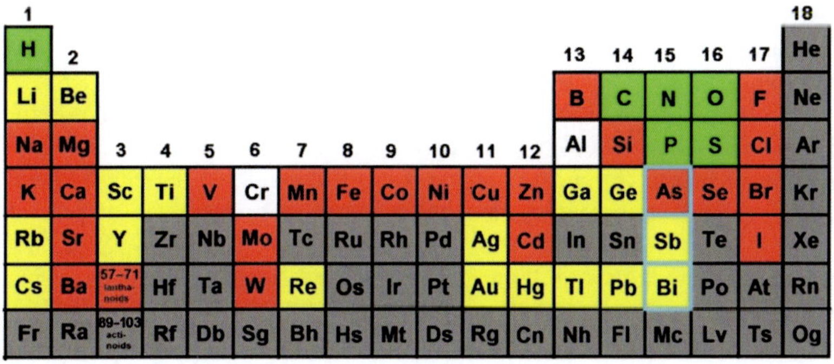

15.1 Arsenic

Arsenic is mostly known as an element with poisonous compounds, but here we shall deal mainly with other aspects of arsenic biology. Arsenic compounds are poisonous mostly because they can be converted to arsenate, which is sufficiently similar to phosphate to be used by enzymes taking part in energy conversion. Adenosine triphosphate is an important "energy currency" in biochemical reactions. But arsenate is sufficiently different from phosphate to cause disaster in the handling of energy. Also arsenite is poisonous. We shall return to the toxicity of arsenic, but first deal with other aspects of the element.

For previous reviews of the role of arsenic in the biosphere I refer to Stolz et al. (2006) and Zhu et al. (2014). A review of arsenic-related genes in various organismal groups is provided by Dunivin et al. (2019).

Some bacteria use arsenate (containing pentavalent arsenic) as oxidant (e.g., Oremland & Stolz 2003) to make energy available, while others (e.g., Ahmann et al. 1994) use arsenite (containing trivalent arsenic) as reductant for the same purpose. In oxygen-free parts of the ocean there are bacteria respiring arsenate instead of oxygen (Saunders et al. 2019, Shi et al. 2020). The reason that there can be so much arsenate in comparison to arsenite in these areas, despite the reducing environment, is not known.

Some bacteria have evolved mechanisms for using methylated arsenic compounds as antibiotics to acquire competitive advantage over other organisms (Chen et al. 2020). Giovannoni et al. (2019) describe an interesting case of parasitism, in which arsenic is involved. Unicellular algae of the genus *Prochlorococcus*, very common in the sea, defend themselves (as many other organisms do) against the danger of arsenate by reducing it to arsenite and then methylate it to monomethyl arsenite and dimethylarsenite, which they can excrete. Bacteria of the genus *Pelagibacter* consume the methyl groups and oxidize the remains back to arsenate, so *Prochlorococcus* has to reduce it and put the methyl groups back again.

It is possible that in the past arsenic has played a more important role for life on Earth than it does today. There are reasons to think that 2.7 billion years ago, before oxygenic photosynthesis had rendered the environment oxidizing, the biological cycling of arsenic was greater than at present (Sforna et al 2014). The "Great oxidation event" (GOE) caused by photosynthesis about 2.4 billion years ago caused great changes in the living world, expressed, e.g., in the appearance of novel enzyme systems for handling oxidized (pentavalent) arsenic (Chen et al. 2020). That it took so long for GOE to occur is, in part, due to the fact that oxidation of water is a very complicated and difficult process, in part because of the presence of large amounts of reduced compounds. Prior to the evolution of oxygen-evolving photosynthesis there were other forms of photosynthesis, e.g., those using arsenite instead of water as reductant. Such forms of photosynthesis are still taking place (Hoeft McCann et al. 2017, Hernandez-Maldonado 2017).

In general, however, arsenate is poisonous to plants, because it is mistaken for phosphate. Arsenate is taken up through the phosphate transporters (DiTusa et al. 2016), and arsenite and undissociated methylated As species through the nodulin 26-like intrinsic (NIP) aquaporin channels. Plants have several ways of protecting themselves; As(V) is reduced to the less toxic As(III) (Xu et al. 2007).

So much is published about arsenic accumulation in the fern *Pteris vittata* that I shall make no attempt to review all of it, but the plant is so interesting that a few more words are warranted. As already mentioned, arsenate is taken up via phosphate transporters (DiTusa et al. 2016), and even more may be accumulated via a mycorrhiza-forming fungus (Sun et al. 2022). Selenium increases the uptake of arsenic in *P. vittata* (Dai et al.

2022). The arsenic-accumulating capacity was first noted in ferns growing at a site contaminated with chromated copper arsenate, and there it was the only one among 14 species studied that accumulated large amounts of arsenic (Ma et al. 2001). When grown for 20 weeks on a soil contaminated with chromated copper arsenate containing 97 ppm As, the As concentration in the fronds was over 2500 ppm and still increasing. The concentration in the roots was only about 200 ppm (Ma et al. 2001). *P. vittata* has been successfully grown together with maize in order to decrease the uptake of arsenic in the maize. Very little As accumulated in maize grains, resulting in their meeting food-grade standard (Ma et al. 2018). In attempts to prevent arsenic uptake in white mulberry (*Morus alba*), it was found that phosphate fertilization increased the uptake in *P. vittata* (Wan et al. 2018). The importance of a good nutrient supply to *P. vittata* for efficient arsenic uptake was also demonstrated by Matzen et al. (2022a), who found that insufficient nutrition lead to reduced growth and leaching of arsenic from the soil. The soil texture has a great influence on phytoextraction of arsenic, and the *P. vittata* plants should be replaced after 2–3 years (Matzen et al. 2022b). *P. vittata* is most successfully used in South China or USA. In North-Central China (Henan Province, Zhang et al. 2017) and Japan (Northern Honshu, Kohda et al. 2022) the climate renders it less successful, but the As accumulated in in *P. vittata* was higher than that of As-hyperaccumulator fern *Pteris cretica*, the native fern in the Japanese field trial area. Yang et al. (2022) found that high soil moisture is important for good As extraction, and that under field conditions As concentration increased to a peak on day 90 and subsequently declined until day 180.

There has not been much experimentation to find out if accumulation of arsenic is of any advantage to *P. vittata*. Moths feeding on the plant have been found to take up a large amount of arsenic, but also excrete much of it (Wan et al. 2021). Several other ferns, such as, *Pteris longifolia*, *Pityrogramma calomelanos* and the already mentioned *Pteris cretica* are known to accumulate large amounts of arsenic (Zang et al. 2022).

Arsenic is poisonous also to animals and to humans. One of the main crops used as human food worldwide, but in particular in South and South-East Asia, is rice. It grows in water in conditions under which arsenic is mobilized, and the rice plant readily accumulates the toxic element (Zhao et al. 2010).

Worldwide 160 million (Li et al. 2019), 94 million to 220 million (Podgorski & Berg 2020), or more than 200 million people (Muzaffar et al. 2022) are estimated to be exposed to unsafe levels of arsenic, mostly due to arsenic in groundwater, the vast majority (94%) being in Asia. Several articles show maps of worldwide or regional distribution of arsenic, arsenic mobilization under reducing and oxidizing conditions, and of arsenic in groundwater (Amin et al. 2008, Brammer & Ravenscroft 2009, Rodriguez-Lado et al. 2013, Podgorski & Berg 2020, Bundschuh et al. 2022, Irunde et al. 2022, Nath et al. 2022, Muzaffar et al. 2022).

Bangladesh is one of the countries worst affected by arsenic contamination in groundwater. Most of the people in the country depend on groundwater for domestic and irrigation purposes. Currently, 61 districts out of 64 districts of Bangladesh are affected by arsenic contamination (Ivy et al. 2022).

One of the deleterious effects of arsenic poisoning is induction of various cancer forms, e.g., prostate cancer (Yang et al. 2022), but arsenic is also used for treatment of some cancer forms (Chin et al. 2022). At environmental relevant concentrations, arsenic acts mainly as a co-carcinogen enhancing the carcinogenic effects of other carcinogens. One of the most important mechanisms of arsenic co-carcinogenesis is the inhibition of DNA repair (Zhou et al. 2021).

15.2 Antimony

Antimony is mostly bad for animals as well as for plants, but there are exceptions. Although solid evidence is lacking, it is very likely that plants that hyperaccumulate antimony protect themselves against herbivores and pathogens by this poisonous element, similar to, for instance, nickel hyperaccumulators (see chapter on Nickel, section "Nickel hyperaccumulation" and Hörger et al. 2013). The best known example of an antimony hyperaccumulator is the fern *Pteris vittata* L. (Chinese brake or ladder brake), that is also known as an arsenic accumulator (Tisarum et al. 2014, Wan et al. 2017). It is able to absorb trivalent, but hardly any pentavalent antimony. Most of it is retained in the root, but the fraction that reaches the leaves is mostly oxidized to the pentavalent form (Tisarum et al. 2014). *Pteris cretica* behaves similarly (He et al. 2022, Xi et al. 2022), but it was shown for this plant that only with mycorrhiza (with fungi of the genus *Glomus*) could *P. cretica* tolerate antimony (Xi et al. 2022). *P. cretica* translocates a smaller fraction of the antimony taken up by the roots to the shoot than do some other ferns, such as *Microlepia hancei* (Feng et al. 2009). These differences should be taken into account when considering the possible role of antimony in defense.

Trivalent antimony (Sb^{3+}, as antimony potassium tartrate) is a more efficient inhibitor of human glutathione S-transferase (50% inhibition by 50 μM Sb^{3+}) than Hg^{2+}, Cu^{2+}, Cd^{2+}, Cr^{3+} and Fe^{2+} (Poon & Chu 2000). There are several types of glutathione S-transferase both in animals and plants;

Salix babylonica, for instance, has 22 different full-length genes for it (Zhuge et al. 2020). Generally glutathione S-transferases act by catalyzing the conjugation of reduced glutathione with electrophilic, often hydrophobic toxic compounds, to form derivatives that can be secreted from the cell or catabolized.

Trivalent antimony also has several disruptive effects on the genetic system (Litwin et al. 2021): In the yeast *Saccharomyces cerevisiae* it causes, even at low doses, replication and topoisomerase I-dependent DNA lesions as well as oxidative stress and DNA breaks independently of replication. Furthermore, trivalent antimony can cause telomere dysfunction and distort actin and microtubules, and inhibit repair of double strand breaks in DNA.

Generally speaking, antimony is considered more toxic than arsenic; for instance the United States Environmental Protection Agency has set the maximum contaminant level in drinking water for arsenic to 10 µg/L=130 nanomolar (as of May 2009; before 2006 t it was 50 µg/L). For antimony the limit is 6 µg/L=49 nanomolar. In EU, as well as in China, the limit is 5 µM. Due to their higher affinity for thiol groups of proteins, As(III) and Sb(III) are more toxic than As(V) and Sb(V), respectively (Filella et al. 2002). Thus *Bacillus sp.* S3 was shown to exert a protective effect on plants growing in the presence of As(III) by oxididizing the arsenic to the pentavalent form (Gu et al. 2020).

Some bacteria are quite resistant to trivalent antimony. Thus *Klebsiella aerogenes* strain X (isolated from mine tailings) protects itself by forming extracellular polymeric substances and antioxidant enzymes, and by oxidizing the antimony to the less poisonous pentavalent form (Rong et al. 2022). Other antimony oxidizing bacteria have also been found, mostly in antimony mines or mine tailings; examples are the Gram-negative *Stenotrophomonas sp.* (Hamamura et al. 2020), *Bosea sp.* AS-1 (Wu et al. 2022). *Acidithiobacillus ferrooxidans* oxidizes stibnite (antimonite, Sb_2S_3) under extremely acidic conditions (Wang et al. 2022). In the latter case pH may drop to 1.9 due to production of sulfuric acid. Li et al. (2016) list many antimony-oxidizing bacteria belonging to Alphaproteobacteria, Betaproteobacteria, Gammaproteobacteria, and Actinobacteria. About half of the species belong to Actionobacteria, the most common ones in the genera *Pseudomonas* and *Acinetobacter*.

Other bacteria reduce antimony. Species of *Desulfovibrio*, abundant in antimony mine tailings, reduce Sb^{V} to Sb^{III} at the same time as they oxidize sulfur to sulphate (Sun et al. 2022). Another species capable of this, *Desulfuribacillus stibiiarsenatis* MLFW-2T, has been isolated from sediment of Mono lake, California (Abin & Hollibaugh 2014). Since compounds with trivalent antimony are generally less soluble than those with pentavalent antimony, Sb-reducing microbes are considered for rendering antimony less dangerous in the environment (Yang et al. 2021). Even *Escherichia coli* W3110, without known Sb(V)-reducing genes, is able to reduce Sb(V) to Sb(III) by extracellular polymeric substances (Zhang et al. 2021).

Some bacteria rid their environment from antimony by methylating it and thereby making it volatile. Methylation may result in mono-, di-, or trimethyl antimony (mono-, di-, or tri-methyl stibine; all volatile compounds. Also stibine, SbH_3, and other volatile antimony compounds may be formed biochemically.

The reader may have noticed that many of the author names of the references in this chapter are Chinese. This has a cause. The Xikuang-shan Sb mine, located in Lengshui-jiang City ("World Capital of Antimony") in Hunan Province of China, has the largest Sb deposit in the world (Wang et al. 2023). Antimony-mining there is associated with considerable environmental problems. Despite its toxicity and use in many electronic devices, and for hardening lead in firearm bullets, and due to technical difficulties, hardly any antimony is recycled (Karlsson et al. 2018), and its accumulation in the environment is an increasing problem (Herath et al. 2017).

15.3 Bismuth

Bismuth is not very toxic to humans, but very toxic to some bacteria. Of particular interest is the toxicity to *Helicobacter pylori*, which causes various gastric diseases, such as gastritis, peptic ulcerations and gastric cancer. Many treatments of *Helicobacter pylori* infections are based on bismuth. Bismuth is inhibitory to several enzymes, including urease, alcohol dehydrogenase, fumarate reductase, fumarase and proteases such as pepsin and phospholipase C and A2 (literature reviewed by Ge et al. 2012). Bismuth (as colloidal bismuth subcitrate, $C_{12}H_{10}BiK_3O_{14}$ as potassium salt, Fig, 15.3.1) decreases the ATP level *Helicobacter pylori* cells to zero.

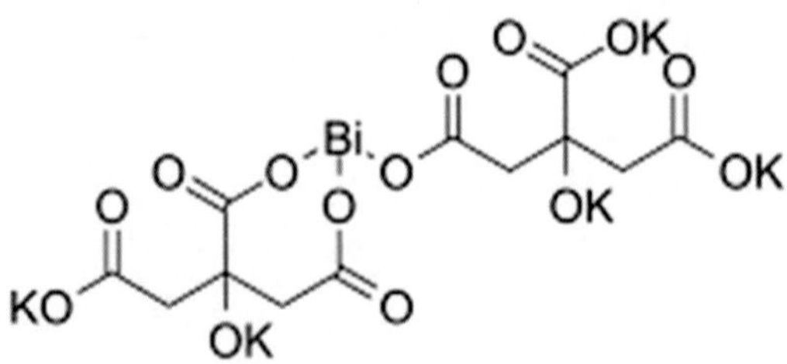

Fig. 15.3.1. Bismuth subcitrate potassium salt

Alcohol dehydrogenase catalyzes the oxidation of alcohol to acetaldehyde, and acetaldehyde mediates mucosal damage by irreversibly binding to the host's phospholipids and cysteine residues of proteins. It appears that bismuth can use certain iron-transport pathways to enter *Helicobacter pylori* cells (Tsang et al. 2011) and that iron and bismuth compete for these pathways (Bland et al. 2004).

Although the most important medical use of bismuth is to fight *Helicobacter pylori* infections, it is also used against stomach infection by *Desulfovibrio desulfuricans* (Barton et al, 2019). This organism causes reduction of sulfate to poisonous hydrogen sulfide. Bi^{3+} ions precipitate the hydrogen sulfide as Bi_2S_3, and can also oxidize them back to sulfate by being reduced to metallic bismuth, Bi^0.

A fine mesh gauze called Xeroform, containing 3% bismuth tribromophenate, has been successfully used for dressings for wounds and burns for inhibiting pathogenic bacteria (Barillo et al. 2017).

References

Abin, C.A. & Hollibaugh, J.T. 2014. Dissimilatory antimonate reduction and production of antimony trioxide microcrystals by a novel microorganism. Environ. Sci. Technol. 48, 681–688.

Ahmann, D., Roberts, A.L., Krumholz, L.R. & Morel, F.M.M. 1994. Microbe grows by reducing arsenic. Nature 371, 750–750.

Amini, M., Abbaspour, K.C., Beerg, M., Winkel, L., Hug, S.J., Hoehn, E., Yang, H. & Johnson, C.A. 2008. Statistical modeling of global geogenic arsenic contamination in groundwater. Environ. Sci. Technol. 42, 3669–3675. https://doi.org/10.1021/es702859e

Barillo, D.J., Barillo, A.R., Korn, S., Lam, K. & Attar, P.S. 2017. The antimicrobial spectrum of Xeroform. Burns 43, 1189–1194. http://dx.doi.org/10.1016/j.burns.2016.10.023

Barton, L.L., Granat, A.S., Lee, S., Xu, H., Ritz, N.L., Hider, R. & Lin, H.C 2019. Bismuth(III) interactions with *Desulfovibrio desulfuricans*: inhibition of cell energetics and nanocrystal for-mation of Bi_2S_3 and Bi^0. Biometals 32, 803–811. https://doi.org/10.1007/s10534-019-00213-4

Bland, M.V., Ismail, S., Heinemann, J.A. & Keenan, J.I. 2004. The action of bismuth against *Helicobacter pylori* mimics but is not caused by intracellular iron deprivation. Antimicrobial Agents and Chemotherapy 48, 1983–1988. https://doi.org/10.1128/AAC.48.6.1983–1988.2004

Brammer, H. & Ravenscroft, P. 2009. Arsenic in groundwater: A threat to sustainable agriculture in South and South-east Asia. Environ. Int. 35, 647–654.

Bundschuh, J., Niazi, N.K., Alam, M.A., Berg, M., Herath, I., Tomaszewska, B., Maity, J.P. & Ok, Y.S. 2022. Global arsenic dilemma and sustainability. J. Hazardous Mat. 436: 129197 (11 pp.). https://doi.org/10.1016/j.jhazmat.2022.129197

Chen, S.-C., Sun, G.-X., Yan, Y., Konstantinidis, K.T., Zhang, S.-Y., Deng, Y., Li, X.-M., Cui, H.-L., Musat, F., Popp, D., Rosen, B.P. & Zhua, Y-G. 2020. The Great Oxidation Event expanded the genetic repertoire of arsenic metabolism and cycling. Proc. Natl Acad. Sci. USA 117, 10414–10421. https://doi.org/10.1073/pnas.2001063117

Chen, J. & Rosen, B.P. 2020. The arsenic methylation cycle: How microbial communities adapted methylarsenicals for use as weapons in the continuing war for dominance. Front. Environ. Sci. 8:43 (14 pp.). doi: https://doi.org/10.3389/fenvs.2020.00043

Chin, L., Kumana, C.R., Kwong, Y.-L. & Gill, H. 2022. The development and clinical applications of oral arsenic trioxide for acute promyelocytic leukaemia and other diseases. Pharmaceutics 14: 1945 (18 pp.). https://doi.org/10.3390/pharmaceutics14091945

Dai, Z.-H., Peng, Y.-J., Ding, S., Chen, J.-Y., He, S.-X., Hu, C.-Y., Cao, Y., Guan, D.-X. & Ma, L.Q. 2022. Selenium increased arsenic accumulation by upregulating the expression of genes responsible for arsenic reduction, translocation, and sequestration in arsenic hyperaccumulator *Pteris vittata*. Environ. Sci. Technol. 56, 14146−14153. https://doi.org/10.1021/acs.est.2c03147

DiTusa, S.F., Fontenot, E.B., Wallace, R.W., Silvers, M.A., Steele, T.N., Elnagar, A.H., Dearman, K.M. & Smith, A.P. 2016. A member of the phosphate transporter 1 (Pht1) family from the arsenic-hyperaccumulating fern *Pteris vittata* is a high-affinity arsenate transporter. New Phytol. 209, 762–772.
https://doi.org/10.1111/nph.13472

Dunivin, T.K., Yeh, S.Y. & Shade, A. 2019. A global survey of arsenic-related genes in soil microbiomes. BMC Biology 17:45 (17 pp.). https://doi.org/10.1186/s12915-019-0661-5

Feng, R., Wei, C., Tu, S., Wu, F. & Yang, L. 2009. Antimony accumulation and antioxidative responses in four fern plants. Plant Soil 317, 93–101. https://doi.org/10.1007/s11104-008-9790-2

Filella, M., Belzile, N., Lett, M.C. 2007. Antimony in the environment: a review focused on natural waters. III. Microbiota relevant interactions. Earth Sci. Rev. 80, 195–217.
https://doi.org/10.1016/j.earscirev.2006.09.003

Ge, R., Chen, Z. & Zhou, Q. 2012. The actions of bismuth in the treatment of *Helicobacter pylori* infections: an update. Metallomics 4, 239–243.

Giovannoni, S.J., Halsey, K.H., Saw, J., Muslin, O., Suffridge, C.P., Sun, J., Lee, C.-P., Moore, E.R., Temperton, B. & Noell, S.E. 2021. A parasitic arsenic cycle that shuttles energy from phytoplankton to heterotrophic bacterioplankton. mBio 10:e00246-19 (12 pp.). https://doi.org/10.1128/mBio.00246-19

Hernandez-Maldonado, J., Sanchez-Sedillo, B., Stoneburner, B., Boren, A., Miller, L., McCann, S., Rosen, M., Oremland, R.S. & Saltikov, C.W. 2017. The genetic basis of anoxygenic photosynthetic arsenite oxidation. Environ Microbiol. 19, 130–141. https://doi.org/10.1111/1462-2920.13509

He, S.-H., Chen, J.-Y., Hu, C.-Y., Han, R., Dai, Z.-H., Guan, D.-X. & Ma, L.Q. 2022. Uptake, speciation and detoxification of antimonate and antimonite in As-hyperaccumulator *Pteris cretica* L. Environm. Pollut. 308: 119653 (7 pp.). https://doi.org/10.1016/j.envpol.2022.119653

Herath, I., Vithanage, M. & Bundschuh, J. 2017. Antimony as a global dilemma: Geochemistry, mobility, fate and transport. Environm. Pollut. 223, 545–559. http://dx.doi.org/10.1016/j.envpol.2017.01.057

Hoeft McCann, S., Boren, A., Hernandez-Maldonado, J., Stoneburner, B., Saltikov, C.W., Stolz, J.F. & Oremland, R.S. 2017. Arsenite as an electron donor for anoxygenic photosynthesis: Description of three strains of *Ectothiorhodospira* from Mono Lake, California and Big Soda Lake, Nevada. Life 7, 1 (14 pp.). https://doi.org/10.3390/life7010001

Hörger, A.C., Fones, H.N. & Preston, G.M. 2013. The current status of the elemental defense hypothesis in relation to pathogens. Front. Plant Sci. 13:395 (11 pp.). https://doi.org/10.3389/fpls.2013.00395

Irunde, R., Ijumulana, J., Ligate, F., Maity, J.P., Ahmad, A., Mtamba, J., Mtalo, F. & Bhattacharya, P. 2022. Arsenic in Africa: potential sources, spatial variability, and the state of the art for arsenic removal using locally available materials. Groundwater for Sustainable Developm. 10: 100746 (18 pp.). https://doi.org/10.1016/j.gsd.2022.100746

Ivy, N., Mukherjee, T., Bhattacharya, S., Ghosh, A. & Sharma, P. 2022. Arsenic contamination in groundwater and food chain with mitigation options in Bengal delta with special reference to Bangladesh Environ. Geochem. Health, online 2022. https://doi.org/10.1007/s10653-022-01330-9

Karlsson, T., Forsgren, C. & Steenari, B.-M. 2018. Recovery of antimony: A laboratory study on the thermal decomposition and carbothermal reduction of Sb(III), Bi(III), Zn(II) oxides, and antimony compounds from metal oxide varistors. J. Sust. Metallurgy 4, 194–204. https://doi.org/10.1007/s40831-017-0156-y

Kohda, Y. H.-T., Endo, G., Kitajima, N., Sugawara, K., Chien, M.-F., Inoue, C. & Miyauchi, K. 2022. Arsenic uptake by *Pteris vittata* in a subarctic arsenic-contaminated agricultural field in Japan: An 8-year study. Sci. Total Environm. 831: 154830 (8 pp.). http://dx.doi.org/10.1016/j.scitotenv.2022.154830

Litwin, I., Mucha, S., Pilarczyk, E., Wysocki, R. & Maciaszczyk-Dziubinska, E. 2021. Complex mechanisms of antimony genotoxicity in budding yeast involves replication and topoisomerase I-Associated DNA lesions, telomere dysfunction and inhibition of DNA repair. Int. J. Mol. Sci. 22: 4510 (30 pp.). https://doi.org/10.3390/ijms22094510

Ma, J., Lei, E., Lei, M., Liu, Y. & Chen, T. 2018. Remediation of Arsenic contaminated soil using malposed intercropping of *Pteris vittata* L. and maize. Chemosphere 194, 737–744. https://doi.org/10.1016/j.chemosphere.2017.11.135

Ma, L.Q., Komar, K.M., Tu, C., Zhang, W., Cai, Y. & Kennelley, E.D. 2001. A fern that hyperaccumulates arsenic. Nature 409, 579–579.

Matzen, S.L., Lobo, G.P., Fakra, S.C., Kakouridis, A., Nicod, P.S. & Pallud, C.E. (2022a) Arsenic hyperaccumulator *Pteris vittata* shows reduced biomass in soils with high arsenic and low nutrient availability, leading to increased arsenic leaching from soil. Sci. Total Environment 818: 151803 (14 pp.). http://dx.doi.org/10.1016/j.scitotenv.2021.151803

Matzen, S.L., Olson, A.L. & Pallud, C.E. 2022b. Soil texture and climate limit cultivation of the arsenic hyperaccumulator *Pteris vittata* for phytoextraction in a long-term field study. J. Hazardous Materials 436: 129151 (11 pp). https://doi.org/10.1016/j.jhazmat.2022.129151

Muzaffar, S., Khan, J., Srivastava, R., Gorbatyuk, M.S. & Athar. M. 2022. Mechanistic understanding of the toxic effects of arsenic and warfare arsenicals on human health and environment. Cell Biol. Toxicol. Online 2022. https://doi.org/10.1007/s10565-022-09710-8

Nath, B., Chowdhury, R., Ni-Meister, W. & Mahanta, C. 2022. Predicting the distribution of arsenic in groundwater by a geospatial machine learning technique in the two most affected districts of Assam, India: The public health implications. GeoHealth, 6: e2021GH000585 (15 pp). https://doi.org/10.1029/2021GH000585

Oremland, R.S. & Stolz, J.F. 2003. The ecology of arsenic. Science 300, 939–941.

Podgorski, J. & Berg, M. 2020. Global threat of arsenic in groundwater. Science 368, 845–850. https://www.science.org/doi/10.1126/science.aba1510

Poon, R & Chu, I. 2000. Effects of trivalent antimony on human erythrocyte glutathione-S transferases. J. Biochem. Mol. Toxicol. 14, 169–176.

Rodriguez-Lado, L., Sun, G., Berg, M., Zhang, Q., Xue, H., Zheng, Q. & Johnson, C.A. 2013. Groundwater arsenic contamination throughout China. Science 341, 866–868.

Rong, Q., Ling, C., Lu, D., Zhang, C., Zhao, H., Zhong, K., Nong, X. & Qin, X. 2022. Sb(III) resistance mechanism and oxidation characteristics of *Klebsiella aerogenes* X. Chemosphere 293: 133453 (9 pp.). https://doi.org/10.1016/j.chemosphere.2021.133453

Saunders, J.K., Fuchsman, C.A., McKay, C. & Rocap, G. 2019. Complete arsenic-based respiratory cycle in the marine microbial communities of pelagic oxygen-deficient zones. Proc. Natl Acad. Sci. USA 116, 9925–9930. https://doi.org/10.1073/pnas.1818349116

Sforna, M.C., Philippot, P., Somogyi, A., van Zuilen, M.A., Medjoubi, K., Schoepp-Cothenet, B., Nitschke, W. & Visscher, P.T. 2014- Evidence for arsenic metabolism and cycling by microorganisms 2.7 billion years ago. Nature Geosci. 7, 811–815. https://doi.org/10.1073/pnas.181834911610.1038/NGEO2276

Shi, K., Qian Wang, Q. & Wang, G. 2020. Microbial oxidation of arsenite: Regulation, chemotaxis, phosphate metabolism and energy generation. Front. Microbiol. 11:569282 (10 pp.). https://doi.org/10.1073/pnas.181834911610.3389/fmicb.2020.569282

Stolz, J.F., Basu, P., Santini, J.M. & Oremland, R.S 2006. Arsenic and selenium in microbial metabolism. Annu. Rev. Microbiol. 60, 107–130. https://doi.org/10.1073/pnas.181834911610.1146/annurev.micro.60.080805.142053

Sun, D., Zhang, X., Liao, D., Yan, S., Feng, H., Tang, Y., Cao,Y., Qiu, R. & Ma, L.Q. 2022. Novel mycorrhiza-specific P transporter PvPht1;6 Contributes to As accumulation at the symbiotic interface of As-hyperaccumulator *Pteris vittata*. Environ. Sci. Technol. 56, 14178−14187. https://doi.org/10.1021/acs.est.2c04367

Sun, X., Kong, T., Li, L., Häggblom, M.M., Kolton, M., Lan, L., Lau Vetter, M.C.Y., Dong, Y., Gao, P., Kostka, J.E., Li, B. & Sun, W. 2014. Desulfurivibrio spp. mediate sulfur-oxidation coupled to Sb(V) reduction, a novel biogeochemical process. ISME J. (10 pp., under publication 2022). https://doi.org/10.1038/s41396-022-01201-2

Tisarum, R., Lessl, J.T., Dong, X., de Oliveira, L.M., Rathinasabapathi, B, & Ma, L.Q. 2014. Antimony uptake, efflux and speciation in arsenic hyperaccumulator *Pteris vittata*. Environm. Pollut. 186, 110–114. http://dx.doi.org/10.1016/j.envpol.2013.11.033

Tsang, C.-N., Ho, K.-S., Sun, H. & Chan, W.-T. 2011. Tracking bismuth antiulcer drug uptake in single *Helicobacter pylori* cells. J. Am. Chem. Soc. 133, 7355–7357. https://doi.org/10.1021/ja2013278

Wan, X., Lei, M. & Yang, J. 2017. Two potential multi-metal hyperaccumulators found in four mining sites in Hunan Province, China. Catena 148, 67–73. http://dx.doi.org/10.1016/j.catena.2016.02.005

Wan, X., Lei, M. & Chen, T. 2021. Concentration and speciation of arsenic in an insect feeding on the leaves of *Pteris vittata*. Soil Ecol. Lett. 3, 279–287. https://doi.org/10.1007/s42832-021-0091-5

Wan, X. & Yang, J. 2018. The soil amendments to Improve the eEfficiency of the intercropping system of *Pteris vittata* and *Morus alba*. Water Air Soil Pollut. 229: 149 (11 pp.). https://doi.org/10.1007/s11270-018-3805-2

Wang, A., Liu, S., Xie, J., Ouyang, W., He, M., Lin, C. & Liu, X. 2023. Response of soil microbial activities and ammonia oxidation potential to environmental factors in a typical antimony mining area. J. Environm. Sci. 127, 767–779. https://doi.org/10.1016/j.jes.2022.07.003

Xu, X.Y., McGrath, S.P. & Zhao, F.J. 2007. Rapid reduction of arsenate in the medium by plant root. New Phytol. 176, 590–599.

Yang, J., Yan, Y., Lu, N., Wan, X., Yang, J., Shi, H., Chen, T. & Lei, M. 2022. The key nodes and main factors influencing accumulation of soil arsenic in *Pteris vittata* L. under field conditions. Sci. Total Environm. 807: 150787 (9 pp.). https://doi.org/10.1016/j.scitotenv.2021.150787

Yang, Y., McDonald, A.C., Wang, X., Pan, Y. & Wang, M. 2022. Arsenic exposures and prostate cancer risk: A multilevel meta-analysis. J. Trace Elem. Medicine Biol. 72: 126992 (7 pp.). https://doi.org/10.1016/j.jtemb.2022.126992

Yang, Z., Hosokawa, H., Kuroda, M., Inoue, G. & Ike, M. 2021. Microbial antimonate reduction and removal potentials in river sediments. Chemosphere 266: 129192 (10 pp.). https://doi.org/10.1016/j.chemosphere.2020.129192

Zhang, E., Deng, X. & Zhou, G. 2022. Research progress on biotreatment of arsenic pollution. J. Biobased Materials Bioenergy 16, 1–14. https://doi.org/10.1166/jbmb.2022.2111

Zhang, L., Ye, L., Yin, Z., Xiao, K. & Jing, C. 2021. Mechanistic study of antimonate reduction by *Escherichia coli* W3110. Environm. Pollut. 291: 118258 (7 pp.). https://doi.org/10.1016/j.envpol.2021.118258

Zhang, Y., Wan, X. & Lei, M. 2017. Application of arsenic hyperaccumulator *Pteris vittata* L. to contaminated soil in Northern China. J. Geochem. Explor. 182, 132–137. http://dx.doi.org/10.1016/j.gexplo.2016.07.025

Zhao, F.-J., McGrath, S.P. & Meharg, A.A. 2010. Arsenic as a food chain contaminant: Mechanisms of plant uptake and metabolism and mitigation strategies. Annu. Rev. Plant Biol. 61, 535–59. https://doi.org/10.1146/annurev-arplant-042809-112152

Zhou, X., Speer, R.M., Volk, L., Hudson, L.G. & Liu, K.J. 2021. Arsenic co-carcinogenesis: Inhibition of DNA repair and interaction with zinc finger proteins. Seminars in Cancer Biology 76, 86–98. https://doi.org/10.1016/j.semcancer.2021.05.009

Zhu, Y.-G., Yoshinaga, M., Zhao, F.-J. & Rosen, B.P. 2014. Earth abides arsenic Biotransformations. Annu. Rev. Earth Planet. Sci. 42, 443–67. https://doi.org/10.1146/annurev-earth-060313-054942

Zhuge, X.-L., Xu, H., Xiu, Z.-J. & Yang, H.-L. 2020. Biochemical functions of glutathione S-transferase family of *Salix babylonica.* Front Plant Sci. 11: 364 (11 pp.). https://doi.org/10.3389/fpls.2020.00364

CHAPTER SIXTEEN

SELENIUM

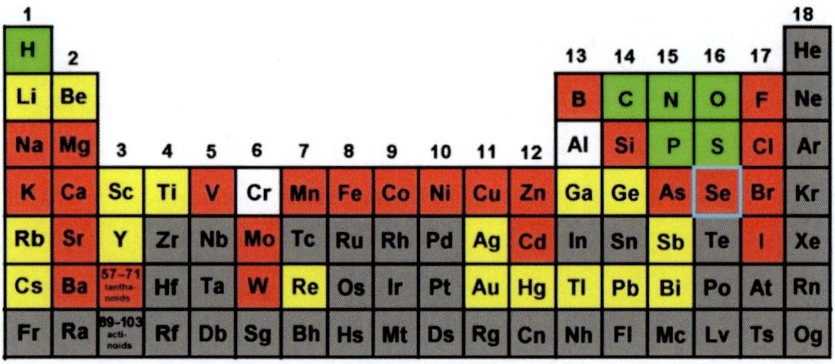

16.1 Introduction

Selenium, in the periodic system between sulfur and tellurium, can in the pure form occur as a grey metal, or in a red non-metallic form. It forms vitally important organic compounds, but is toxic if consumed in too large amounts.

In his famous travel account from end of the thirteenth century, Marco Polo recounts how he was traveling along the Silk Road in the Suzhou region of western China (the present Hexi Corridor, perhaps Sunan County of Gansu Province). He noticed that his horses lost hair and hooves because they had eaten poisonous grass. The toxic symptoms described by Marco Polo were supposed to be typical of selenosis caused by having eaten selenium-accumulating vegetation.

It has long been thought that this is the first report of selenium poisoning. It is a good story, quite recently repeated (Tsuji et al. 2022), but Shao & Zheng (2008) doubt that it is correct. After carefully surveying the vegetation and analyzing different plants as well as the soil in the area in question, the so-called Hexi Corridor, which is part of ancient Suzhou in China, they arrive at the conclusion that the concentration of selenium in the area is far lower than that in a typical seleniferous area, with soils that are unable to support seleniferous plants. Selenium levels in various plants are around 0.1 mg/kg, much lower than the toxic threshold of 3 mg/kg for most farm livestock. One of the grasses in the area is a species of *Oxytropis*, and Shao & Zheng (2008) assume that this is what poisoned Marco Polo's horses. They point out that this grass contains the alkaloid swainsonine, i.e. (1*S*,2*R*,8*R*,8a*R*)-octahydroindolizine-1,2,8-triol, which gives intoxication symptoms similar to those of selenium.

On the other hand, it seems that the American cavalry (especially in Dakota and Nebraska), in their attempts of "civilizing" the "Wild West" during the 19th century had their horses exposed to real selenium poisoning. It has been claimed that General Custer's defeat against the Sioux Indians at Little Bighorn in June 1876 that inspired movies like "They died with their boots on" (Fig. 16.1.1) and "Custer's last stand", https://www.youtube.com/watch?v=lPG-O0DEaVI , took place because Custer's horses were poisoned by selenium.

Fig. 16.1.1. They died with their boots on. Levels of dietary selenium which may cause chronic toxicity are 5 mg/kg or more for cattle and 2 mg/kg or more for sheep (WHO 1987).

Kieliszek (2019) states "Selenium is a micronutrient that is essential for the proper functioning of all organisms", and Bao et al. (2020) write "Selenium (Se) is an essential micronutrient in living organisms", and Mojadadi et al. (2021): "Selenium (Se) is a micronutrient essential for life.". As we shall see further on, these statements are a bit misleading. At least selenium is not directly essential for most plants. But it is essential for vertebrates. Since the element is very unevenly distributed over our planet (and even within a single country: China), both toxic and deficiency occurs; deficiency is more common. It was first observed in humans and pigs In the 1930s as Se deficiency associated cardiomyopathy under the name Keshan disease, and was described for the first time in China (Oropeza-Moe et al. 2017). It was often fatal.

16.2 Selenium in plants

Selenium alleviates the toxicity to plants of several metals: Cadmium toxicity to durum wheat (*Triticum durum*) (Bayat et al. 2021), chromium toxicity to Chinese cabbage (*Brassica campestris* L. ssp. Pekinensis) (Zhao et al. 2019), lead toxicity to *Aechmea blanchetiana* (Bromeliaceae) (Rodrigues Martins et al. 2021), and arsenic toxicity to rice (Bhadwal & Sharma 2022).

Selenium also helps plants to endure drought (Han et al. 2022, Vega-Ravello et al. 2022) and cold (Yang et al. 2021, de Sousa et al. 2022).

While there is no evidence that selenium is an indispensable nutrient element for higher plants, several reports show that when it is added at low concentrations, selenium exerts beneficial effects on plant growth, even if the plants are not subjected to any of the stresses mentioned above (El-Ramady et al. 2016). In addition to the examples mentioned by El-Ramady et al. (2016) we cite Khalofah et al. (2021), who grew quinoa (*Chenopodium quinoa* Willd) plants for 30 days in a sand-peat mixture. They found that there was a linear increase dry matter with Na_2SeO_3 supplementation from zero to 5 mg/L of Na_2SeO_3, and a decline with further increase. For the roots the dry matter increased from 15.0 to 20.2 mg per plant, and for the shoots from 13.3 to 19.5 mg per plant. Malik et al. (2011) grew mung bean (*Phaseolus aureus* Roxb.) hydroponically for 10 days. They do not report the dry matter, but the root length increased from 5.3±0.12 cm without added selenium to 6.30±0.09 cm at 0.75 ppm. The corresponding increase for the shoot was from 3.60±0.11 cm to 4.70±0.12 cm. For tea trees grown in soil, fertilization with selenate was more effective than fertilization with selenite, and 75 Se g/ha increased the spring harvest of leaves from 4.97 to 10.24 g/m^2 (Hu et al. 2003). For some (but not all) varieties of safflower

(*Carthamus tinctorius* L.) grown in soil the oil yield as well as the "biological yield" (kg/hectare) was increased by selenium supplementation (Sher et al. 2022). In such cases (there are several described in the literature), when the plant is grown in soil, it is not clear if there is a direct effect of selenium on the plant, or if the selenium addition releases some other nutrient from the soil. Cunha et al. (2022) grew peanut plants (*Arachis hypogaea* L.) hydroponically with and without supplementation with sodium selenate, and found that the the selenate could increase the peanut yield from 55.5 to 70.3 g per pot. The yield of tartary buckwheat (*Fagopyrum tataricum*) was increased by spraying the leaves with sodium selenite solution. The optimum amount was 2.32 g sodium selenite per 100 m^2 (Liu et al. 2022). In conclusion it can be said that, although higher plants do not have any absolute requirement for selenium, many beneficial effects of the element have been demonstrated. The yeast *Rhodotorula mucilaginosa* CAM4 increased uptake of selenium (from sodium selenite) in spinach (*Spinacea oleracea*) plants, and thereby increased growth (Silambarasan et al. 2022).

There are also plants that hyperaccumulate selenium, particularly in the family Brassicaceae. Several are mentioned by El-Mehdawi et al. (2017). One example is *Stanleya pinnata*, which accumulates selenium to 0.5% of its dry mass (Freeman et al. 2010). Other examples are *Cardamine hupingshanensis*, which can accumulate 1 g selenium per kg (White 2016, Xiao et al. 2022), and *Cardamine violifolia* (Both et al. 2020), but there are also *Cardamine* species that are not selenium hyperaccumulators (e.g., *Cardamine pratensis*. A hyperaccumulating genus in another family is *Astragalus* (Fabaceae). Examples of hyperaccumulating species are *Astragalus bisulcatus*, *A. praelongus*, *A. racemosus*, and *A. rafaelensis*, but there are also several species in this genus that do not hyperaccumulate selenium (Alford et al. 2012).

All plants do take up selenium salts when available, and also produce amino acids, and proteins, including enzymes, containing selenium. Seleno-amino acids that occur are γ-glutamyl-selenocysteine, γ-glutamyl-methyl-selenocysteine, methylselenocysteine, seleno-homocysteine, and methyl-selenomethionine. Other occurring organic selenium-containing compounds are Se-cystathionine, Se-dimethyldiselenide, dimethylselenide, and adenosine phosphoselenate (Hu et al. 2022). In the green alga *Chlamydomonas reinhardtii*, the cDNA-cloned sequence of a glutathione peroxidase homologue codon for selenocysteine was found (Fu et al. 2002), and *Chlorella vulgaris* R117 forms methyl-selenocysteine, selenocysteine, and selenomethionine when selenium (as sodium selenite) is available (Mylenko et al. 2020). Martins Alves et al. (2020) found a glutathione peroxidase in the coacoa plant (*Theobroma cacao*) that has affinity with seleno-methionine in the active site.

Neither for algae can selenium be classified as an element that is absolutely required, although there are cases in which an alga would hardly survive without selenium in its usual environment. An example of this is provided by *Haematococcus pluvialis*, which provides the colour of "red snow". It lives exposed to very strong sunlight, and is provided with red carotenoids for protection, but in addition it has selenium-containing methionine sulfoxide reductase that plays a crucial role in cellular defense against environmental stress (Zheng et al. 2022). Algae are able to rid themselves of excess selenium by converting it to dimethylselenide that evaporates (Zhang et al. 2023, and literature cited by them).

16.3 Selenium in fungi

Selenium has not been shown to be an essential element for any fungus, but fungi take up selenium and form organic selenium compounds when the element is available, e.g., synthesis of Se-methyl-seleno-L-cysteine by *Lentinula edodes* (Klimaszewska et al. 2016). Mycorrhiza-forming fungi may help plants to take up selenium (Zhang et al. 2022). The ascomycetes *Paraconiothyrium sporulosum* and *Stagonospora* sp. reduce Se^{IV} and Se^{VI} aerobically to elemental Se and Se^{-II}, while simultaneously the opposite redox biomineralization process of mycogenic Mn^{II} oxidation to Mn oxides takes place (Rosenfeld et al. 2020).

16.4 Selenium in bacteria and archaea

Bacteria that are able to reduce selenite (SeO_3^{2-}) or selenate (SeO_4^{2-}) to elemental selenium are of interest both for cleaning the environment from these ions, and for producing useful selenium nanoparticles. Of several bacterial strains studied, Won et al. (2021) recommend strain NDSe-7, belonging to the genus *Lysinibacillus*, for reduction of selenite since it can do that at pH 6 – 10 and at a salinity of 0.1 – 7%. Another selenite-reducing bacterium was isolated by Borah et al. (2022). It can be active at selenite concentrations up to 1 *M*. Another bacterial species that is able to reduce selenite to selenium nanoparticles is *Proteus mirabilis* QZB-2, which is active in 300 mM selenite (Huang et al. 2022). The reaction can also be made to proceed in the opposite direction: Luo et al. (2022) report that four strains of bacteria, namely *Dyella* spp. LX-1 and LX-66, and *Rhodanobacter* spp. LX-99 and LX-100, can oxidize seleno-methionine, selenocystine, selenourea and Se(0) to selenite.

Patel et al. (2018) review how various bacteria (species of *Enterobacter*, *Stenotrophomonas*, *Pseudomonas*, and *Bacillus*) living in the rhizosphere of wheat, as well as some living inside the roots (species of *Acinetobacter*, *Bacillus*, and *Klebsiella*) can increase the uptake of selenium by the plants.

Zhang et al. (2022b) provide a broad review of the bacterial selenium biochemistry, but also include a section on selenium in some Archaea: Methanococcales, Methanopyrales, and Lokiarchaeota. Most archaeal selenoproteins are involved in methanogenesis. Lokiarchaeota seem to have the largest archaeal selenoproteome known (12 selenoprotein genes).

16.5 Selenium and human and animal health

The medical literature on selenium effects is very large, and we must limit ourselves to a selective pick and focus on recent literature. Selenium—and more specifically the amino acid selenocysteine—is present in at least 25 human selenoproteins involved in a wide variety of essential biological functions (Barchielli et al. 2022, Huang et al. 2022; a list is provided by Minich 2022). They are all expressed in the brain, which seems to be a prioritized organ in case of selenium deficiency: While most organs lose up to 90% of the selenium during deficiency, the mouse brain retains >75% of it (Nakayama et al. 2007). Deletion of the selenoprotein P gene leads to atrophy of the central nervous system, in particular the cerebellum in dogs (Christen et al. 2021). Low blood concentration of selenium is associated with risk of schizophrenia (Deng et al. 2022).

Already Schwarz & Foltz (1957, 1958) showed that selenium (as pure element or as any of several compounds tested) is required for survival of rats. Al-Mubarak et al. (2022) found that high selenium levels are associated with reduced mortality and reduced new-onset heart failure in non-smokers. Selenium deficiency in patients with COVID- 19 correlated with disease severity and mortality risk (Demircan et al. 2022). Significantly reduced selenoprotein P (a selenium-rich glycoprotein that contains most of the selenium in plasma) concentrations were found in association with intellectual disability in children (Görlich et al. 2022). Schweizer & Fabiano (2022) provide an overview of neurological problems, including epilepsy, associated with the lack of various selenoproteins.

Selenoproteins needed for triiodothyronine synthesis also protect thyroidal and extrathyroidal tissues from hydrogen peroxide produced to kill invading microorganisms. Disbalanced iodine-selenium explains relations between thyroid autoimmune disease and cancer of thyroid, stomach, breast, and prostate (Dijck-Brouwer et al. 2022).

Selenium deficiency causes impaired male and female fertility in humans and other animals (review by Mojadadi et al. 2022). It impairs sperm motility, and severe deficiency stops sperm formation. For further details see the comprehensive review by Barchielli et al. (2022).

Experimental and epidemiological studies have suggested the potential of low selenium intake as a risk factor for type 2 diabetes (Gorini & Vassalle 2022, Zhao et al. 2022). da Fonseca Grili et al. (2022) found a positive correlation between selenium consumption and bone mineral density. According to Yu et al. (2022) also this is related to decrease of the selenoprotein P (see above).

References

Alford, E.R., E.A.H., Fakra, S.C. & M.W 2012. Selenium hyperaccumulation by *Astragalus* Fabaceae does not inhibit root nodule symbiosis. Am. J. Bot. 99, 1930–1941. https://doi.org/10.3732/ajb.1200124.

Amaral, D., Brown, P., Chalfun-Junior, A. & Guilherme, L.R.G. 2022. Selenium enhances chilling stress tolerance in coffee species by modulating nutrient, carbohydrates, and amino acids content. Front. Plant Sci. 13:1000430 (15 pp.). https://doi.org/10.3389/fpls.2022.1000430

Bao, P., Li, G.-X., He, Y.-Q. & Ren, H.-Y. 2020. Selenium nanovirus and its cytotoxicity in selenite-exposed higher living organisms. Biochem. Biophys. Reps 21: 100733. https://doi.org/10.1016/j.bbrep.2020.100733

Barchielli, G., Capperucci, A. & Tanini, D. (2022) The role of selenium in pathologies: An updated review. Antioxidants 11: 251 (49 pp.). https://doi.org/10.3390/antiox11020251

Bayat, S., Dalir, N., Mokhtassi-Bidgoli, A., Malakouti, M.J. & Shahbazi, K. 2021. Selenium alleviates cadmium-induced stress in durum wheat (*Triticum durum*) by enhancing the accumulation of cadmium in the roots and by modulating of photosynthesis parameters. J. Plant Nutr. (in press) (17 pp.). https://doi.org/10.1080/01904167.2022.2105713

Bhadval, S. & Sharma, S. 2022. Selenium alleviates carbohydrate metabolism and nutrient composition in arsenic stressed rice plants. Rice Sci. 29, 385−396. http://dx.doi.org/10.1016/j.rsci.2021.11.008

Borah, S.N., Goswami, L., Sen, S., Sachan, D., Sarma, H., Montes, M., Peralta-Videa, J.R., Pakshirajan, K. & Narayan, M. 2021. Selenite bioreduction and biosynthesis of selenium nanoparticles by *Bacillus paramycoides* SP3 isolated from coal mine overburden leachate. Environm. Pollut. 285: 117519 (12 pp).

https://doi.org/10.1016/j.envpol.2021.117519
Both, E.B., Stonehouse, G.C., Lima, L.W., Fakra, S.C., Aguirre, B., Wangeline, A.L., Xiang, J., Yin, H., Jókai, Z., Soós, A., Dernovics, M. & Pilon-Smits, E.H.A. 2020. Selenium tolerance, accumulation, localization and speciation in a *Cardamine* hyperaccumulator and a non-hyperaccumulator. Sci. Tot. Environm. 703: 135041 (12 pp.). https://doi.org/10.1016/j.scitotenv.2019.135041

Christen, M., Högler, S., Kleiter, M., Leschnik, M., Weber, C., Thaller, D., Jagannathan, V. & Leeb, T. 2021. Deletion of the SELENOP gene leads to CNS atrophy with cerebellar ataxia in dogs. PLoS Genet 17 :e1009716 (12 pp.). https://doi.org/10.1371/journal.pgen.1009716

Cunha, M.L.O., de Oliveira, L.C.A., Silva, V.M., Montanha, G.S. & dos Reis, A.R. 2022. Selenium increases photosynthetic capacity, daidzein biosynthesis, nodulation and yield of peanuts plants (*Arachis hypogaea* L.). Plant Physiol. Biochem. 190, 231–239. https://doi.org/10.1016/j.plaphy.2022.08.006

da Fonseca Grili, P.P., Vilarinho Vidigal, C., Ferreira da Cruz, G., Albergaria, B.H., Marques-Rocha, J.L., Silva Pereira, T.S. & Guandalin, V.R. 2022. Dietary consumption of selenium inversely associated with osteoporosis in postmenopausal women. Front. Nutrit. 9:997414. (9 pp.). https://doi.org/10.3389/fnut.2022.997414

Deng, M.-G., Cui, H.-T., Nie, J.-Q., Liang, Y. & Chai, C. 2022. Genetic association between circulating selenium level and the risk of schizophrenia in the European population: A two-sample Mendelian randomization study. Front. Nutr. 9:969887 (8 pp.). https://doi.org/10.3389/fnut.2022.969887

de Sousa, G.F., Silva, M.A., de Morais, E.G., Van Opbergen, G.A.Z., Van Opbergen, G.G.A.Z., de Oliveira, R.R., Amaral, D., Brown, P., Chalfun-Junior, A. & Guilherme, L.R.G. 2022. Selenium enhances chilling stress tolerance in coffee species by modulating nutrient, carbohydrates, and amino acids content. Front. Plant Sci. 13:1000430 (15 pp.). https://doi.org/10.3389/fpls.2022.1000430

Dijck-Brouwer, D.A.J., Muskiet, F.A.J., Verheesen, R.H., Schaafsma, G., Schaafsma, A. & Geurts, J.M.W. 2022. Thyroidal and extrathyroidal requirements for iodine and selenium: A combined evolutionary and (patho)physiological approach. Nutrients 14, 3886 (29 pp.). https://doi.org/10.3390/nu14193886

El Mehdawi, A.F., Jiang, Y., Guignardi, Z.S., Esmat, A., Pilon, M., Pilon-Smits, E.A.H., Schiavon, M. 2017. Influence of sulfate supply on selenium uptake dynamics and expression of sulfate/selenate transporters in selenium hyperaccumulator and nonhyperaccumulator Brassicaceae. New Phytol. 217, 194–205. https://doi.org/10.1111/nph.14838

El-Ramady, H., Abdalla, N., Taha, H.S., Alshaal, T., El-Henawy, A., Faizy, S. E.-D.A., Shams, M.S. et al. 2016. Selenium and nano-selenium in plant nutrition. Environ. Chem. Lett. 14, 123–147. https://doi.org/10.1007/s10311-015-0535-1

Freeman, J.L., Tamaoki, M., Stushnoff, C., Quinn, C.F., Cappa, J.J., Devonshire, J., Fakra, S.C., Marcus, M.A., McGrath, S.P., Van Hoewyk, D. & Pilon-Smits, E.A.H. 2010. Molecular mechanisms of *Selenium* tolerance and hyperaccumulation in *Stanleya pinnata.* Plant Physiol. 153, 1630–1652. www.plantphysiol.org/cgi/doi/10.1104/pp.110.156570

Fu, L.-H., Wang, X.-F., Eyal, Y., She, Y.-M., Donald, L.J., Standing, K.G. & Ben-Hayyim, G. 2002. A selenoprotein in the plant kingdom. J. Biol. Chem. 277, 25983–25991.

Gorini, F. & Vassalle, C. 2022. Selenium and selenoproteins at the intersection of type 2 diabetes and thyroid pathophysiology. Antioxidants 11, 1188 (22 pp.). https://doi.org/10.3390/antiox11061188

Han, D., Tu, S., Dai, Z., Huang, W., Jia, W., Xu, Z. & Shao, H. 2022. Comparison of selenite and selenate in alleviation of drought stress in *Nicotiana tabacum* L. Chemosphere 287: 132136 (11 pp.). https://doi.org/10.1016/j.chemosphere.2021.132136

Hu, J., Wang, Z., Zhang, L., Peng, J., Huang, T., Yang, X., Jeong, B.R. & Yang, Q. 2022. Seleno-amino acids in vegetables: A review of their forms and metabolism. 13: 804368 (13 pp.). https://doi.org/10.3389/fpls.2022.804368

Hu, Q., Xu, J. & Pang, G. 2003. Effect of selenium on the yield and quality of green tea leaves harvested in early spring. J. Agric. Food Chem. 51, 3379–3381.

Khalofah, A., Migdadi, H. & El-Harty, E. 2021. Antioxidant enzymatic activities and growth response of quinoa (*Chenopodium quinoa* Willd) to exogenous selenium application. Plants 10: 719 (13 pp.). https://doi.org/10.3390/plants10040719

Kieliszek, M. 2019. Selenium–fascinating microelement, properties and sources in food. Molecules 24, 1298 (14 pp.). https://doi.org/10.3390/molecules24071298

Klimaszewska, M., Górska, S., Dawidowski, M., Podsadni. P. & Turło, J. 2016. Biosynthesis of Se-methyl-seleno-lcysteine in Basidiomycetes fungus *Lentinula edodes* (Berk.) Pegler. SpringerPlus 5:733 (8 pp.). https://doi.org/10.1186/s40064-016-2498-5

Liu, P., Song, L., Hao, S., Qin, J., Yang, C., Yang, W., Feng, M., Zhang, M., Wang, C. & Song, X. 2022. Effects of selenium application concentration, period and method on the selenium content and grain yield of Tartary buckwheat of different varieties. J. Sci. Food Agric. 102, 6868–6876. https://doi.org/10.1002/jsfa.12048

Luo, X., Wang, Y., Lan, Y., An, L., Wang, G., Li, M. & Zheng, S. 2022. Microbial oxidation of organic and elemental selenium to selenite. Sci. Total Environm.833: 155203 (8 pp.). http://dx.doi.org/10.1016/j.scitotenv.2022.155203

Malik, J.A., Kumar, S., Thakur, P., Sharma, S., Kaur, N., Kaur, R.P., Pathania, D., Bhandhari, K., Kaushal, N., Singh, K., Srivastava, A., & Nayyar, H. 2011. Promotion of growth in mungbean (*Phaseolus aureus* Roxb.) by selenium is associated with stimulation of carbohydrate metabolism. Biol. Trace Elem. Res. 143, 530–539. https://doi.org/10.1007/s12011-010-8872-1

Martins, J.P.R., Conde, L.T., Falqueto, A.R. & Gontijo, A.B.P.L. 2021. Selenium biofortified *Aechmea blanchetiana* (Bromeliaceae) can resist lead-induced toxicity during in vitro culture. Acta Physiolog. Plantarum 43:149 (16 pp.). https://doi.org/10.1007/s11738-021-03323-0

Martins Alves, A.M., Pereira Menezes Reis, S., Peres Gramacho, K. & Micheli, F. 2020. The glutathione peroxidase family of Theobroma cacao: Involvement in the oxidative stress during witches' broom disease. Int. J. Biol. Macromolec. 164, 3698–3708. https://doi.org/10.1016/j.ijbiomac.2020.08.222

Minich, W.B 2022. Selenium metabolism and biosynthesis of selenoproteins in the human body. Biochemistry (Moscow) 87, S168–S177 (transl. from Uspekhi Biolog. Khimii 62, 369–390).

Mojadadi, A., Au, A., Salah, W., Witting, P. & Ahmad, G. 2021. Role for selenium in metabolic homeostasis and human reproduction. Nutrients 13: 3256 (21 pp.). https://doi.org/10.3390/nu13093256

Mylenko, M., Vu, D.L., Kuta, J., Ranglová, K., Kubáč, D., Lakatos, G., Grivalský, T., Caporgno, M.P., da Câmara Manoel, J.A., Kopecký, J., Masojídek, J. & Hrouzek, P. 2020. Selenium incorporation to amino acids in *Chlorella* cultures grown in phototrophic and heterotrophic regimes. J. Agric. Food Chem. 68, 1654−1665. https://dx.doi.org/10.1021/acs.jafc.9b06196

Nakayama, A., Hill, K.E., Austin, L.M., Motley, A.K. & Burk, R.F. 2007. All regions of mouse brain are dependent on selenoprotein P for maintenance of selenium. J. Nutrit. 137, 690–693.

Oropeza-Moe, M., Wisløff, H. & Bernhoft, A. 2015. Selenium deficiency associated porcine and human cardiomyopathies. J. Trace Elements Med. Biol. 31, 148–156. http://dx.doi.org/10.1016/j.jtemb.2014.09.011

Patel, P.J., Trivedi, G.R., Shah, R.K. & Saraf, M. 2018. *Selenorhizobacteria*: As biofortification tool in sustainable agriculture. Biocatalysis Agricult. Biotechnol. 14, 198–203. https://doi.org/10.1016/j.bcab.2018.03.013

Rosenfeld, C.E., Sabuda, M.C., Hinkle, M.A.G., James, B.R. & Santelli, C.M. 2020. A fungal-mediated cryptic selenium cycle linked to manganese biogeochemistry. Environ. Sci. Technol. 54, 3570−3580. https://dx.doi.org/10.1021/acs.est.9b06022

Schwarz, K. & Foltz, C.M. 1957. Selenium as an integral part of factor 3 against dietary necrotic liver degeneration. J. Am. Chem. Soc. 79, 3292–3293, reprinted in Nutrition 15, 338–340 (1978).

Schwarz, K. & Foltz, C.M. J. 1958. Factor 3 activity of selenium compounds. J. Biol. Chem. 233, 245–251.

Schweizer, U. & Fabiano, M. 2022. Selenoproteins in brain development and function. Free Radical Biol. Med. 190, 105–115. https://doi.org/10.1016/j.freeradbiomed.2022.07.022

Sher, A., Ul-Allah, S., Sattar, A., Ijaz, M., Ahmad, W., Bibi, Y. & Qayyum, A. 2022. The Effect of selenium concentration on the quantitative and qualitative yield of four safflower (*Carthamus tinctorius* L.) genotypes. J. Soil Sci. Plant Nutr. 22, 2663–2669. https://doi.org/10.1007/s42729-022-00834-8

Silambarasan, S., Logeswari, P., Vangnai. A.S. & Cornejo, P. 2022. *Rhodotorula mucilaginosa* CAM4 improved selenium uptake in *Spinacia oleracea* L. and soil enzymatic activities under abiotic stresses. Environm. Sci. Pollut. Res., online 2022. https://doi.org/10.1007/s11356-022-21935-y

Tsuji, P.A., Santesmasses, D., Lee, B.J., Gladyshev, V.N. & Hatfield, DL. 2022. Historical roles of selenium and selenoproteins in health and development: The good, the bad and the ugly. Int. J. Mol. Sci. 23: 5 (20 pp.). https://doi.org/10.3390/ijms23010005

Vega-Ravello, R., Romero-Poma, M.B., de Oliveira, C., Guimarães Guilherme, L.R. & Lopes, G. 2022. Soil selenium addition for producing Se-rich quinoa and alleviating water deficit on the Peruvian coast. J. Soil Sci. Plant Nutr. (in press). https://doi.org/10.1007/s42729-022-01004-6

White, P.J. (2016) Selenium accumulation by plants. Ann. Bot. 117, 217–235. https://doi.org/10.1093/aob/mcv180

WHO 1987. Environmental health criteria 58: selenium. Geneva: World Health Organisation.

Won, S., Ha, M.-G., Nguyen, D.D. & Kang, H.Y. 2021. Biological selenite removal and recovery of selenium nanoparticles by haloalkaliphilic bacteria isolated from the Nakdong River. Environm. Pollut. 280: 117001 (11 pp.). https://doi.org/10.1016/j.envpol.2021.117001

Xiao, Z., u, Y., Zou, Y., Zhang, C., Ding, L., Luo, K., Tang, Q. & Zhou, Y. 2022. Gene Identification, expression analysis and molecular docking of ATP sulfurylase in the selenization pathway of *Cardamine hupingshanensis*. Plant Biology 22:491 (17 pp.). https://doi.org/10.1186/s12870-022-03872-7

Yang, N., Sun, K., Wang, X., Wang, K., Kong, X., Gao, J. & Wen, D. 2021. Melatonin participates in selenium-enhanced cold tolerance of cucumber seedlings. Front. Plant Sci. 12:786043 (11 pp.). https://doi.org/10.3389/fpls.2021.786043

Yu, R., Wang, Z., Ma, M., Xu, P., Liu, L., Tinkov, A.A., Lei, X.G. & Zhou, J.-C. 2022. Associations between circulating SELENOP level and disorders of glucose and lipid metabolism: A meta-analysis. Antioxidants 11: 1263 (20 pp.). https://doi.org/10.3390/antiox11071263

Zhang, G., Chen, X., Li, F., Que, W., Qian, J., Fang, J. & Ding, T. 2023. Effects of environmental factors on selenite volatilization by freshwater microalgae. Sci. Total Environm. 854: 158539 (10 pp.). http://dx.doi.org/10.1016/j.scitotenv.2022.158539

Zhang, Z., Li, B., Liu, Y.;,He, L.; Pang, T., Chen, Z., Shohag, M.J.I., Miao, X., Li, X.; Gu, M. & Wei, Y. 2022a. Arbuscular mycorrhizal fungal inoculation increases organic selenium accumulation in soybean (*Glycine max* (Linn.) Merr.) growing in selenite-spiked soils. Toxics 10: 565 (13 pp.). https://doi.org/10.3390/toxics10100565

Zhang, Y., Jin, J., Huang, B., Ying, H., He, J. & Jiang, L. 2022b. Selenium metabolism and selenoproteins in prokaryotes: A bioinformatics perspective. Biomolecules 12: 917 (14 pp.). https://doi.org/10.3390/biom12070917

Zhao, J., Zou, H., Huo, Y., Wei, X. & Li, Y. 2022. Emerging roles of selenium on metabolism and type 2 diabetes. Front. Nutr. 9:1027629 (11 pp.). https://doi.org/10.3389/fnut.2022.1027629

Zhao, Y., Hua, C., Wang, X., Qing, X., Wang, P., Zhang, Y., Zhang, X. & Zhaoa, X. 2019. Selenium alleviated chromium stress in Chinese cabbage (*Brassica campestris* L. ssp. Pekinensis) by regulating root morphology and metal element uptake. Ecotox. Environm. Safety 173, 314–321. https://doi.org/10.1016/j.ecoenv.2019.01.090

Zheng, Y., Wang, Z., ue, D., Tao, M., Jiang, F., Jia, B., Li, Y., Huang, G. & Hu, Z. 2022. Characterization of a new selenoprotein methionine sulfoxide reductase from *Haematococcus pluvialis* and its antioxidant activity in response to high light intensity, hydrogen peroxide, glyphosate, and cadmium exposure. Ecotoxicol. Environm. Safety 242: 113903 (11 pp.). https://doi.org/10.1016/j.ecoenv.2022.113903

CHAPTER SEVENTEEN

HALOGENS

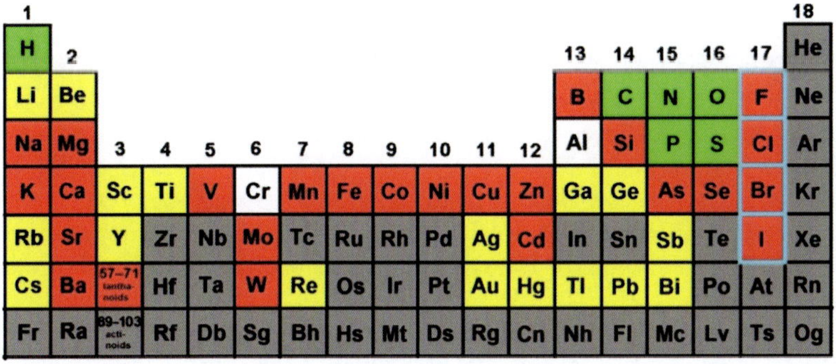

17.1 Fluorine

When we hear the word fluorine (or fluoride) we probably first think of toothpaste or teeth. The outermost, most resistant layer of our teeth, the enamel, contains fluoride ions. It consists of fluorapatite, in which some of the hydroxygroups of hydroxyapatite have been changed to fluoride. This is the only known function for fluorine in our own bodies.

The fluorine in fluorapatite forms hydrogen bonds to neighbouring hydroxyl groups, thereby increasing the dissociation energy barrier of the small anions (i.e., OH^- and F^-), and therefore, protecting the fluorine anion and the hydroxyl groups in the inner layer (Wang et al. 2022).

There is consensus about the tooth-protecting effects of fluoride, but has been much discussion about the mode of action. One issue has been whether it is the ingestion of fluoride, i.e. systemic effects, that are important, or the "topical" effects, i.e. the direct contact of teeth with fluoride on the tooth surface. The evidence now seems to be in favour of topical effects; see ten Cate & Buzalaf (2019) for literature. Another issue has been whether fluoride prevents or heals caries. The correct opinion is probably "both" (e.g., ten Cate 2004, Gao et al. 2016). While most researchers have focused on effects on biomineral (hydroxyapatite) crystals (e.g., Wang et al. 2022), Zhang et al. (2022) have focused on effects on the microbe flora.

Fluorine has unusual properties. It has (1) the smallest atomic radius of all non-hydrogen and non-noble-gas elements, (2) the largest electronegativity of all open-shell atoms, and (3) the highest reduction potential of all elements (2.87 V) (Dehnen et al. 2021).

But fluorine biology extends beyond our teeth. Fluoroacetate is the most common natural fluorinated organic substance. It was first isolated by Marais (1943) from a very poisonous African plant, *Dichapetalum cymosum*, see also Meyer (2019). It has been found in more than 40 plant species, mostly belonging in the families of Malpighiaceae, Fabaceae, Dichapetalaceae, Rubiaceae, and Bignoniaceae (Grobbelaar and Meyer 1990; Lee et al. 2014 as cited by Cheng & Ma 2021). Fluoroacetate and other fluorinated organic compounds are produced by various microorganisms: *Streptomyces cattleya*, *Streptomyces sp.* MA37, *Streptomyces xinghaiensis* NRRL B24674, *Norcardia brasiliensis*, *Actinoplanes sp.* N902-109 and *Actinopolyspora mzabensis* (Deng et al. 2014; Huang et al. 2014, Ren et al. 2020, Sanada et al. 1986, Sooklal et al. 2020). The synthesis of fluoroacetate is mediated by the enzyme fluorinase (Fig. 17.1.1).

As mentioned above, fluorinated organic compounds are produced by several microorganisms, but the best investigated one is *Streptomyces cattleya* (Hamilton et al. 1997, Moss et al. 2000, O'Hagan et al. 2002, Cadicamo et al. 2004, Onega et al. 2007, Cobb et al. 2015). It produces also 5′-fluoro-5-deoxyfluoroadenosine (O'Hagan et al. 2002).

Some steps in fluorine metabolism, ending in the formation of fluoroacetaldehyde, are described in Fig. 17.1.1.

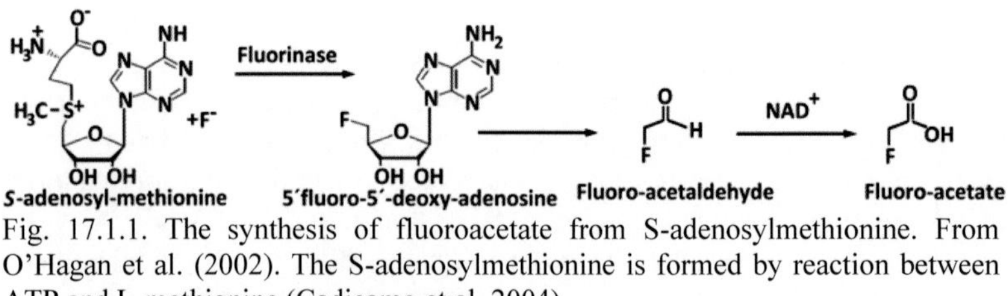

Fig. 17.1.1. The synthesis of fluoroacetate from S-adenosylmethionine. From O'Hagan et al. (2002). The S-adenosylmethionine is formed by reaction between ATP and L-methionine (Cadicamo et al. 2004).

From fluoroacetate 4-fluorothreonine is formed by the action of the enzyme 4-fluorothreonine transaldolase; alternatively fluoroacetate results from the action of fluoroacetaldehyde dehydrogenase (Deng et al. 2008).

An important medical product produced by *Streptomyces calvus* and some other bacteria is the fluorine-containing antibiotic nucleocidin (Zhu et al. 2015, Chen et al. 2021, Ngviprom et al. 2021, Fig. 17.1.2).

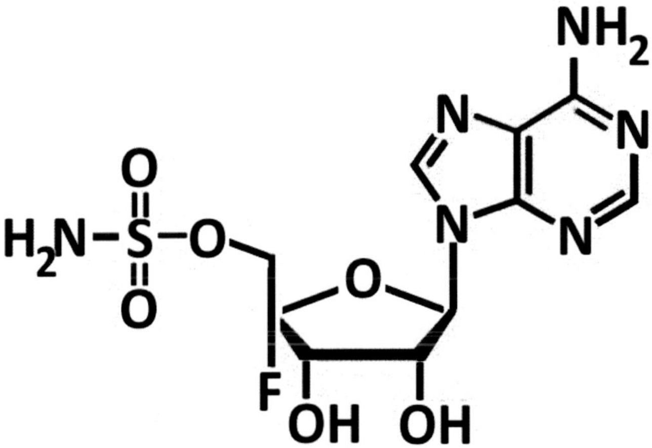

Fig. 17.1.2. Nucleocidin.

Fluorinated compounds have been found in animals, but they probably derive from the animal feed (Cheng & Ma 2021).

In plants several fluoro-fatty acids have been found: ω-fluorooleic acid, ω-fluoropalmitic acid,

ω-fluorocapric acid, ω-fluoromyristic acid, and ω-fluoro derivatives of palmitoleic, stearic, tlinoleic, arachidic, and eicosenoic acids and threo-18-fluoro-9,10-dihydroxystearic acid. In addition the following compounds have been found in plants: Fluoroacetate, fluoroacetone, and (2R,3R)-2-fluorocitrate (see Harper & Hagan 1994 for literature). Baunthiyal & Pandey (2012) report that the fluoro-fatty acids in plants derive from the replacement of acyl-CoA by fluoroacyl-CoA in their synthesis, and fluorocitric acid is synthesized in the corresponding way (Fig. 17.1.3).

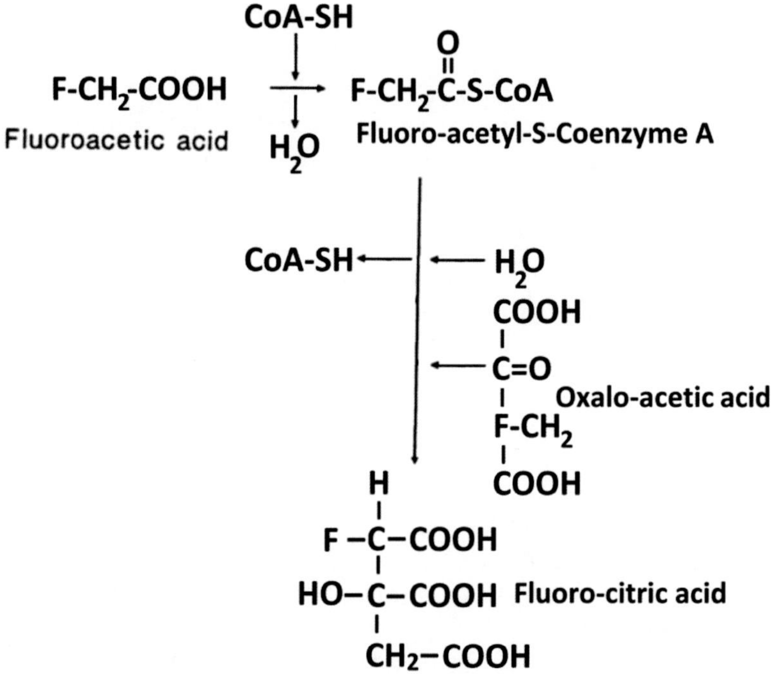

Fig. 17.1.3. Synthesis of fluorocitric acid in *Dichapetalum cymosum*. From Meyer (1991).

17.2 Chlorine

One important role of chlorine is carried out as the free chloride ion. In our blood plasma the concentration is, at 103 m*M*, the dominating negative ion, helped by bicarbonate, on second place at 27 m*M*, to balance the dominating positive ion, Na^+ (142 m*M*). It is probably because of its high concentration in seawater that the chloride ion has got this dominating position, because in the solid Earth crust there is more fluoride than chloride.

Chlorine also occurs in organic biomolecules, such as the antibiotics aureomycin, chloramphenicol (Fig. 17.2.1, Fig. 17.2.2), and vancomycin (from the soil bacterium *Amycolatopsis orientalis*).

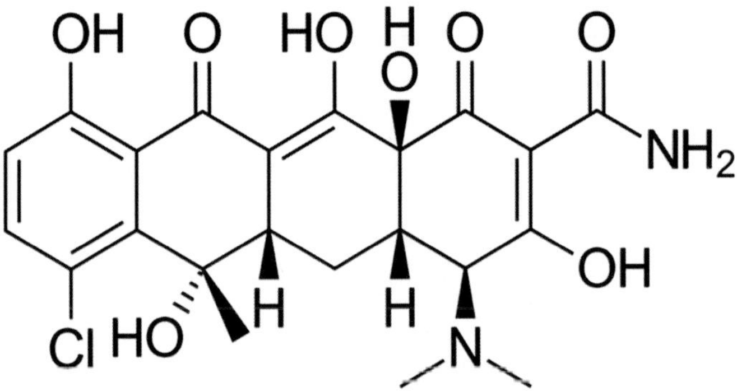

Fig. 17.2.1. Aureomycin (chlortetracycline) from the actinomycete *Streptomyces aureofaciens,*

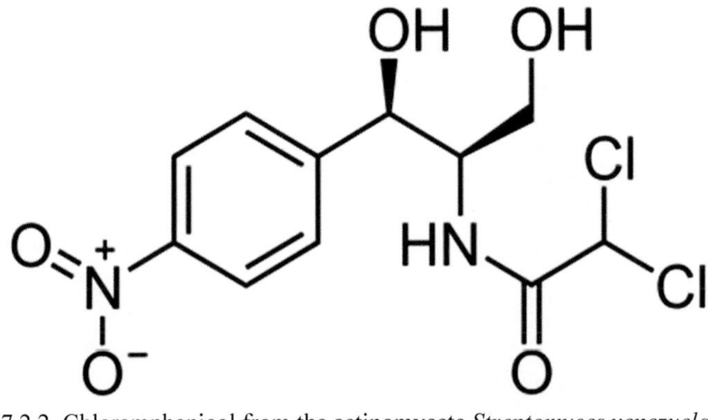

Fig. 17.2.2. Chloramphenicol from the actinomycete *Streptomyces venezuelae.*

A general review of chlorine in organisms and their environment is provided by Winterton (2000). Naturally formed chlorinated organic compounds in soil frequently exceed the abundance of chloride (Svensson et al. 2021). The roles of chlorine in biology are too many to enumerate here, and I refer to the recent review by Barnum & Coates (2022). Here I shall just stress the importance of one role of chlorine that is often neglected, without which most present life on our planet, including our own, would be impossible.

I am thinking of the two chloride ions present in the oxygen-evolving center of the water-oxidizing photoenzyme (photosystem II) of cyanobacteria, algae, and plants (see, e.g., Imaizumi & Ifuku 2020 or Brahmachari et al. 2021). Without this, there would be at most traces of life-giving molecular oxygen. The rôles of these chloride ions are described as presently understood by Imaizumi & Ifuku (2020).

The chloride ions are located near the Mn_4CaO_5 cluster at the oxygen-evolving center of PSII. They are held in place by the surrounding protein. In the absence of chloride, the $S_1 \rightarrow S_2$ transition takes place as in its presence, but the $S_2 \rightarrow S_3$ and $S_3 \rightarrow S_0$ transitions are blocked (Wincencjusz 1997).

17.3 Bromine

Surprisingly late was it realized that bromine is an essential element for humans and animals (McCall et al. 2014), with the possible exception of sponges (Fidler et al. 2014), although some sponges are reported to accumulate bromopyrrole alkaloids (Rua et al. 2018), and at least those of the genus *Aplysina* harbour tryptophan halogenases due to bacteria (Gutleben et al. 2019). Bromine is required for properly crosslinking the collagen IV scaffold of basement membranes, and the enzyme, a peroxidase, executing this construction is called peroxidasin (Péterfi & Geiszt 2014). It is a heme peroxidase which oxidizes bromide to hypobromous acid (HOBr), which oxidatively creates sulfilimine crosslinks between the sulfur of methionine-93 and the amine nitrogen of hydroxy-lysine-211 in collagen IV (Vanacore et al. 2009, Bathish et al. 2020). Bromine deficiency has been shown to be lethal in the fruitfly (Ceko et al. 2015).

Bromine compounds occur also in some algae, especially red algae (Fries 1975, Küpper et al. 2014), and bromide in their environment can be necessary for their growth (Fries 1975). But bromine compounds occur also in brown algae, e.g., as bromophenol in *Leathesia nana* (Xu et al. 2004) and ethyl methyl 2-bromobenzene 1,4-dioate in *Dictyopteris hoytii* (Rehman et al. 2019). The brown alga *Ectocarpus siliculosus* accumulates both iodide and bromide in the cell wall, and a range of volatile halogenated compounds are emitted. The oxidation of bromide ions by bromoperoxidase II from *Ascophyllum nodosum* is described by Radlow et al. (2018). The largest of all brown algae, kelp (*Macrocystis pyrifera*) forms dibromomethane, CH_2Br_2 and bromoform, $CHBr_3$ (Goodwin et al. 1997). For a red alga it was shown that the brominated compounds decrease the growth of epiphytic bacteria (Paul et al. 2006).

Bromine compounds, or the enzymes that produce them, have been found also in many other algae, e.g., Chlorarachniophyta (Abe et al. 2020), Chlorophyta (Lim et al. 2019), diatoms (Moore et al. 1996, Hill & Manley 2009, Hughes & Sun 2016, Nguvava et al. 2016, Lim et al. 2019, Abe et al.

2020). They are also emitted by cyanobacteria of the genus *Synechococcus* (Johnson et al. 2015), Lim et al. 2019).

Halogenated organic compounds are generated by bromoperoxidases and other haloperoxidases (halogenating enzymes). The haloperoxidases are, based on phylogeny, classified into six families (Xu & Wang 2016).

Production of volatile organic bromine compounds is detrimental to the ozone. In the troposphere the compounds are stable, but after reaching the stratosphere, where most of the ozone is, they are decomposed by ultraviolet-C radiation, resulting in free bromine atoms, and oxygen molecules are also split to produce free oxygen atoms. The following reactions result: $Br+O_3 \rightarrow BrO+O_2$; $BrO+O \rightarrow Br+O_2$. In other words the bromine atoms are not consumed, they act as a catalyst that destroys ozone. Although the role of algae for this process was small compared to anthropogenic release of ozone-destroying compounds at its peak, it is not negligible. And there is uncertainty about how the production of bromine organics will be affected by the ongoing global warming (Keng et al. 2020).

It was mentioned above that bromine is an essential element for animals with the exception of sponges, but in most cases only very small amounts are required. Some animals contain larger amounts. Ohyoshi et al. (2022) found bromoditerpenes with an α-methylene carbonyl in the sea hare *Aplysia kurodai*, and in *Aplysia dactylomela* Binh et al. (2022) found brominated sesquiterpenes. In the same animal Palaniveloo et al. (2020) found several brominated and chlorinated organic compounds, and concluded that they were derived from the food.

Mammals: During inflammation, leukocyte derived eosinophil peroxidase catalyses the formation of hypobromous acid, which can brominate tyrosine residues in proteins to form bromotyrosine (Mani et al. 2016), and Arnhold & Malle (2022) described the halogenation activity of mammalian heme peroxidases.

According to Cochereau et al. (2022) 133 bromine compounds are known from fungi. They stress that fungi living in the sea, where bromide is more accessible than on land, are not yet well explored.

17.4 Iodine

For humans iodine is important, since it is part of important hormones, triiodothyronine and thyroxine, that are synthesized in cells called thyrocytes in the thyroid. Deficiency can result not only from lack of consumed iodide, but also from disturbance or lack of transporters of iodide into the thyrocytes. At least in mice the most important iodide transport protein is Slc26a7 (Yamaguchi et al. 2022). Disturbed thyroxine and triiodothyronine production can lead to a number of problems (https://medlineplus.gov/genetics/condition/congenital-hypothyroidism/) including goiter, dwarfism (cretinism) and intellectual disability.

But thyroxine is important not only for vertebrates and lancelets, but also for other animals, such as echinoderms (Chino et al. 1994), arthropods (Dvoretsky et al. 2022), and molluscs (Li et al. 2020), and not only for animals, but also for some other organisms, such as the slime mold *Dictyostelium discoideum* (Singh et al. 2014).

Iodine (I) is one of the least abundant elements on Earth's surface; soils have only about 3 mg kg^{-1} of total iodine (Mohiuddin et al., 2019).

Iodine is not an essential element for land plants, but can have beneficial effects, such as accelerating flowering (Kiferle 2020). Plants, algae, and phytoplankton are capable of volatilizing I in the form of iodomethane (also known as methyl iodide, CH_3I), and this reaction is catalyzed by enzymes with methyltransferase activity dependent on S-adenosyl-L-methionine (Itoh et al., 2009). More than 110 iodine-containing metabolites are known, and a general survey of the older literature is provided by Dembitsky (2006). Thus the statement by Holzer & Laudet (2015): "…thyroid hormones… are the only known biological iodinated compounds" is not correct,

Scientists have been interested in iodine not only for its role in biochemistry and physiology. A second reason is the generation of radioactive iodine by fission of uranium in power stations and warfare, and the consequences of warfare and accidents with power stations, as has happened in Harrisburg, Fukushima, and Chernobyl. There are 37 known isotopes of iodine, and only one is not radioactive, ^{127}I. Four radioactive iodine nuclides result from uranium fission. The yield of ^{129}I from ^{235}U fission is about one percent, and the half-life 15.7 million years, so it does not represent any immediate great danger; the activity is low. In the long run, however, its accumulated effect may be important, since it has become widely distributed in the environment. The yields of ^{133}I and ^{135}I are both between 6 and 7%, and their half-lives 21 h and 6.6 h, respectively, and do not present any great danger. But the iodine nuclide to really worry about is ^{131}I. The yield from ^{235}U fission is about 3%, and the half-life 8.0 days, long enough for finding its way into people and other living beings, and short enough to deliver its radiation dose before being excreted. In humans and other mammals it accumulates in the thyroid and gives rise to thyroid cancer. If people consume enough stable iodine, ^{127}I, some of the ^{131}I is "washed out", and the effect decreased (Fig. 17.4.1).

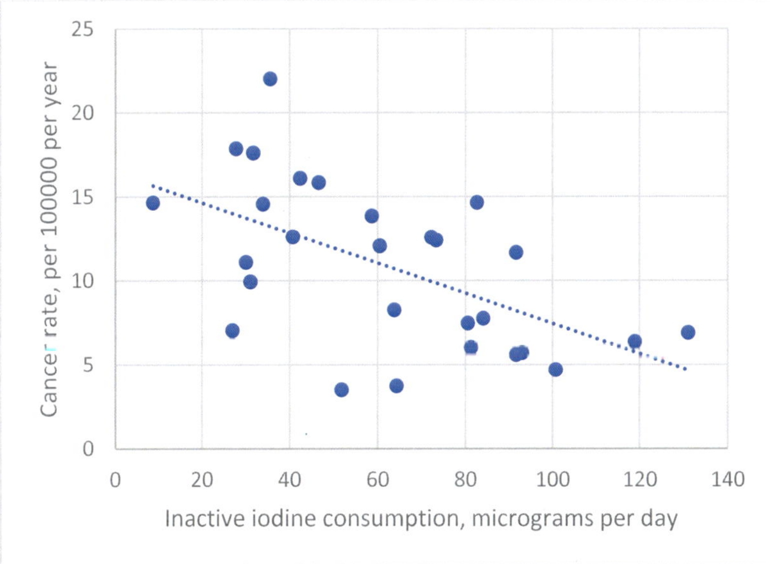

Fig. 17.4.1. Relation between consumption of (stable) iodine and incidence of thyroid cancer. Data from Korobova et al. (2022).

Most of the iodine in the biosphere is in the ocean, and if there were no way for it to get unto land, many terrestrial animals would be in great trouble, and in the past people would also have been. Fortunately there are in the ocean organisms that form volatile iodine compounds: methyliodide (CH_3I), diiodomethane (CH_2I_2), chloroiodomethane (CH_2ClI), iodocthanc (C_2H_5I), and iodopropane (C_3H_7I); for literature see Amachi (2008). CH_3I is the most important of these (Amachi et al. 2001) and is important for spreading iodine in the environment. It is estimated that in the regions between 40°N and 40°S up to 174 Gg of methyliodide is produced annually via biological and photochemical processes (Stemmler et al. 2014). According to Bell et al. (2002), almost 70% of the annual flux of methyl-iodide originates from the open oceanic waters, but some is produced also in rice-fields (e.g., Lee-Taylor & Redeker 2005, Carlessi et al. 2021), as well as from many other plants (Itoh et al. 2009). Methyl-iodide forms from inorganic iodide and S-adenosyl-L-methionine by halide/thiocyanate methyl transferase, which can also catalyze corresponding reactions with CN^-, SCN^-, Cl^-, and Br^- (Lankau et al. 2022). Also some fungi produce methyl-iodide, the basidiomycete *Lentinula edodes* showing the highest rate of those tested (Ban-nai et al. 2006). Other fungal species, e.g., *Alternaria alternata* and *Cladosporium cladosporioides* accumulate iodine in their hyphae.

Although the soil is severely iodine-deficient in many parts of the world (Fig. 17.4.2), the situation for the human nutrition is gradually being improved (Fig. 17.4.3), to a large extent by fortification of household salt (NaCl) with iodide (NaI or KI).

There exist bacteria in the genus *Pseudomonas* that carry out a kind of anaerobic oxidation with iodate (IO_3^-) as terminal electron acceptor (Yamazaki et al. 2020). Two of the proteins involved in this process are related to subunits of respiratory arsenite oxidase, two other distantly related to cytochrome c peroxidase.

Bacteria of the genus *Iodidimonas* oxidize iodide, I^-, to elemental iodine, I_2 (Amachi & Iino 2022). These bacteria were isolated from a brine from a gas well (Muramatsu et al. 2001) containing 1 mM iodide, which is 2000 times greater than that of natural seawater.

Brown algae are efficient accumulators of iodine (Küpper & Carrano 2019). The genus *Laminaria* are the strongest accumulators of iodine known and are major emitters of molecular iodine as well as iodinated organic compounds into the atmosphere (Leblanc et al. 2006).

Fig. 17.4.2. Distribution of land areas with severely suboptimal iodine content in the soil. From Medrano-Macías et al. (2016), adapted from Burlingame (2013).

1993

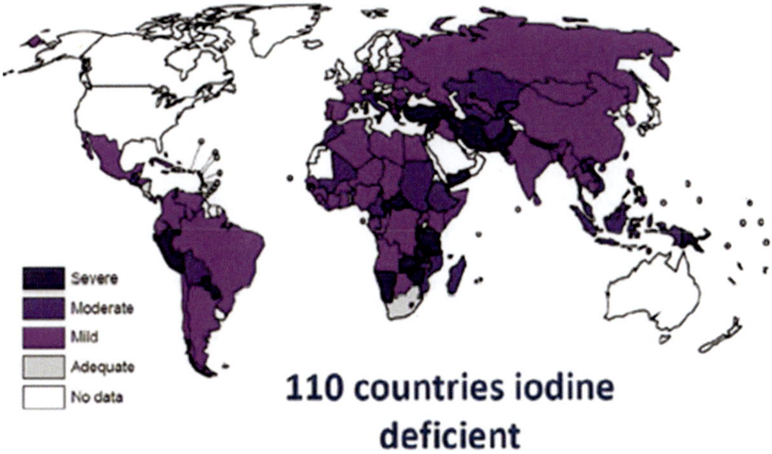

2013

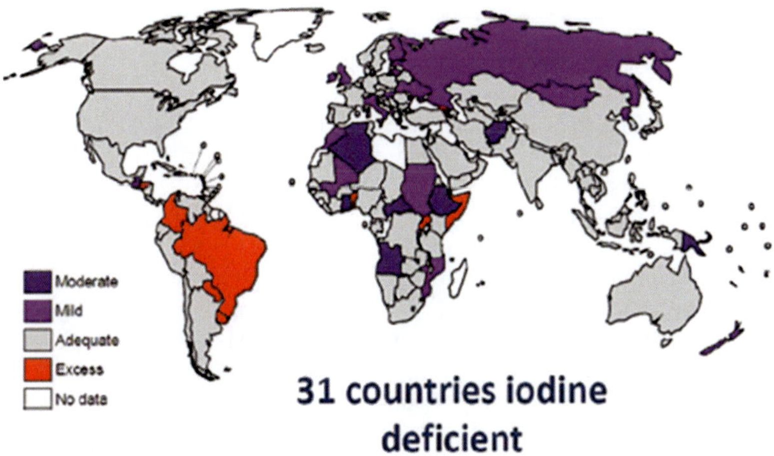

Fig. 17.4.3. Iodine deficiency in the world is decreasing: Deficiency in 110 countries in 1993, in 31 countries in 2013. From "Salt reduction and iodine fortification strategies in public health". Report of a joint technical meeting convened by World Health Organization and the George Institute for Global Health in collaboration with the International Council for the Control of Iodine Deficiency Disorders Global Newwork, Australia, March 2013.

Some red algae, e.g., *Polysiphonia urceolata*, but not all, need iodine for growth (Fries 1966). Red algae produce organic iodine compounds. Romarís-Hortas et al. (2014) found several iodine-containg proteins in *Porphyra umbilicalis*. Jesus et al. (2019) give examples of aromatic compounds from red algae containing iodine or bromine, or both.

The animals having the highest content of iodine are the "thorn corals", Antipatharia. The older parts of their skeletons can contain more than contain more than 20% iodine by weight (Goldberg 1978). The halogen content of *Antipathes salix* constituted about 5% of the skeletal weight and consisted primarily of iodine in both tip and base (Goldberg et al. 1994).

Since the amount of iodine reaching the stratosphere has been low compared to the amounts of chlorine and bromine, its effect on stratospheric ozone depletion has been judged negligible. But new calculations indicate that iodine can significantly impact the lower part of the Antarctic ozone hole, contributing, on average, 10% of the lower stratospheric ozone loss during spring (Cuevas et al. 2022).

References

Amachi, S. (2008) Microbial contribution to global iodine cycling: Volatilization, accumulation, reduction, oxidation, and sorption of iodine. Microbes Environ, 23, 269–276. https://doi.org/10.1264/jsme2.MR08548

Amachi, S., Kamagata, Y., Kanagawa, T. & Muramatsu, Y. (2001) Bacteria mediate methylation of iodine in marine and terrestrial environments. Appl. Environm. Microbiol. 67, 2718–2722. https://doi.org/10.1128/AEM.67.6.2718-2722.2001

Ban-nai, T., Muramatsu, Y. & Amachi, S. (2006) Rate of iodine volatilization and accumulation by filamentous fungi through laboratory cultures. Chemosphere 65, 2216–2222. https://doi.org/10.1016/j.chemosphere.2006.05.047

Barnum, T.P & Coates, J.D. (2022) The biogeochemical cycling of chlorine. Geobiol. 20, 634–649. https://doi.org/10.1111/gbi.12513

Bathish, B., Paumann-Page, M., Paton, L.N., Kettle, A.J. & Winterbourn, C.C. (2020) Peroxidasin mediates bromination of tyrosine residues in the extracellular matrix. J. Biol. Chem. 295, 12697–12705. https://doi.org/10.1074/jbc.RA120.014504

Baunthiyal, M. & Pandey, A. (2012) Organofluorine metabolism in plants. Fluoride 45, 78–85.

Bell, N., Hsu, L., Jacob, D. J., Schultz, M. G., Blake, D. R., Butler, J. H., King, D.B., J. M. Lobert, J.M. & Maier-Reimer, E. (2002) Methyl iodide:

atmospheric budget and use as a tracer of marine convection in global models. J. Geophys. Res. 107:4340 (12 pp.).

Binh, P.T., Trang, D.T., Thao, N.P., Mai, N.C., Cuong, N.X., Nam, N.H. & Thanh, N.V. (2022) Structure elucidation of new brominated sesquiterpenes from the sea hare *Aplysia dactylomela* by experimental and DFT computational methods. J. Molecul. Struct. 1259: 132744 (7 pp.). https://doi.org/10.1016/j.molstruc.2022.132744

Brahmachari, U., Gonthier, J.F., Sherrill, C.D. & Barry, B.A. (2017) Chloride maintains a protonated internal water network in the photosynthetic oxygen evolving complex. J. Phys. Chem. B 121, 10327−10337. https://doi.org/10.1021/acs.jpcb.7b08358

Carlessi, M., Mariotti, L., Giaume, F., Fornara, F., Perata, P. & Gonzali, S. (2021) Targeted knockout of the gene OsHOL1 removes methyl iodide emissions from rice plants. Sci. Rep. 11:17010 (13 pp.). https://doi.org/10.1038/s41598-021-95198-x

Ceko, M.J., Hummitzsch, K., Hatzirodos, N., Bonner, W., James, S.A., Kirby, J.K., Rodgers, R.J. & Harris, H.H. (2015) Distribution and speciation of bromine in mammalian tissue and fluids by X-ray fluorescence imaging and X-ray absorption spectroscopy. Metallomics 7, 756–765. https://doi.org/10.1039/c4mt00338a

Cheng, X. & Ma, L. /2021) Enzymatic synthesis of fluorinated compounds. Appl. Microbiol. Biotechnol. 105, 8033–8058. https://doi.org/10.1007/s00253-021-11608-0

Chance, R., Malin, G., Jickells, T., and Baker, A. R. (2007) Reduction of iodate to iodide by cold water diatom cultures. Mar. Chem. 105, 169–180.

Chino, Y., Saito, M., Yamasu, K., Suyemitsu, T. & Ishihara, K. (1994) Formation of the adult rudiment of sea urchins is influenced by thyroid hormones. Dev. Biol. 161, 1–11. http://dx.doi.org/10.1006/dbio.1994.1001

Cochereau, B., Meslet-Cladière. L., Pouchus, Y.F., Grovel, O. & Roullier, C. (2022) Halogenation in Fungi: What Do We Know and What Remains to Be Discovered? Molecules 27: 3157 (45 pp.). https://doi.org/10.3390/molecules27103157

Cosimo D. Cadicamo, C.D, Courtieu, J., Deng, H., Meddour, A. & O'Hagan, D. (2004) Enzymatic fluorination in *Streptomyces cattleya* takes place with an inversion of configuration consistent with an S_N2 reaction mechanism. ChemBioChem 5, 685 –690. https://doi.org/10.1002/cbic.200300839

Cuevas, C.A., Fernandez, R.P., Kinnison, D.E., Lia, Q., Lamarque, J.-F., Trabelsf, Y., Francisco, J.S., Solomong , S. & Saiz-Lopeza, A. (2022)

The influence of iodine on the Antarctic stratospheric ozone hole. Proc. Natl Acad. Sci. USA 119: e2110864119 (10 pp.). https://doi.org/10.1073/pnas.2110864119

Dehnen, S., Schafer, L.L., Lectka, T. & Togni, A. (2021) Fluorine: A very special element and its very special impacts on chemistry. Inorg. Chem. 60, 17419−17425. https://doi.org/10.1021/acs.inorgchem.1c03509

Dembitsky, V.M. (2006) Biogenic iodine and iodine-containing metabolites. Natural Product Commun. 1, 139 – 175.

Deng, H., Cross, S.M., McGlinchey, R.P. & Hamilton, J.T.G. O'Hagan, D. (2008) In vitro reconstituted biotransformation of 4-fluorothreonine from fluoride ion: Application of the fluorinase. Chemistry & Biology 15, 1268–1276

Dvoretsky, A.G., Tipisova, E.V., Alikina, V.A., Elfimova, A.E. & Dvoretsky, V.G. (2022) Thyroid hormones in hemolymph of red king crabs from the Barents Sea. Animals 12, 379 (11 pp.). https://doi.org/10.3390/ani12030379

Fidler, A.L., Vanacore, R.M., Chetyrkin, S.V., Pedchenko, V.K., Bhave, G., Yin, V.P., Stothers, C.L., Rose, K.L., McDonald, W.H., Clark, T.A., Borza, D.-B., Steele, R.E., Ivy, M.T., Aspirnauts, T., Hudson, J.K. & Hudson, B.G. (2014) A unique covalent bond in basement membrane is a primordial innovation for tissue evolution. Proc. Natl Acad. Sci. USA 111, 331–336. https://doi.org//10.1073/pnas.1318499111

Fries, L. (1966) Red algae in axenic culture. Physiol. Plantarum 19, 800–808.

Gao, S.S., Zhang, S., Mei, M.L., ChinLo, E.C.M. & Chu, C.-H. (2016) Caries remineralisation and arresting effect in children by professionally applied fluoride treatment – a systematic review. BMC Oral Health 16:12 (9 pp.). https://doi.org/10.1186/s12903-016-0171-6

Goldberg, W.M. (1978) Chemical changes accompanying maturation of the connectivetissue skeletons of gorgonian and antipatharian corals. Mar. Biol. 49, 203–210.

Goldberg, W.M., Hopkins, T.L., Holl, S.M., Schaefer, J., Kramer, K.J., Morgan, T.D. & Kim, K. (1994) Chemical composition of the sclerotized black coral skeleton (Coelenterata: Antipatharia): a comparison of two species. Comp. Biochem. Physiol. 107B, 633-643.

Goodwin, K.D., North, W.J. & Lidstrom, M.E. (1997) Production of bromoform and dibromomethane by Giant Kelp: Factors affecting release and comparison to anthropogenic bromine sources. Limnol. Oceanogr: 42, 1725–I734.

Gutleben, J., Koehorst, J.J., McPherson, K., Pomponi, S., Wijffels, R.H., Smidt, H. & Sipkema, D. (2019) Diversity of tryptophan halogenases in sponges of the genus *Aplysina*. FEMS Microbiol. Ecol., 95, fiz108 (15 pp.). https://doi.org/10.1093/femsec/fiz108

Harper, D.B. & O'Hagan, D. (1994) The fluorinated natural products. Nat. Prod. Rep. 11, 123–133. https://pubs.rsc.org/en/content/articlelanding/1994/NP/np9941100123

He, C., Song, W., Weston, T.A., Tran, C., Kurtz, I., Zuckerman, J.E., Guagliardo, P., Miner, J.H., Ivanov, S.V., Bougoure, J., Hudson, B.G., Colon, S., Voziyan, P.A., Bhave, G., Fong, L.G., Young, S.G. & Jiang, H. (2020) Peroxidasin-mediated bromine enrichment of basement membranes. Proc. Natl Acad. Sci. USA 117, 15827–15836. www.pnas.org/cgi/doi/10.1073/pnas.2007749117

Holzer, G. & Laudet, V. (2015) Thyroid hormones: A triple-edged sword for life history transitions. Curr. Biol. 25, R328–R347. https://doi.org/10.1016/j.cub.2015.02.026

Imaizumi, K. & Ifuku, K. (2022) Binding and functions of the two chloride ions in the oxygen-evolving center of photosystem II. Photosynth. Res. publ. online June 13, 2022. https://doi.org/10.1007/s11120-022-00921-y

Itoh, N., Toda, H., Matsuda, M., Negishi, T., Taniguchi, T. & Ohsawa, N. (2009). Involvement of S-adenosylmethionine-dependent halide/thiol methyltransferase (HTMT) in methyl halide emissions from agricultural plants: isolation and characterization of an HTMT-coding gene from *Raphanus sativus* (daikon radish). BMC Plant. Biol. 9:116 (10 pp.). https://doi.org/10.1186/1471-2229-9-116

Jesus, A., Correia-da-Silva, M., Afonso, C., Pinto, M. & Cidade, H. (2019) Isolation and potential biological applications of haloaryl secondary metabolites from macroalgae. Mar. Drugs 17: 73 (19 pp.). https://doi.org/10.3390/md17020073

Keng, F.S.-L., Phang, S.-M., Rahman, N.A., Elvidge, E.C.L., Malin, G. & Sturges, W.T. (2020) The emission of volatile halocarbons by seaweeds and their response towards environmental changes. J. Appl. Phycol. 32, 1377–1394. https://doi.org/10.1007/s10811-019-02026-x

Kiferle, C., Martinelli, M., Salzano, A.M., Gonzali, S., Beltrami, S., Salvadori, P.A., Hora, K., Holwerda, H.T., Scaloni, A. & Perata, P. (2020) Evidences for a nutritional role of iodine in plants. Front. Plant Sci. 12:616868 (18 pp.). https://doi.org/10.3389/fpls.2021.616868

Korobova, E.M., Baranchukov, V.S., Kurnosova, I.V. & Silenok, A.V. (2022) Spatial geochemical differentiation of the iodine-induced health risk and distribution of thyroid cancer among urban and rural population of the Central Russian plain affected by the Chernobyl NPP accident. Environ. Geochem. Health 44, 1875–1891.

https://doi.org/10.1007/s10653-021-01133-4
Küpper, F-C. & Carrano, C.J. (2019) Key aspects of the iodine metabolism in brown algae: a brief critical review. Metallomics 11, 756 –764. https://doi.org/10.1039/c8mt00327k
Lankau, T., Ken, H.C., Chang, H.M. & Yu, C.H. (2022) A Computational study of the promiscuity of the SAM-dependent methyltransferase AtHTMT1. ACS Omega 7, 12753−12764. https://doi.org/10.1021/acsomega.1c07327
Lee, S.T., Cook, D., Pfister, J.A., Allen, J.G., Colegate, S.M., Riet-Correa, F. & Taylor, C.M. (2014) Monofluoroacetate-containing plants that are potentially toxic to livestock. J. Agric. Food Chem. 62, 30, 7345–7354. https://doi.org/10.1021/jf500563h
Lee-Taylor, J. & Redeker, K.R. (2005) Reevaluation of global emissions from rice paddies of methyl iodide and other species. Geophys. Res. Lett. 32: L15801 (5 pp.). https://doi.org/10.1029/2005GL022918
Li, Y.F., Cheng, Y.-L., Chen, K., Cheng, Z.-Y., Zhu, X., Cardoso, J.C.R., Liang, X., Zhu, Y.-T., Power, D.M. & Yang, J.-L. (2020) Thyroid hormone receptor: A new player in epinephrine-induced larval metamorphosis of the hard-shelled mussel. Gen. Comp. Endocrinol. 287: 113347 (9 pp.). https://doi.org/10.1016/j.ygcen.2019.113347
Marais, J.S.C. (1943) The Isolation of the toxic Principle "Potassium Cymonate" from "Gifblaar" *Dichapetalum cymosum* (Hook) Engl. Onderstepoort J. Vet. Sci. Animal Industry 18, 203–206.
McCall, A.S., Cummings, C.F., Bhave, G., Vanacore, R., Page-McCaw, A. & Hudson, B.G. (2014) Bromine Is an essential trace element for assembly of collagen IV scaffolds in tissue development and architecture. Cell 157, 1380–1392. http://dx.doi.org/10.1016/j.cell.2014.05.009
Medrano-Macías, J., Leija-Martínez, P., González-Morales, S., Juárez-Maldonado, A. & Benavides-Mendoza, A. (2016) Use of iodine to biofortify and promote growth and stress tolerance in crops. Front. Plant Sci. 7:1146 (20 pp.). https://doi.org/10.3389/fpls.2016.01146
Meyer, J.J.M. (2019) Fluoroacetate production by *Dichapetalum cymosum*. PhD thesis, Univ. Pretoria.
Mohiuddin, M., Irshad, M., Ping, A., Hussain, Z. & Shahzad, M. (2019). Bioavailability of iodine to mint from soil applied with selected amendment. Env. Poll. Bioaval. 31, 138–144. https://doi.org/10.1080/26395940.2019.1588077
Ngivprom, U., Kluaiphanngam, S., Ji, W., Siriwibool, S., Kamkaew, A., Cairns, J.R.K., Zhang, Q. & Lai, R.-Y. (2021) Characterization of

NucPNP and NucV involved in the early steps of nucleocidin biosynthesis in *Streptomyces calvus*. RSC Adv. 11, 3510–3515. https://doi.org/10.1039/d0ra10878b

O'Hagan, D., Schaffrath, C., Cobb, S.L., Hamilton, J.T.G. & Murphy, C.D. (2002) Biosynthesis of an organofluorine molecule. Nature 416, 279.

ten Cate, J.M. (2004) Fluorides in caries prevention and control: Empiricism or science. Caries Res. 38, 254–257. https://doi.org/10.1159/000077763

Ohyoshi, T., Zhao, Y. & Kigoshi, H. (2022) Isolation and synthesis of azuriaplysins A and B, bromoditerpenes with an α-methylene carbonyl from the sea hare *Aplysia kurodai*. J. Nat. Prod. 85, 2082–2089. https://doi.org/10.1021/acs.jnatprod.2c00476

Palaniveloo, K., Rizman-Idid, M., Nagappan, T. & Razak, S.A. (2020) Halogenated metabolites from the diet of *Aplysia dactylomela* Rang. Molecules 25: 815 (18 pp.). https://doi.org/10.3390/molecules25040815

Paul, N.A., de Nys, R. & Steinberg, P.D. (2006) Chemical defence against bacteria in the red alga *Asparagopsis armata*: linking structure with function. Mar. Ecol. Prog. Ser. 306, 87–101. https://doi.org/10.3354/meps306087

Péterfi, Z, & Geiszt, M. (2014) Peroxidasins: novel players in tissue genesis. Trends Biochem. Sci. 39, 305–307. http://dx.doi.org/10.1016/j.tibs.2014.05.005

Radlow, M., Czjzek, M., Jeudy, A., Dabin, J., Delage, L., Leblanc, C. & Hartung, J. (2018) X-ray diffraction and density functional theory provide insight into vanadate binding to homohexameric bromoperoxidase II and the mechanism of bromide oxidation. ACS Chem. Biol. 13, 1243−1259. https://doi.org/10.1021/acschembio.8b00041

Rehman, N.U., Rafiq, K., Khan, A., Halim, S.A., Ali, L., Al-Saady, N., Al-Balushi, A.H., Al-Busaidi, H.K. & Al-Harrasi, A. (2019) a-glucosidase inhibition and molecular docking studies of natural brominated metabolites from marine macro brown alga *Dictyopteris hoytii*. Mar. Drugs 17: 666 (15 pp.). https://doi.org/10.3390/md17120666

Romarís-Hortas, V., Bianga, J., Moreda-Piñeiro, A., Bermejo-Barrera, P. & Szpunar, J. (2014) Speciation of iodine-containing proteins in Nori seaweed by gel electrophoresis laser ablation ICP–MS. Talanta127, 175–180. http://dx.doi.org/10.1016/j.talanta.2014.04.003

Rua, C.P.J., de Oliveira, L.S., Froes, A., A. Tschoeke, D.A., Soares, A.C., Leomil, L., Gregoracci, G.B., Coutinho, R., Hajdu, E., Thompson, C.C., Berlinck, R.G.S. & Thompson, F.L. (2018) Microbial and functional biodiversity patterns in sponges that accumulate bromopyrrole alkaloids suggest horizontal gene transfer of halogenase genes. Microbial Ecol. 76, 825–838. https://doi.org/10.1007/s00248-018-1172-6

Singh, S.P., Dhakshinamoorthy, R., Jaiswal, P., Schmidt, S., Thewes, S. & Baskar, R. (2014) The thyroxine inactivating gene, typeIII deiodinase, suppresses multiple signaling centersin *Dictyostelium discoideum*. Developm. Biol. 396, 256–268. http://dx.doi.org/10.1016/j.ydbio.2014.10.012

Stemmler, I., Hense, I., Quack, B. & Maier-Reimer, E. (2014) Methyl iodide production in the open ocean. Biogeosci. 11, 4459–4476. https://doi.org/10.5194/bg-11-4459-2014

Svensson, T., Kylin, H., Montelius, M., Sandén, P. & Bastviken, D. (2021) Chlorine cycling and the fate of Cl in terrestrial environments. Environm. Sci. Pollut. Res. 28, 7691–7709. https://doi.org/10.1007/s11356-020-12144-6

Taylor, E. & Heyland, A. (2017) Evolution of thyroid hormone signaling in animals: Non-genomic and genomic modes of action. Molecul. Cellul. Endocrin. 459, 14–20. http://dx.doi.org/10.1016/j.mce.2017.05.019

ten Cate, J.M. & Buzalaf, M.A.R. (2019) Response to the Letter to the Editor: "Fluoride Mode of Action". J. Dental Res. 98, 1549. https://doi.org/10.1177/0022034519882663

Vanacore, R., Ham, A.-J.L., Voehler, M., Sanders, C.R., Conrads, T.P., Veenstra, T.D., Sharpless, B., Dawson, P.E. & Hudson, BG. (2009) A Sulfilimine Bond Identified in Collagen IV. Science 325, 1230–1234. https://www.science.org/doi/10.1126/science.1176811

Wang, M., Zhang, H.-Y., Xiang, Y.-Y., Qian. Y.-P., Ren, J.-N. & Jia, R. (2022) How does fluoride enhance hydroxyapatite? A theoretical understanding. App. Surface Sci. 586: 152753 (9 pp.). https://doi.org/10.1016/j.apsusc.2022.152753

Wincencjusz, H., van Gorkom, H.J. & Yocum, C.F. (1997) The photosynthetic oxygen evolving complex requires chloride for its redox state S_2→S_3 and S_3→S_0 transitions but not for S_0→S_1 or S_1→S_2 transitions. Biochemistry 36, 3663–3670.

Winterton, N. (2000) Chlorine: the only green element – towards a wider acceptance of its role in natural cycles. Green Chemistry 2, 173–225. https://doi.org/10.1039/b003394o

Xu, X.L., Fan, X., Song, F.H., Zhao, J.L., Han, L.J. & Shi, J.G. (2004) A new bromophenol from the brown alga *Leathesia nana*. Chines Chem. Lett. 15, 661–663.

Xu, G. & Wang, B.-G. (2016) Independent evolution of six families of halogenating enzymes. PLoS ONE 11(5): e0154619 (13 pp.). https://doi.org/10.1371/journal.pone.0154619

Yamaguchi, N., Suzuki, A., Yoshida, A., Tanaka, T., Aoyama, K., Oishi, H., Hara, Y., Ogi, T., Amano, I., Kameo, S., Koibuchi, N., Shibata, Y., Ugawa, S., Mizuno, H. & Saitoh, S. (2022) The iodide transporter Slc26a7 impacts thyroid function more strongly than Slc26a4 in mice. Sci. Rep. 12:11259 (9 pp.). https://doi.org/10.1038/s41598-022-15151-4

Yamazaki, C., Kashiwa, S., Horiuchi, A., Kasahara, Y., Yamamura, S. & Amachi, S. (2020) A novel dimethylsulfoxide reductase family of molybdenum enzyme, Idr, is involved in iodate respiration by *Pseudomonas sp*. SCT. Environm. Microbiol. 22, 2196–2212. https://doi.org/10.1111/1462-2920.14988

Zhang, Q., Guan, L., Guo, J., Chuan, A., Tong, J., Ban, J., Tian, T., Jiang, W. & Wang, S. (2022) Application of fluoride disturbs plaque microecology and promotes remineralization of enamel initial caries. J. Oral Microbiol. 14: 2105022 (14 pp.). https://doi.org/10.1080/20002297.2022.2105022

Genus and Species Index

SUBJECT INDEX